THE MOLECULAR BASIS OF CANCER

THE MOLECULAR BASIS OF CANCER

Edited by
Peter B. Farmer
and
John M. Walker

A WILEY-INTERSCIENCE PUBLICATION

JOHN WILEY & SONS
New York

Published in the USA by
Wiley-Interscience, a Division of John Wiley & Sons, Inc.,
New York

ISBN 0-471-82755-X

Printed in Great Britain

CONTENTS

PREFACE

This book aims to describe the current state of knowledge and possible future developments in a number of major areas of research into the nature, causes and treatment of cancer. The contributing authors have been encouraged to discuss their subjects at the molecular level. It will become apparent to the reader that considerable developments in the understanding of the fundamental nature of cancer, in molecular terms, are constantly being made. This is particularly the case in the area of oncogene research where differences between tumour and normal cells can now be defined in terms of altered expression of DNA sequences. An understanding of the methods available for detecting cancer, of the process of carcinogenesis and of the means available for treating cancer can only be achieved with a precise knowledge of the basic biochemical and molecular processes involved.

Since it is all to easy for the research scientist to become totally absorbed within the specialised area of research in which he is involved, the first chapter is an attempt to encourage a broader field of vision by introducing the clinician's view of the cancer problem, which illustrates the broad spectrum of basic problems that need to be solved by the cancer researcher. This is followed by a systematic description of the mode of action of the various causative agents in cancer (radiation, viruses and chemicals), methods for detecting cancer and carcinogens and three of the major forms of cancer treatment available (hormone therapy, radiation therapy and chemotherapy). Although the book does not attempt to cover all aspects of cancer treatment and research (areas such as cell kinetics and the surgical treatment of tumours are not covered), this volume is intended to encompass the basic biochemical, chemical and molecular biological areas of cancer study. Consequently it should be of particular interest to undergraduate science students, medical students and postgraduate students entering the field of cancer research. Because of the proposed readership a basic knowledge of biochemistry and molecular biology has been assumed. However, since much of the language of the clinician is foreign to the scientist, and vice versa, a glossary of some of the more commonly used terms has been included.

1 THE CLINICAL PROBLEM

M.D. Vincent

Contents

1.1 Introduction

The relief of human suffering and the intellectual challenge of a problem almost unrivalled in breadth and depth are the two major yet distinct factors motivating cancer research. Thus clinicians (whose job it is to cure people) are primarily concerned with the control of medical events, whereas scientists aim to achieve theoretical understanding of the problems. Failure to cure disseminated cancer is still depressingly frequent and any solution to this problem will require intensive collaboration between clinicians and scientists.

One hopes, particularly, that knowledge of how a cell becomes malignant will lead to rationally designed effective means of cure and prevention. This is what we in the clinic demand of the scientist. However, this would not be in the tradition of medical therapeutics, where most successes stemmed from empirical observations, often accidental, and the theoretical understanding usually followed afterwards.

The situation is now changing. Increasingly, our knowledge permits drug design from theoretical principles. Many believe that, in cancer, this rational approach will achieve more than largely empirical attempts, exemplified by the random screening of thousands of compounds for cytotoxicity. Of course, every age has 'explained' cancer and devised 'cures' in terms of the available ideas; there are many exotic examples of this.[1] But there is now reason to believe that our concepts are at last beginning to prove the equal of reality. Unfortunately, we must also accept that ultimate knowledge may reveal that no perfect treatment is possible, because it may be the case that no exploitable fundamental differences exist between normal and cancerous cells. If this is true, cancer treatments will always be more or less poisonous and clinicians will always be engaged in damage-limiting exercises. The complexity of these issues further emphasises the need for communication between clinicians and scientists. Awareness of each other's practical difficulties, tasks and language, is vital. This chapter attempts to deal with this by providing a clinician's perspective.

It is worth re-emphasising that communication problems may also arise because of the difference in the ways clincians and scientists think.

Clinicians have to be more interested in applied factual knowledge. A deep theoretical grasp of how a drug is thought to work is of little importance at the bedside. Rather, whether or not it works is much more critical. This difference in interest may cause clinicians and scientists to talk past or avoid each other and is an example of the ancient dichotomy of empiricism and rationalism which permeates the conceptual history of cancer.[1] The best remedy is to spend time in each other's working environment.

Whether fully understood or not, some effective treatments do exist for cancer. Their successful use depends entirely on recognition of the appropriate clinical context, which demands some knowledge of pathology.

1.2 Pathology

Pathology, the study of disease, progresses from the naked eye (macroscopic) appearance through succeeding levels of resolution (light and electron microscopy) to the molecular and submolecular levels of chemical pathology. Apart from describing and classifying the changes of disease, pathologists are concerned with pathogenesis (the sequence of events in time and space) and identifying aetiological relations of cause and effect.

The body is disturbed by myriad diseases. We understand them as well as we do only because they are essentially reducible to a few basic pathological processes. One of these is the abnormal proliferation of cells; obviously, if this occurs faster than cells are removed, growth will result. Growth (i.e. enlargement) may also occur as a result of increase in size of existing cells (hypertrophy) without increase in their numbers, as occurs in repeatedly exercised muscle. Accumulation of various non-cellular materials (e.g. glycogen, oedema fluid) can also result in 'growth'.

Cellular proliferation, whether normal or pathological, is traditionally subdivided into three further categories: hyperplasia, malformation with excess tissue and neoplasia. The last two are always abnormal, as are some forms of hyperplasia; however, hyperplastic growth includes such perfectly normal phenomena as growth from fetus to adulthood and the changes of puberty.

Compensatory hyperplasia, as when one kidney enlarges after removal of the other, is an appropriate response to an abnormal situation. Administration of female sex hormones to males causes hyperplastic growth of breast cells; whether or not this is 'abnormal' depends on the context; sex-change patients welcome it, men with prostatic cancer do not. (Oestrogens are used to cause temporary regression of prostatic cancer.) Hyperplasia may also be the response to a noxious agent, as when bronchial epithelium is exposed to tobacco smoke. Wound repair and inflammatory cell proliferations are closely allied to hyperplasia. The hallmark of hyperplastic growth is that it is dependent absolutely on the continuing presence of some external stimulus and does not occur actively in the absence of one.

This book is concerned with neoplasia. Tumours are strictly defined as neoplasms, although the word 'tumour' may be loosely applied to any swelling. In neoplasia growth occurs because cells are not removed fast enough from the dividing pool of stem cells. This process is normally in equilibrium. Neoplasms are a broad group which ranges from the slowly growing benign (innocent) (see later) to the rapidly growing malignant forms which are invasive, disseminating and lethal. Cancer is defined as a malignant neoplasm.

The formulation of a concise and correct definition of neoplasia is impossible. This is because we do not know at a molecular level what constitutes its unique irreducible essence. We have to rely on outward descriptive criteria that are second best. However, the variability of neoplasms is such that no single criterion is both constant and exclusive. We have to use several criteria so that overall requirements are met.

Most existing definitions can be faulted: they may beg the question, they may contain ambiguities, they may depend on concepts beyond our knowledge or they may be simply incorrect in terms of new knowledge. A significant problem is that outward differences between normal and neoplastic cells may be more relative than absolute. However, the 'successful treatment of cancer necessitates the detection and exploitation of consistent and significant biological differences between cancer cells and their normal counterparts'[2] and an adequate description of the overt differences is an essential prerequisite to discovering the biochemical differences. Thus those seeking to define neoplasia and those designing treatments for it are in pursuit of exactly the same goal, namely the unique and consistent differences which explain the tumour cell.

A widely quoted definition is that of Willis[3] 'A tumour is an abnormal mass of tissue, the growth of which exceeds, and is unco-ordinated with, that of normal tissues, and persists in the same excessive manner after cessation of the stimuli which evoked the change'.

By excessive growth, Willis means that the mitotic rate of all tumours is greater than that of the parent tissue; we now know that this is not a general property of neoplasms. The inco-ordination criterion is probably valid but is imprecise and difficult to recognise in early cases. Because we know little of what co-ordinates normal cell growth, the inco-ordination of neoplasia has presumably to be inferred from the destruction and disruption of normal tissue as well as the uselessness of the tumour. But the smaller and more benign a tumour, the less are these effects recognisable. Furthermore inco-ordination is admitted by Willis to be relative and not absolute. Even if co-ordination could be quantitated, no-one has identified a clear point beyond which cells are sufficiently inco-ordinated to be neoplastic. Finally, inco-ordination is not unique to neoplasia: thyroid hyperplasia (goitres), for example, may attain debilitating proportions whose inco-ordination is more obvious than that of small dormant neoplasms. This non-specificity is a consequence of the fuzziness of the concept.

Willis[4] claims that the inco-ordinate nature of neoplastic growth is partly revealed by its 'unlimited . . . indefinitely progressive . . . continuous' nature. But although proliferation of tumour cells is usually relentless, the phenomena of regression and dormancy, described later, invalidate this as a general property of neoplasms. Further, even the wildest tumours do not grow in a limitless fashion in all directions. They are constrained within the capacities of the supporting tissue, especially blood vessels, to nourish them; they cannot breach hard barriers like cartilage. Although these are not the factors limiting normal growth, the concept of limitless growth clearly has to be qualified.

Finally, persistence '. . . after cessation of the stimuli which evoked the change' must be disputed. While this is true for irradiation and crude chemical carcinogens, it does not generally accord with our present knowledge of oncogenes, especially now that viruses are the established first causes of some tumours (see Chapter 4). Indeed, not only the persistence of the oncogene, but its persistent activity, may well be required for the continued expression of the malignant phenotype. It is currently not possible, and may never be, to make universally valid statements about the first causes of neoplasms; therefore references to causality are out of place in a definition.

Despite the difficulties, there is agreement that spontaneous tumours consist of host-derived neoplastic cells and variable numbers of other cells, not neoplastic themselves, but incorporated into the mass in reaction or response, or by accident. Another variable component is the acellular stroma in between the cells. The neoplastic cell population is not entirely homogeneous; only the tumour stem cells are self-regenerating (clonogenic) and able to form new tumours. There are other tumour cells derived from the stem cells that, for a variety of reasons not entirely understood, are unable to divide further.

Thus some of the problems of definition may be circumvented by distinguishing the neoplasm as a whole from the component cells and both from neoplastic growth (proliferation). Tumour stem cells usually (but not always) exhibit neoplastic growth; this occurs in the absence of any sufficient, continuing, known, external stimuli (even if some external stimuli are necessary to sustain the proliferation) and is not limited by the factors which usually limit cell growth. It results in an increasing distortion of the pattern and proportion of cells characterised by an ongoing accumulation of tumour cells which is inevitably damaging unless checked. In so far as the neoplastic cells combine uselessness, an open-ended life-span within the host and an ability to compete successfully with the host for space and nutrition, it is inferred that they have escaped from at least some of whatever control mechanisms exist and can thus be said to have achieved a degree of autonomy. Because the rest of the cells behave as before (unless affected by the tumour), it is evident that the tumour cells, being derived from host cells, are changed internally in some stable and heritable way and in this sense are abnormal.

In this definition, I have had to impose conditions of necessity and sufficiency on the 'continuing external stimuli' to cater for the (important) hormone-dependent tumours, like some breast cancers, and for any tumour that may require the supply of basal levels of growth factors in the circulation (this latter may in fact apply to many tumours). The point is that, notwithstanding any necessity for these basal levels of hormones or growth factors, they are insufficient to cause progressive proliferation of the generality of cells sensitive to their effects and therefore by themselves are not enough to explain the neoplastic growth. The full explanation has to be sought within the tumour cells, as the last sentence of the definition articulates. All this makes for a rather unwieldy descriptive definition, but I feel that the readership would be misled by an uncritical repetition of the classical definitions.

1.3 Metaplasia and Dysplasia

↑ derranged cellular proliferation

Hyperplasia, metaplasia, atypical metaplasia, dysplasia, carcinoma *in situ*, invasive cancer and anaplastic cancer are all on a continuum, in that order, representing increasingly deranged cellular proliferation. From atypical metaplasia onward there is a progressive escape from normal control mechanisms associated with a decreasing likelihood of reversibility. Carcinoma *in situ* may be 'spontaneously' reversible but invasive and anaplastic cancers are not, for practical purposes. [It is convenient to illustrate these issues with epithelial examples. A 'carcinoma' is an epithelial cancer, '*in situ*' implies that it has not (yet) invaded through the basement membrane. The basic principles apply to connective tissue as well.]

Metastasis

Metaplasia is the orderly substitution of one type of differentiated adult cell by another, in response to a long-standing stimulus. Metaplasia occurs because the undifferentiated stem cells redirect their line of differentiation, not because mature cells metamorphose. Exposure to harmful chemical or physical agents, inflammation, or even vitamin deficiencies,[5] suffice. The change is considered protective: for example, the delicate mucus-secreting ciliated columnar epithelium is replaced by tougher stratified squamous epithelium in the respiratory tracts of smokers. The price for protection is that differentiated functions are sacrificed and there is an increased risk of cancer supervening in the metaplastic field of cells. Whether metaplasia itself is premalignant is controversial. The commonest form of lung cancer is a squamous carcinoma; as there is not normally squamous epithelium in the respiratory tract, metaplasia must have occurred either before or after transformation.

Proceeding along the continuum to dysplasia, the cells become more and more different from the norm and from each other (atypia). These differences apply to the appearances of the individual cells and to their architectural relations with one another. The cell nuclei become larger and more darkly staining (hyperchromatic). The cells become larger and more variable in size and shape; mitoses become more frequent and occur away from the basal layer. Disorderly architecture, with loss of progressive maturation from base to surface and random orientation of the cells are additional dysplastic features. Common sites of dysplasia are the mouth, cervix uteri, gall bladder and respiratory tract.

Dysplasia is potentially reversible, but this cannot be predicted in the individual case. There is a high incidence of cancer within dysplastic fields of cells; the initially localised 'carcinoma *in situ*' phase may intervene. Why some dysplasias progress to cancer and others are static or reverse to normality is not well understood. Metaplasia and especially dysplasia increase the risk of malignant transformation; but the development of cancer can also occur without the cells going through these 'premalignant' phases.

1.4 Benign and Malignant Tumours

Ideally, all tumours fall into one of two categories: benign (innocent) or malignant. This extremely important division is based fundamentally on the actual or projected behaviour of the tumour. In fact, perfect innocence and absolute malignancy are end points of a whole spectrum of behaviour; although most neoplasms are obviously one or the other, the middle ground is well represented and one has often to talk of degrees of malignancy.

Unfortunately, attempts to quantify these degrees of malignancy in precise terms have proven unworkable. This is partly because not one but two behavioural criteria determine malignancy; these are local invasion and distant (discontinuous) spread (metastases; see later and Chapter 2). These two properties are not always present (or absent) together; the basal-cell carcinoma of the skin, for example, is locally destructive because of direct infiltration, but never spreads to distant sites. Furthermore morphological appearances known collectively as anaplasia correlate quite well, but not perfectly, with malignant behaviour and therefore morphology also influences the decision as to where on the spectrum a particular tumour should lie; but the relevant microscopic parameters do not lend themselves to accurate and reproducible measurement.

The word cancer derives from the Latin for crab and suggests its capacity to reach out and cling tenaciously to adjacent tissues. While this direct infiltration may be referred to as 'local metastasis', the term metastasis more strictly refers to the process of spread, by various routes, to distant discontinuous sites where separate colonies are established called 'metastases' or secondaries. The initial tumour is then known as the primary. The outcome, without effective treatment, is nearly always the death of the host.

This capacity to kill the host justifies the word 'malignant'. It would seem therefore logical to suppose that benign tumours were never fatal. This is not true; they may have a variety of adverse effects relating to the progressive occupation of vital space (like inside the skull). Also, benign tumours of endocrine origin produce biochemically active substances which affect well-being and sometimes cause death. These hormones are the same as those normally produced by such endocrine tissues: for instance benign adrenal tumours produce excessive amounts of adrenal corticosteroids (hormone production is not a general property of benign tumours). The word 'benign' refers only to a form of behaviour characterised by non-invasiveness and inability to metastasise. 'Innocence' is an equally misleading synonym.

While behaviour is the paramount interest and criterion, often the pathologist has to rely heavily on morphology. Once biopsied and under the microscope cells are not able to behave anymore. (If, however, they are in sites where they ought not to be, it is impossible to infer past misbehaviour.) Fortunately the correlation between the morphological appearances and benign or malignant behaviour is sufficiently good to enable a confident decision to be made in the vast majority of cases. The light microscope identifies a collection of features, known as anaplasia, which are empirically associated with malignant behaviour; both the appearances of individual cells and their architectural relations to one another are important.

These features are an extension of those of dysplasia and further include a complete loss of polarity of cellular orientation, more variation in cell size and shape (pleomorphism) and the emergence of larger and even compound giant cells. A more florid increase in the number of mitotic structures and the appearance of abnormally situated and shaped mitoses as well as a deterioration of the nuclear features described under dysplasia (nuclear aberration) are also characteristic of anaplasia.

Consequently, anaplastic cells do not manifest the differentiated structural and functional properties of the normal mature cells of their tissues or origin. Some authors use the term anaplasia to denote this absence of differentiation, rather than the associated atypical features described above. This goes to the heart of one of the oldest unresolved questions concerning the nature of cancer: whether malignant cells have failed to differentiate or whether they have dedifferentiated, i.e. lost differentiation characteristics. The question as to whether the 'target cell of origin' is a stem cell or a mature differentiated cell addresses the same issue.

Although some malignant cells retain differentiated functions, there is a strong association between highly malignant behaviour, anaplasia and lack of differentiation. Normal stem cells (especially in the fetus) are also undifferentiated and this suggests that malignancy may be the consequence of a differentiation block. Others feel that malignant behaviour depends on entirely new properties in the cell, not found in normal stem cells. The discovery in tumour cells of many proteins [e.g. carcinoembryonic antigen (CEA) in colon cancer and α-fetoprotein (AFP) in liver cancer] characteristic of normal immature, especially fetal, cells (oncofetal proteins), has fuelled this debate (see Chapter 7). The regulatory growth and inhibitory factors and their receptors that appear to control normal growth and differentiation are also generating much interest in this context.

Other features of malignant neoplasms are lack of a capsule, difficulty in defining the edge, a variable host response and a rapid rate of enlargement. Necrosis (death) of tumour cells and haemorrhage in and around the malignancy are frequently associated due to disturbance of blood supply. (These factors explain why it is difficult to obtain uncontaminated viable cancer cells from a natural tumour specimen.)

Benign tumours, by contrast, are composed of well-differentiated cells. They are encapsulated, well defined, have a low mitotic rate and grow slowly by expansion, not by infiltration (or metastasis). The capsule is a consequence, not a cause, of the slowly expansive non-invasive growth pattern of benign tumours. It is not a barrier to cell spread but may facilitate surgical removal. All the features of benign neoplasm point to their curability by local excision. Occasionally, a benign neoplasm transforms further into a malignancy; however, most malignancies do not arise in this way.

The reactive host cells and acellular material within the tumour (the stroma) contribute to the texture, a feature of some limited help in distinguishing benign from malignant growths. Cancer, in situations like the breast, may provoke a lot of fibrosis which imparts a firmness to the tumour, consequently described as 'scirrhous'. This is not a feature of benign breast neoplasms.

1.5 Histochemistry

The microscopic diagnosis of cancer depends on the delineation of micro-anatomy which requires histochemical techniques. The routine haematoxylin-eosin stain usually suffices. But beyond this, newer 'functional' uses for histochemistry have emerged in tumour pathology: identification (in doubtful cases) of the tissue of origin (histogenesis); extension of classifications; diagnosis of the neoplastic or malignant state by appropriate markers; performance of genetic analysis. Our improved ability to identify not only the gross shapes but the chemical constituents of organelles has allowed us to address these pressing clinicopathological issues. Indeed it is a feature of histochemistry that chemical analysis and structural delineation can occur simultaneously, advancing our knowledge of structure–function relations in an unique way.

The complex physico-chemistry of the multiple steps (fixation, dehydration, wax impregnation and staining) cannot be detailed here. All the major biochemical classes, and many individual members thereof, can be recognised by routine and special stains whose biochemical basis is now largely understood. Enzymes are identifiable by their substrate specificity; immunocytochemistry exploits the antigenicity of cellular constituents and relies on the fact that it is possible to join tracer molecules to antibodies without compromising antigen binding. Lectins, able to distinguish various carbohydrate moieties of membrane macromolecules, contribute to the analysis of the cell surface (see Chapter 2). All these newer reactions, like the traditional methods, result in coloured or fluorescent products, visible down the microscope.

Resolving the origins of undifferentiated anaplastic cancers is critical and can be achieved by revealing preserved normal features, e.g. enzymes like acid phosphatase (prostate) and dopa oxidase (melanoma). Appropriate antisera can identify a lymphoid or epithelial origin. Neuroendocrine tumours (e.g. carcinoid, phaeochromocytoma) can be recognised with silver stains. The classification of particularly the leukaemias and lymphomas has been facilitated and extended by a combination of enzymic and immunological methods.[10] The role of histochemistry in diagnosing the malignant, or even neoplastic, state *per se* has been limited by the (to date) non-availability of suitable indicators. However, at least with B-lymphocytes, histochemistry can establish the clonality and by inference the nature of cell proliferation. These cells produce either κ- or λ-type light chains, identifiable immunocytochemically. Reactive proliferations always contain both cell types; all the cells of a B-lymphoma by contrast, are uniformly one or the other.

The existence and expression of genes within identified cells can be studied microscopically by the technique of hybridisation *in situ*, whereby specific radiolabelled complementary DNA (cDNA) probes combine with cellular DNA or RNA if homologous. The tissue preparations are then autoradiographed and counterstained. This system complements Southern blotting, where the identity of the individual cell is necessarily lost. Hybridisation of chomosome preparations *in situ* can be combined with banding techniques to increase the precision of gene mapping, a topic of outstanding importance in cancer. For example, the genetic significance of the consistent translocations in Burkitt's lymphoma[6] and chronic myeloid leukaemia[7] is now emerging, due to these methods (see Chapter 4).

1.6 Tumour-cell Heterogeneity

It has been widely assumed in the past that all the cells of a neoplasm are derived from a single transformed 'target cell of origin', and (therefore) are carbon copies of one another, i.e. are genetically and phenotypically identical. This is the dogma of monoclonality. Strong biochemical and cytogenetic evidence supports this; many markers, that would be expected to vary from cell to cell if the tumour were multicellular in origin, are in fact constantly of the same type.

However, it is obvious that in many solid tumours there are, incorporated either by accident or 'reaction', normal cells like endothelium, fibroblasts, macrophages, etc., which cause practical difficulties to anyone trying to obtain 'pure' neoplastic cells from the tumour mass. However, this does not threaten the supposed monoclonality of the genuine tumour-cell compartment.

Occasionally, as in some lung cancers, there are two morphologically distinct neoplastic types, e.g. squamous and glandular. A possible explanation (preserving a monoclonal origin) is of metaplasia after transformation, which has affected one of the descendents of the original cell, and which results in a coexisting, but different, subclone. Indeed, this concept can be generalised; the genetic instability of the neoplastic state may couple with mutagenic therapies to produce mutant subclones which are both genetically and phenotypically different. These phenotypic differences are possibly critical to our understanding of the biology and the treatment of cancer; for, while many markers in tumours are the same, increasingly differences are being documented among neoplastic cells of the same tumour, with respect to such important characteristics as karyotype, expression of surface antigens, drug sensitivities and presence of hormone receptors. The concept of progression, whereby a tumour becomes more malignant with time, may be explicable in terms of the evolutionary emergence of different subclones.

There are other possible explanations for tumour-cell heterogeneity. Obviously the tumour may have arisen not from one but from several original cells (polyclonal origin). Secondly, tumour cells may hybridise with each other, or with normal cells, with resulting chromosome loss. There is evidence that both of these mechanisms may operate.[8] Thirdly, cellular variation may be due to 'trivial' factors like being in different stages of the cell cycle or different surroundings may induce superficial epigenetic difference, e.g. when metastases in different organs are compared.

Developmental tumours may be truly mixed. Teratomas comprise cells of all three germ layers (ecto- , endo- and meso-derm) to variable extents. These tissues are foreign to the area in which the tumour arises; this distinguishes teratomas from other embryonal tumours and hamartomas. Teratomas may be benign or malignant; their obscure origins probably involve primitive germ cells with pluripotential differentiating capacities (they usually arise in the testis or ovary). Monoclonality is still likely.

Neurofibromatosis exemplifies the inherited conditions in which multiple tumours occur. Presumably all cells inherit a predisposition to neoplastic transformation, so the documented tumour polyclonality here is expected. These are rare conditions; however, the clonality of tumours in general is still a realm of legitimate debate. In particular, genetic heterogeneity is liable to prove an obstacle for specific and rational gene therapy, when that becomes possible.

The clonality issue could be resolved in the near future as the problems of tracing cell lineage yield to new developments in molecular biology.

1.7 Kinetics

Kinetics is the study of tumour growth rates. For ethical reasons hard data on human neoplasms are difficult to obtain. This is unfortunate because of the important implications for radio- and chemo-therapy (see Chapters 9 and 11) and because kinetic information may tell us much about intrinsic tumour biology and host–tumour interactions.

Several factors determine growth rate: the growth fraction (proportion of cells dividing), the frequency of their divisions and rate of cell loss. Cells may be lost after treatment, from host defences, by emigration (the beginnings of metastasis), by 'maturation' out of the growth fraction into a postmitotic, possibly differentiated, state and from lack of nutrition. This usually occurs in the centres of large tumour masses, where blood supply is inadequate. These centres are often necrotic. Attempts to describe growth in refined mathematical terms meet with many difficulties; a commonly used concept is the 'doubling time', which is the time taken for any convenient parameter to double. In the clinic it refers to the linear dimensions of the tumour (as measured, e.g., on the X-ray), from which only crude inferences about kinetic fundamentals can be drawn. However, it may also refer to tumour volume or mass, or cell number.

A tumour, beginning with one cell, has to divide about 30 times to reach a mass of 1 g (diameter about 1 cm): this is roughly the current level of detectability. This period from inception to detection is the latent or silent period. Most tumours kill their hosts when they achieve a mass of 500 g–1 kg, i.e. 10^{11} or 10^{12} cells, after only seven to ten more doublings. Thus three-quarters of the generation are silent and growth rates are difficult to investigate.

In the 'visible' macroscopic range growth rates tend to slow as the nutritive capacity of the vascular infrastructure is exceeded. This may approach an idealised mathematical form (known as 'Gompertzian' growth) as the initially rapid growth rates diminish exponentially with time, represented by the latter part of a sigmoid curve. The constancy of growth rates in the silent period is debatable; our scanty information suggests that some tumours grow slowly or not at all (dormancy); a proportion of these seem suddenly to accelerate, perhaps with the appearance of more anaplastic cells (progression). Either the growth rate genuinely increases or it may only appear to increase while in fact remaining constant. But if growth rates do indeed increase (i.e. dormancy and progression are true phenomena) then either the nature of the tumour cells alters or there are external constraining influences that may be shrugged off at some critical moment. Differentiation, clonal heterogeneity, host defences and tumour vascularisation may be implicated; clarification of these uncertainties may yield much therapeutic benefit. Spontaneous, as opposed to treatment-induced, regression of cancer is a true but rarely documented phenomenon. The interest it excites does not relate to the (actually negligible) chance of its saving patients, but rather to the biological insights it might provide and which may be exploited for therapy in all cancer patients. The true incidence of spontaneous regression in microscopic ('silent') cancers is unknown and probably unknowable. The possible mechanisms may well be the same as those underlying progression and dormancy. Regression has been documented in melanoma, renal cancers, neuroblastomas and a variety of other cancers.[9]

1.8 Classification

Cancer is said to comprise not one but 'over 200' diseases. Our main concern is what they have in common; but tumours arising from different organs and tissues behave in ways sufficiently distinct to warrant talk of different diseases.

Classifying these different tumours produces especially practical advantages and many schemes based on various criteria have been advanced. Attempts to classify by causation have been confounded by ignorance and might anyway be of little practical value. New concepts continue to influence these endeavours; however, three ancient criteria, used in Egypt 3 500 years ago, are still valid, i.e. site (organ) of origin, gross appearance and behaviour.

Light microscopy added two more: histogenetic (tissue of origin) and histological (cellular appearance and arrangement). The most important are the histogenetic and behavioural criteria. Broadly, the body comprises two main tissues, epithelial and connective, along with subsidiary and miscellaneous types. The behavioural criterion (innocence and malignancy) has been discussed.

A tumour is denoted by the suffix '-oma', e.g. a surface epithelial tumour is a papilloma and one of solid (glandular) epithelium is an adenoma. These are benign, as are the connective-tissue tumours osteoma (bone), chondroma (cartilage), fibroma (from fibroblasts), etc. Malignant epithelial tumours are carcinomas and are further characterised by their direction of differentiation, e.g. adenocarcinoma (glandular) and squamous (epidermoid or keratinising) carcinoma. Malignant connective-tissue tumours are known as sarcomas. Simple histologic descriptive terms may be added, e.g. large cell, small cell, clear cell, etc. The organ of origin is named if it is known and gross description terms of shape and texture may be appended, e.g. fungating vs sessile, or scirrhous (hard) vs medullary (soft).

The occurrence of specialised tissues, not fitting readily into the epithelial/connective-tissue divide, necessitates expansion of this simple scheme. Tumours of the adrenal medulla, whether benign or malignant, are phaeochromocytomas. Pigment-cell tumours are naevi (benign) or malignant melanomas. There are other exceptions: the nervous system is so specialised that a different terminology is required.

Haemopoetic neoplasms (leukaemias) are different in that they rarely form discrete masses but spread diffusely in the marrow (and usually blood stream) *ab initio*. They are all regarded as malignant although some may 'grumble on' slowly for years; these are therefore called 'chronic' as opposed to their more rapidly growing and quickly lethal ('acute') counterparts. Thus there are acute and chronic forms of both lymphocytic and myeloid leukaemia. Plasma-cell tumours are myelomas: these tend to spread within bone marrow and seldom spill over into peripheral blood. Germ cells give rise to teratomas and seminomas. In children primitive embryonal cells generate embryonal tumours, e.g. nephroblastomas, retinoblastomas and neuroblastomas.

Classifications should reconcile the different aims of scientific accuracy, clinical usefulness, logicality and easy reproducibility. Logicality requires the avoidance of arbitrariness and eponyms (personal names) which habit has entrenched, e.g. 'hepatoma' sounds benign but is in fact the name for a malignant liver tumour. Scientific accuracy demands constant revision, e.g. new subclassifications of lymphomas/leukaemias based on enzymes, antigenic surface markers[10] or even nucleic acid content.[11]

Classifiers have been hampered by uncertainty and misinformation, particularly regarding histogenesis. The pleurae give rise to the mesothelioma, but is this an epithelial or connective-tissue tumour? There are several important examples of doubtful histogenesis. All these problems plague the classification of the lymphomas. 'Lymphoma' suggests innocence but they are malignant. The fundamental division into Hodgkin's and non-Hodgkin's lymphomas is eponymous. The cell of origin of Hodgkin's lymphoma may not even be a lymphocyte. New knowledge is spawning new subclassifications[12] and there is much confusion.

If, as seems likely, the oncogene hypothesis is validated, tumours will be subclassified according to which of the various oncogenes or group of oncogenes is activated.

1.9 Grading

The ability to predict the future greatly facilitates appropriate clinical management. The type and site of origin of the tumour, as well as the general nutritional and immune condition of the patient, give us useful prognostic information. Two other factors are important: the extent of the disease (the stage) and the grade.

The tumour grade is a semiquantitative composite of three histological criteria, which taken together give some idea of the degree of malignancy by virtue of their empirical correlation with malignant behaviour. These are the degrees of differentiation and pleomorphism, and the number of mitoses; in other words, how anaplastic a tumour is and how fast it is growing. Grading is usually in terms of low, intermediate or high; more precise schemes have not been workable.

The predictive power of grading is limited by the imperfect correlation of morphology with behaviour and the occurrence of mixed grades in one tumour. However, high-grade neoplasms do tend to be more aggressive; but they also tend to be more sensitive to chemo- and radio-therapy. Overall, stage is more important than grade in prognosis.

1.10 Staging

Cancers start in one place and grow by expansion, infiltration and metastasis. The extent of this process has enormous influence on choice of treatment and prognosis; it is therefore useful to demarcate certain stages of disease progression. This staging information, obtained before, during and after treatment, tells us how well the patient is doing. Responses are classed as progressive disease, no change and partial or complete remission.

Proper management of the individual patient, and progress in cancer research in general, depends absolutely on unambiguous communication. Internationally accepted schemes of grading, classification and staging provide the basis for this.[13] Various staging systems have been proposed, culminating in the 'TNM' system which refers to size of primary tumour (T), involvement of draining lymph nodes (N) and presence or absence of other metastases (M). Older systems, often still used, employ the five stages 0-IV, starting with carcinoma *in situ* and progressing likewise through defined stages of primary tumour size and fixity, lymph nodal status and other metastasis. Stage IV implies inoperability either because of extensive local disease or the presence of metastases. These schemes are applicable to a wide variety of malignancies, with appropriate modifications.

Staging information derives from the bedside physical examination as well as a series of 'special investigations' of varying invasiveness, expense and sophistication. The high priority accorded to developing better ways of attaining this knowledge has resulted in many spectacular bioengineering feats, like ultrasonography, computer-assisted tomography, nuclear magnetic resonance and positron emission tomography (using ultrasound waves, computer integrated X-rays, the magnetic dipole properties of protons and the positron-emitting capacities of certain isotopes respectively).

These have not displaced older investigations like radiographs (X-rays), with or without various radio-opaque dyes which improve delineation and isotope scans with various radiolabelled-pharmaceuticals which may either localise in tumours or show displacement of normal structures by 'space occupying' tumour masses. Surgical biopsy provides staging as well as diagnostic information. In Hodgkin's disease, for example, surgical removal of the spleen is the best way of knowing whether it is affected or not ('staging laparotomy'). These investigations are reflections of reality, not reality itself and therefore subject to distortions both in performance and interpretation (see later this chapter).

Given the inadequacies of systemic therapies (chemotherapy and whole-body irradiation), the major philosophy underlying staging-scheme design concerns the appropriateness and type of a surgical operation. The chief question is whether the disease is localised and thus resectible by knife (or able to be encompassed within one radiotherapy field). We are thus interested in those factors which predict extension to the next stage; e.g. both size and local infiltration of the primary growth predict lymph-node involvement. Likewise multiple, bulky, nodal disease predicts distant organ metastasis.

The traditional basis of surgical and radiotherapeutic planning assumes that the best balance between successful disease eradication and over-radical treatment is achieved by treating for an additional stage beyond what the stage apparently is, i.e. modest over-treatment. Although quite successful, this policy is controversial. The debate over the wisdom of mastectomy in breast cancer, for instance, is understandable only in these terms.

Finally, note that in some cancers like small-cell carcinoma of the bronchus and obviously the leukaemias, precise staging is irrelevant because the cells have spread widely from very early in the natural history of the disease.

1.11 Aetiology, Premalignancy, High Risk and Prevention

Rational treatment suggests removal of known causes. For tumour patients, this is impossible; some causes are transient (e.g. radiation), some unknown and the viruses/oncogenes still beyond our reach. However, awareness of causes does allow prevention, both in the sense of subtracting carcinogens from the environment as well as offering high-risk groups intense surveillance and early treatment. We are all at risk; this is increased by carcinogen exposure (see Chapters 3 and 5), by genetic susceptibility and by having certain diseases which are not themselves neoplastic.

Tumours probably arise unifocally, but carcinogens act over a wide field of cells within which multiple tumours may arise, synchronously or metachronously. Skin cancer from sun exposure is an example; however, the logical step of surgically removing all the at-risk cells is clearly impossible. Next best here is surveillance with early treatment. Genetic predisposition also operates over a wide field of cells. Polyposis coli is an autosomal, dominant, inherited condition in which the colon is covered with multiple small tumorous polyps. These are benign adenomas, but inevitably at least one becomes malignant. The only rational manoeuvre is to remove the whole at-risk field of cells, i.e. perform a total colectomy, before cancer supervenes.

Certain diseases are associated with the subsequent development of tumours. Ulcerative colitis is a non-neoplastic chronic inflammatory disease (of the colon) which is associated with a high risk of cancer in the severely affected. The only sure way to prevent this lies again in removal of the colon, although frequent surveillance suffices in the early stages of the colitis. Cirrhosis of the liver is another disease with a cancer risk, in this case hepatoma. Prophylactic removal of this organ awaits improvement in organ transplantation.

The mechanisms of carcinogenesis in all these cases are obscure. The concept of 'chronic irritation' has little explanatory power. However, it makes the point that any cause of persistent hyperplasia (except physiological) is associated with a risk of neoplastic transformation.

1.12 **Metastasis** (see also Chapter 2)

Invasive local growth, which occurs by infiltration and destruction of surrounding tissue, is probably an essential precondition for metastasis to distant discontinuous sites. Two major surgical principles are thus deducible: resection should be radical, i.e. include a wide margin of ostensibly normal tissue, and early, i.e. before potential metastases have left the primary growth. The most important factor determining curability is whether or not a tumour can be completely cut out. Spread beyond the draining lymph nodes almost always precludes a surgical cure. A further surgical principle is that resection of cancer and draining lymph nodes should be *en bloc* to avoid malignant cells remaining behind.

Many patients referred for curative surgery with apparently localised cancers already have small undetected micrometastases at distant sites; this is the major cause of treatment failure in this group and has prompted two important successful research efforts. The first, 'adjuvant therapy', involves treatment (usually with drugs) to selected patients where micrometastases are suspected but not proven, in addition to surgical excision of the primary.[14] The second concerns the refinement of staging investigations to detect smaller and smaller metastases. With immunohistochemical techniques, isolated malignant epithelial cells can be identified as such, for instance in slides of bone marrow.

Metastasis may occur by several routes: directly in to body cavities followed by widespread seeding on organ surfaces or growth in the fluid form; lymphatic and haematogenous spread by lymph and blood vessels respectively; finally, by direct implantation after artificial interference (e.g. along needle tracts or in surgical scars).

Body cavities include the peritoneum, pleurae, pericardium, subarachnoid space of the central nervous system and bony sinuses. In the first three, fluid effusions often accumulate, in which small 'spheroids' of tumour cells may grow (the 'ascites' form). The compressive effects on lungs and heart of such large volumes of fluid are serious and often fatal.

Once in lymph vessels, tumour cells travel with the drainage to the first set of regional nodes. The extent to which these nodes constitute a barrier or a conduit to further spread is controversial. The nodes may be the site of an immune response to tumour, as well as an albeit imperfect mechanical barrier, and it may be illogical to resect them. Tumour cells can, however, undoubtedly pass on to further lymphatics and connecting blood vessels and it is usual policy to resect the first regional lymph nodes *en bloc* with the tumour.

Tumour cells enter the blood stream either by direct invasion or via lymphatics. The former is favoured by areas of necrosis and bleeding in the primary as well as the effete nature of new tumour vessels; tumours promote development of their own blood supply (neovascularisation), probably by the production of 'tumour angiogenesis factor'. Capillaries and small veins are more susceptible to penetration by tumour than arterioles; this explains why the liver and lungs, which respectively drain the portal and systemic circulations, are the most common sites for secondaries (after the regional lymph nodes).

If tumour cells traverse the lungs, they are then pumped out with the systemic arterial circulation; one could predict the distribution of subsequent metastases to correlate with the cardiac output received by the various organs. This is partly true, except that the heart and skeletal muscles, which receive large proportions of cardiac output, are rarely the sites of metastases. The full explanation of metastatic distribution is not yet known.

The biological properties of the still obscure organ–tumour-cell interaction also influence the distribution of metastases. Much research is devoted to elucidating this 'soil and seed' interaction. Its significance is exemplified by the much higher frequency of lung-cancer metastases in the adrenals than the blood supply would seem to warrant. The brain and bone marrow are also common destinations for blood-borne metastases.

There are, circulating in the blood of cancer patients, considerably more tumour cells than will ever successfully form metastases. This 'metastatic inefficiency' may be due to the mechanical and immunological hostility of the circulation. Also, not every tumour cell may have the ability to implant and certainly not every one is a stem cell. The cells of the primary do not have equal metastatic potential.

Immunological defences include NK (natural killer)-lymphocytes and macrophages; these are important because of the possibility that medical treatment need not eradicate every last cell.

There is much about the process of metastasis that is unpredictable and unexplained. Its pivotal role as the prime determinant of curability and survival, given that most primary growths can now be controlled, justifies the increasingly intense research effort it is receiving (see Chapter 2).

1.13 Effects of Tumour on Host

A large number of tumour cells can exist within the body without producing symptoms. Thus many cancers spread beyond the bounds of resectability, resulting eventually in a 'late' presentation. Also the first symptoms of cancer are often non-specific, e.g. tiredness, and are ignored or explained away, causing further delay. This explains the desire to screen asymptomatic (well) people for 'early' resectable cancer (see Chapter 7).

Tumour effects are of two sorts, local and distant. Local effects, felt in the region of a tumour deposit, result partly from the sheer physical bulk of space-demanding tumour cells as they proliferate. Locally acting chemical substances also play a role which needs further clarification. What may begin in a purely localised way may have 'knock-on' consequences felt more widely, e.g. bowel obstruction by tumour will prevent food intake, with wide-ranging effects. Similarly, urinary-tract obstruction will cause renal failure, a biochemical derangement eventually affecting every organ system. However, these examples are not what is meant by 'distant', as defined above.

Distant effects are mediated partly by biochemicals which are released from the tumour, and spread all over via the bloodstream, acting at vulnerable sites. (A variety of substances escape from normal cells as they are destroyed by the advancing tumour; these harmless tumour-associated derangements must be distinguished from tumour-derived ones.) Tumour cells, being metabolically active, also consume nutrients at the expense of the host and the resulting deprivations may be felt at distant sites. Thus, although the majority of distant effects are thought to be due to tumour-derived chemicals, few are known to be due to consumption of essential nutrients; some are not understood and may be due to either production or consumption (or both).

Obstruction of hollow organ systems has been mentioned; other examples of local effects include destruction of adjacent normal tissue by compression or invasion and the breaching of barriers like skin, vessel wall, gut and various mucous membranes. This breaching provides entry for infectious agents, causes haemorrhage and disrupts the integrity of separate body compartments. These local effects are common and serious.

The many and varied distant effects, which may antedate detection of the actual tumour mass by a long time, have provoked speculation on the biology of tumours. However, of clinical importance is the fact that it is conceptually possible to treat the distant effects with methods that do not involve killing the tumour cells responsible, although it is preferable to remove the tumour cells entirely. Where this is impossible clinicians strive to palliate their effects, using knowledge of the biochemicals involved and how their excesses or deficiencies translate into pathophysiological derangements. This knowledge need not be complete, but generally the more that is known, the more that can be achieved.

Rapidly dividing cells may overload the capacity of the body to excrete their simple waste products. Thus, some leukaemias and lymphomas are associated with elevations of purine-derived uric acid. Particularly dangerous elevations of urate occur after cytotoxic treatment of these drug-sensitive tumours, when many cells die simultaneously and release their intracellular contents. Uric acid may 'clog up' the kidneys, causing renal failure, unless its formation is prevented by the prior administration of the xanthine oxidase inhibitor allopurinol (see also Chapter 11). Potentially lethal elevations of serum potassium may occur, also derived from dying tumour cells; these may cause heart rhythm disturbances. Urate and potassium toxicity are part of the 'tumour lysis syndrome'.

Most simple metabolic upsets are mediated by release of biologically active proteins from live tumour cells. These proteins may resemble naturally occurring hormones like adrenocorticotropic hormone (ACTH), antidiuretic hormone (ADH) and parathormone (PTH) (see Glossary). ACTH-like proteins are often made by small-cell carcinoma of the lung and cause hypokalaemic alkalosis and high blood pressure via stimulation of adrenal corticosteroids. The syndrome of inappropriate ADH production, also associated with small-cell carcinoma, results in an inability to excrete water with resulting haemodilution. The consequent hyponatraemia may be profound and coma-inducing. (This can be treated, e.g., by water restriction.) Hypercalcaemia is to be expected with benign functional parathyroid tumours which secrete, predictably, PTH to excess. Dangerous hypercalcaemia, causing nausea, vomiting, dehydration, polyuria and coma, is common with breast cancer, squamous (epidermoid) lung cancer and some renal cancers. In the last two a PTH-like substance has been proposed and also disputed;[15] in breast cancer an 'osteoclast-activating factor', prostaglandin E_2 and the mechanical pressure of bony metastases have all been implicated. Irrespective of the mechanism, effective treatment is available.

We expect, and find, that tumours of endocrine glands make inappropriate amounts of the hormones normally produced (eutopic production). What needs explanation is why some cancers arising in non-endocrine organs like the lungs should produce protein hormones (ectopic production). Small-cell carcinoma of the lung, a good example, is a veritable factory; it may also manufacture calcitonin and chorionic gonadotrophin. Perhaps small-cell carcinoma originates within a minority population of lung cells which normally produce these substances in low concentrations or perhaps 'derepression' occurs. The existence of these hormone-like proteins, and other products like the oncofetal proteins, have important implications for our understanding of the process of neoplastic transformation.

The terms 'ectopic' and 'inappropriate' are often used synonymously. This is, in my view, a confusing error; 'ectopic' should refer to the site of origin of the hormone and 'inappropriate' to the physiological needs of the body. Furthermore, these ectopic-production syndromes are by no means a feature of every malignancy.

Apart from explaining hitherto mysterious aspects of the clinical picture, products derived from tumour cells, by virtue of their proportionality to tumour load, may serve as indicators of tumour growth or regression in response to treatment (see Chapter 7). In this role they are called tumour markers. Some products are biochemically inactive and without obvious clinical effects; none the less they are still suitable as tumour markers. They can be measured in any convenient body fluid and also on or in the biopsied tumour cells themselves, with the appropriate immunohistochemical or histochemical techniques (see section 1.5). Indeed some substances do not usually diffuse out of the tumour cells to any extent and have to be identified in this way. They are therefore rarely associated with any distant clinical effects.

None of these well-characterised markers are markers of the neoplastic state *per se*: they are normal biochemicals present out of context. Claims for markers of the neoplastic state are now appearing;[16] if verified, they would represent a major advance with profound diagnostic and biological implications.

Many important distant effects are poorly explained. Some, like fever, are not harmful. Others, like anaemia, muscle weakness, various neurological manifestations and immunodepression have sinister implications. Bizarre skin rashes also occur: some of these are mediated by immune complexes, the antigenic component deriving probably from the tumour. Complexes may also localise in the kidney glomerulus producing a massive protein leak with resulting drop in colloid oncotic pressure and gross body swelling from oedema fluid (the nephrotic syndrome).

Cachexia is a common lethal syndrome of progressive weight loss and anorexia, abnormal taste sensation and often anaemia and immunodepression. In some cases it is easily explained by intestinal obstruction and nausea which cancer patients frequently experience for many reasons. However, it may be caused by very small tumours that have no obvious connection with the intake, digestion and absorption of food. Despite the potential importance of this subject, it would be fair to say that we have no clear appreciation of the exact incidence, mechanisms and true significance of cachexia. We do not know with certainty whether the enormously expensive business of parenteral nutrition (intravenous feeding) reverses the malnutrition of cancer or whether it improves survival.

To start with, rapidly growing tumour cells do not handle glucose normally. They are unusually dependent on glucose as an energy substrate but surprisingly metabolise it by glycolysis to lactate at a high rate even in the presence of adequate oxygen (the 'Warburg effect'). This is an extremely inefficient way of making ATP and consequently the cancer cell is in a constant state of 'glucose hunger'. In return for host glucose, the tumour cell gives up the lactate, which it cannot further metabolise. This is reconverted into glucose in the liver by the energy-consuming Cori cycle; the glucose may then return to the tumour to undergo the whole 'futile cycle' again. This process of glucose/energy wastage is probably responsible in part for cancer cachexia (Gold's theory[17]).

Cancer patients probably exhibit increased rates of gluconeogenesis and this explains why they are normoglycaemic in the face of the continuing hypoglycaemic strain imposed by the tumour. Increased supply of gluconeogenic substrates (lactate and certain amino acids) may be responsible for increased gluconeogenesis. The glucose intolerance demonstrated by many cancer patients in the face of a standard tolerance test is probably due to accelerated gluconeogenesis rather than insulin resistance or lack. (Elevated levels of the gluconeogenic hormone cortisol may also play a part.)

It appears that protein metabolism in both tumour and patient is seriously deranged. Protein synthesis in cancer cells is accelerated beyond their capacity to synthesise the necessary amino acids and thus they 'trap' them from the host (particularly glutamine). The commitment to synthesising the nucleic acid and protein components for new chromatin produced at mitosis accounts for part of the increased amino acid consumption by tumours (glutamine, e.g., is a precursor of purine nucleotides).

Alanine, an important gluconeogenic substrate, may be increasingly removed from the blood of cancer patients by their livers to combat possible hypoglycaemia. This would further aggravate the amino acid imbalance that may partly explain the decreased muscle protein synthesis observed in some cancer patients. It is possible that the muscles of these patients exhibit both decreased protein synthesis and increased proteolysis. Stein,[18] noting that muscle protein synthesis is impaired by amino acid imbalance, proposed that the available (but now useless) amino acids are excreted.

The alterations in lipid metabolism are difficult to summarise concisely. Cancer patients do appear to become fat depleted[19] but the postulated tumour-derived lipolytic factor has not been found. There is some evidence that the mobilisation of depot triglycerides occurs,[20] with elevated levels of serum non-esterified (free) fatty acids.[21] These free fatty acids are not used by the tumour as energy substrates to any degree.

The healthy person responds to starvation by the production of ketones, which act as an energy source and inhibit muscle proteolysis, conserving muscle bulk. In muscle, elevation of ketones inhibits oxidation of branched-chain amino acids, a major source of the gluconeogenic substrate alanine. Malnourished cancer patients by contrast, apparently, do not form ketones to the same extent[22] and consequently are unable to prevent the loss of protein from muscle. Furthermore the partial substitution of ketones for glucose as an energy substrate makes it easier for healthy people to successfully avoid hypoglycaemia during starvation. They therefore need to break down less muscle than cancer patients who only maintain normoglycaemia at the expense of their muscle protein. The failure of the liver of patients with cancer to form ketones from the (possibly increased?) free fatty acids available to it may thus be considered an act of sabotage. It is interesting, and perhaps central, to speculate as to how this comes about. The corollary, that infusion of ketones into cancer patients will prevent cachexia, awaits investigation.

The theories proposed here depend on subtle dynamic interactions between tumour and host, rather than the earlier simplistic notions of 'toxohormones' which derive from the tumour and poison the patient. The issue is far from resolved; it is probable that we already have the knowledge and technology to solve this problem, which is one of the least trumpeted but most important challenges to biochemists of the decade.

The existence of so many different forms of cancer, with their differing natural histories, makes precise overall quantification of the statistical cause of death meaningless. Cancer patients usually die with widespread metastatic disease. Infection, cachexia, haemorrhage, compression of vital tissue and therapeutic toxicities are all common and usually coexist. Most of the metabolic upsets listed are fatal if untreated. Destruction of vital tissue by invasion (as opposed to compression) is a less common cause of death and suicide is surprisingly uncommon.

1.14 **Hormone Dependence** (see also Chapter 8)

Manipulation of the patient's hormonal status can alter the natural history of certain cancers. Although good temporary remissions may result, these endocrine manoeuvres are not curative. Partial hormone dependence demonstrates that absolute autonomy is not a universal property of neoplasms.

Generally, these tumours arise in tissues which, in health, are hormone dependent; thus it is a preserved rather than a newly acquired property. Important examples are cancers of the breast, prostate, thyroid and uterus and the lymphomas. The breast cancer–sex steroid relation merits further discussion.

About one-third of patients with breast cancer are hormone-responsive; much research has been concerned with discovering which category of patients (referring to menopausal status) responds to which of the several possible hormonal manipulations. It is possible to antagonise oestrogens, to increase or decrease oestrogen levels and to alter the progestogenic status; the administration of androgens has been superseded, because of virilising side-effects. Pre-menopausal patients tend to respond to oestrogen withdrawal and postmenopausal patients to oestrogen therapy; paradoxically, oestrogen antagonism is also successful in older postmenopausal women.

The discovery of the oestrogen receptor (ER), a protein found in breast epithelium, has partly rationalised the selection of patients for hormone therapy. Oestrogen, entering the cell, binds to ER and the combination interacts with chromatin to influence gene expression. (The exact intracellular loaction of ER is in doubt.[23]) It is now possible, and should be routine, to measure ER presence/absence in breast-cancer biopsies; the recent availability of monoclonal antibodies to ER enhances our ability to measure and study this receptor.[24] ER + (positivity) is the best predictor of a response to hormonal manipulation, though menopausal status, not ER status will determine which particular manoeuvre should be used.

Just over half of patients with breast cancer are ER +; almost two-thirds of these will respond to some endocrine manipulation. Less than 10 per cent of ER − cases will respond. The higher the titre of ER positivity, the more likely a response. ER tumours grow more slowly and these patients survive longer. Metastases are the same ER status as their primaries in 80 per cent of cases. The minor discrepancy is further evidence of clonal variation in tumours. Response of metastases, the major purpose of endocrine therapy, is site-dependent, being better in lung, lymph nodes, bone and soft tissue than brain and liver. Assay of progesterone receptor (an oestrogen-dependent protein) may lead to further improvements.

Despite the receptor concept, we still lack a full explanation for the successes and failures of endocrine therapy and our predictions for benefit are imperfect. But as hormonal treatments are much better tolerated than the cytotoxic alternatives, this is not a great problem.

Tablets easily achieve increased oestrogen levels; oestrogen withdrawal is obtained by surgical removal or irradiation of the ovaries. The adrenals are indirectly a source of oestrogen, because the adrenal steroid androstenedione is converted in peripheral tissues into oestrogen by the enzyme aromatase. This is the chief source of oestrogen in postmenopausal women, in whom removal of the adrenals (or controlling pituitary) was the only way to obtain oestrogen withdrawal. The drug aminoglutethimide now achieves this by impairing both production of androstenedione in the adrenal (by blocking the cholesterol to pregnenolone step) and its peripheral aromatisation to oestrogen (the 'medical adrenalectomy'). Aminoglutethimide binds to cytochrome P-450, an essential part of these enzyme complexes.

Of note is that aromatisation of androstenedione may well occur within breast cancer itself; given that very small amounts of oestrogen may be sufficient to drive the breast cancer, it is vital that improved aromatase inhibitors are developed. The agent 4-hydroxyandrostenedione is currently under evaluation in this role.

Tamoxifen is the least toxic of the available oestrogen antagonists; these bind to ER and displace oestrogen. The extent to which tamoxifen is a pure antagonist or a partial agonist is uncertain. A widely used progestogen is medroxyprogesterone acetate. A significant minority of patients benefit from medroxyprogesterone acetate, but by which of the many possible mechanisms of action is unknown. Medroxyprogesterone acetate interacts with ER, as well as progesterone, androgen and glucocorticoid receptors. Gonadotropins are inhibited and there may be a direct cytotoxic effect at high dose. Danazol, a synthetic steroid, also inhibits gonadotropins and is used as a second- or third-line agent. However, it must be re-emphasised that our understanding of these issues will remain inadequate until the basic molecular events are revealed. Most hormonal therapies probably exploit physiological growth-control mechanisms rather than non-specific cytotoxicity.

The lympholytic affect of glucocorticoids ensures their prominent place in combination protocols for leukaemia, lymphoma and myeloma. Prostatic cancer, being androgen-dependent, responds to castration and/or oestrogen therapy. Continuous luteinising-hormone releasing factor administration may prove more acceptable than surgical castration. A subset of thyroid cancers respond to measures which result in thyroid-stimulating hormone suppression.

Significantly, hormone therapy provides a way of controlling cancer that does not depend on surgery, irradiation or non-specific poisoning. Some believe that the best hope for cancer treatment lies in the generalisation of the principles and practice of hormone therapy. The putative association of some oncogene products with growth factor/receptor mechanisms[25-27] has added credibility to this view.

1.15 Problem Areas in Management

The major reasons for our frequent failure to cure cancer are, first, that we diagnose it after it has spread, secondly, our drugs are too toxic to normal tissues and, thirdly, we lack the means of changing the behaviour of cancer cells, e.g. by inducing differentiation and thus 'taming' them; we do not, in general, fail because of major diagnostic errors or because we implement the wrong treatment. However, the potential dangers of misinformation and uncertainty are ever present, affect everyone concerned with the cancer problem and must not be underestimated.

An optimal combination of diagnostic information and therapy may achieve the clinician's tasks (ideally, to cure the patient without toxicity, discomfort or unbearable expense; to relieve suffering, to allay anxiety).

Completely effective non-toxic treatment requires little information; unfortunately, maximum high-quality information is needed to maximise the effectiveness of our very imperfect therapies.

Complete diagnosis requires the identification of the neoplastic state, the degree of malignancy, and the tissue/organ of origin; description of the extent of spread and prediction of the timing/nature of future events are also necessary. Some of the uncertainty, attendant at each step, may be reduced by research, but some is inherent.

While tumour markers provide some diagnostic information, the definitive diagnosis is nearly always histological, i.e. the light microscopy of a biopsy. But there is no guarantee that these samples are truly representative of the tumour. Indeed 'blind' biopsies of organs like liver may miss the neoplastic zones altogether (false negative). Considering that natural history may depend on a small percentage of critically different cells, these sampling errors assume significance.

Of the many features characterising cancer histology, none is constantly present and all are occasionally seen in other conditions, i.e. no feature, by itself, is necessary or sufficient. Severe uncertainty may result.

The goal of cancer surgeons is to excise the whole tumour; chemo- and radio-therapists try to kill or sterilise every tumour cell. Unfortunately it is currently impossible to prove that these aims have been achieved. Given that the absence of cancer can not and may never be amenable to proof, the concept of cure needs to be understood as a statistical phenomenon and oncologists prefer to talk about disease-free survival without persisting treatment toxicity.

Other fundamental problems, relating to the imperfect microscopic morphology/behaviour correlations have been alluded to. Granted, certain morphological features imply previous neoplastic behaviour. Tumorous distortions of architecture imply proliferation, mitoses imply that more would have occurred. Misplaced cells suggest spread. But the criteria used to define cancer are behavioural and the diagnostic criteria are morphologic; their equation with each other is more empirical than rational. However good the correlation, atypical features give no clue as to how malignant behaviour is mediated. Malignancy must have an explanation, which will ultimately be found on and around cancer-cell surfaces[28] (see Chapter 2). Within this biochemical milieu exist enzymes (assisting invasiveness), angiogenesis factor (promoting vascularisation), etc. Identification of these critical substances with new technologies (e.g. immunocytochemistry/enzyme histochemistry/hybridisation *in situ*) will give us mechanistic insight into what the tumour is capable of, providing an escape from the weakness of empiricism.

The prediction of drug sensitivities from histology faces exactly the same problems which urgently need solution. Because drug resistance is mediated by identifiable proteins, the new techniques just mentioned hold much promise (see Chapter 11).

Specificity and sensitivity are two concepts necessary for the formal consideration of the uncertainties of the microscope and staging investigations.[29] Sensitivity (the 'pick-up' rate) is the proportion of disease sufferers who are correctly identified as such; it is the true positive rate and is corrupted by false negatives. Specificity is defined as the proportion of people without the disease who are correctly identified as such, i.e. it is the true negative rate and is corrupted by false positives. Specificity is the measure of exclusivity of the relation between a particular disease and a particular parameter. The chest X-ray illustrates these concepts: it may be 'normal' in a patient with lung cancer because the tumour is too small to cast an identifiable shadow; this false negative is a problem of sensitivity. Non-specificity is illustrated by the fact that a definite abnormal shadow on a chest X-ray can be due to cancer, pneumonia or many other diseases. Equating every chest X-ray shadow with lung cancer will result in many false positives.

False negative and positive rates, expressed numerically, quantitate uncertainty and allow rational planning: because we know that some pretreatment evaluations are going to be incorrect we can devise tactics to achieve the best compromise between the conflicting interests of the correctly and incorrectly evaluated. For uncertainty may lead to over- or under-diagnosis; the consequences (excessive or inadequate treatment) do not have to be spelled out but do have to be paid for in human terms and financial ones which are so enormous, even for effective and appropriate treatment, that they need to be borne by the taxpayer.

We are all force-fed a diet of 'breakthroughs' by the media; undiscriminating readers may be surprised that many practising doctors believe that nothing curative can be done for most cancer patients. They believe that, if a tumour is excised and the patient survives normally, then that tumour never had malignant potential anyway, irrespective of the histology, or that if a tumour is truly malignant, that it is impossible to detect it and cut it out before spread.[30] This is therapeutic nihilism and it is based on many disappointments and a realistic appreciation of the uncertainty problem. The only weapon against it is the controlled trial, with high-quality information, and conducted by those who do not have to purchase their subsidies with 'positive' results. It can not be overemphasised that control patients should be strictly comparable; 'historical' controls in particular are a potent source of confusion.

1.16 On and Over the Horizon

Many medical advances have sprung from ostensibly unrelated projects.

This makes predicting the future hazardous. Few, for example, could have predicted that the work which led to the discovery of the 'Australia antigen' would lead to a vaccine that may well prevent liver cancer in Africa and Asia.[31] However, certain endeavours have attained such momentum that results can reasonably be expected soon: the most important are our investigations into the basis of transformation (see Chapters 4 and 5) and the chemical mediation of malignant behaviour, especially metastasis (see Chapter 2). The availability of large amounts of pure monoclonal antibodies, coupled with better understanding of cell surfaces, will allow us to 'target' more selectively on cancer cells. The antibodies could themselves be cytotoxic, or they could carry various 'warheads', radiolabels for localisation (see Chapter 7) or substances that could modify cell behaviour (see Chapter 11). Besides antibody, the hybridoma technique can be modified to produce pure unlimited quantities of immunoregulatory molecules (like lymphokines); these will enable us to dissect and hopefully improve on the immune respone to tumours by, for example, enhancing tumour-cell antigenicity, inhibiting suppressor cells and amplifying defence systems (macrophages/NK-lymphocytes). Other 'biological modifiers' like interferon and thymosin are relevant here.

Modulating the differentiation status of cancer cells, a major new line of potential therapy, has already been achieved *in vitro* by using various chemicals, e.g. phorbol esters and dimethylsulphoxide. Understanding the very complex nature of tumour-cell phenotypic heterogeneity will be essential for this. Beneficial interactions between conventional treatment and differentiation inducers are beginning to emerge. Also, the significance of growth factors and their receptors is increasing; some of these may be made by the tumour cells themselves (autostimulation). The maintenance of the malignant phenotype may depend on their constant production; there are many diagnostic and therapeutic opportunities here.

Flow cytometry is a versatile technique which, apart from improving some of the diagnostic limitations of light microscopy, is able to 'sort' and separate living cells according to various criteria. It provides a rapid, automated, objective, quantitative determination of cell size, nucleic acid content (including ploidy) and surface antigenicity. Nuclear magnetic resonance can be used to study metabolism within living intact host–tumour systems. The problems of organ transplantation are gradually yielding; bone-marrow grafting, already feasible, will expand. Auto-transplantation on one's own marrow, 'laundered' of tumour cells by monoclonal antibodies (if necessary), will permit much higher doses of cytotoxic treatment. New antibiotics, especially anti-viral and antifungal agents, will be critically important.

The modification of existing drugs, and new combinations of them, should produce modest benefit at most. This explains the interest in 'targeting', in radically new drugs and above all in 'biological' therapies, the ultimate of which, gene therapy, is still at the conceptual stage. Gene therapy will require much improved 'targeting', as well as a much better understanding of the workings of the mammalian genome. Finally, the elucidation of the nature and function of oncogene translational products is a matter of the utmost interest and importance; initial therapeutic strategies will probably be directed at these proteins rather than their nucleic acid templates (see Chapter 4).

1.17 Conclusion

Long-term survival, even to a normal life-span, is now a realistic expectation for patients with some embryonal cancers, testicular cancers, malignant gestational tumours (choriocarcinomas) and some forms of leukaemia and lymphoma. It is especially gratifying that most of these patients are young. However, enthusiasm for this considerable achievement is tempered by the significant toxic aftermath, by the inability to guarantee a cure and by the knowledge that the tumours listed above are relatively uncommon.

It has not been possible to extend these benefits to the common killers, exemplified by cancers of the lung, breast, colorectum, pancreas and stomach. Certain generalisations can be made: first, their victims are by no means always elderly; secondly, many, perhaps most, are beyond curative surgery at presentation, even though this may only be recognised in retrospect (with lung, pancreatic and stomach cancers, survival to five years even of patients with ostensibly resectable tumours is uncommon; breast and colorectal tumours are better in this respect); thirdly, the survival of all but a few patients recognised to be surgically incurable is at most a few years and is more commonly measured in months or weeks despite the best available treatment. Thus there is little or no diminution in the appalling toll of the common cancers and for many patients the best we can offer them is an experiment.

I therefore make no apology for emphasising present inadequacies. The nature of the problem is such that we cannot be less than perfectionist. There are many human reasons for investing in fantasy, but students should reconcile the highest standards of objectivity with their creative impulses. Only in this way can the genuine advances be made and recognised.

References

1. Kardinal, C. & Yarbro, J. (1970) A conceptual history of cancer. *Semin. Oncol.*, *6*, 396-408
2. Currie, G. & Currie, A. (1982) *Cancer: the Biology of Malignant Disease*, Edward Arnold, London, p. 97
3. Willis, R. (1967) *Pathology of Tumours*, 4th edn, Butterworths, London, p. 1
4. Ibid., pp. 4, 8, 22
5. Bollag, W. (1983) Vitamin A and retinoids: from nutrition to pharmacotherapy in dermatology and oncology. *Lancet*, *i*, 860-3
6. Editorial (1984) Molecular biology and lymphoma. *Lancet*, *i*, 26
7. Canaani, E., *et al.* (1984) Altered transcription of an oncogene in chronic myeloid leukaemia. *Lancet*, *i*, 593-5
8. Woodruff, M. (1983) Cellular heterogeneity in tumours. *Br. J. Cancer*, *47*, 589-94
9. Everson, T. & Cole, W. (1966) *Spontaneous Regression of Cancer*. W.B. Saunders, Philadelphia
10. Hoffbrand, A. & Janossy, G. (1981) Enzyme and membrane markers in leukaemia: recent developments. *J. Clin. Pathol.*, *34*, 254-62
11. Birnie, G. *et al.* (1983) A new approach to the classification of human leukaemias: measurement of the relative abundance of a specific RNA sequence by means of molecular hybridization. *Lancet*, *i*, 197-200
12. Non-Hodgkins Pathologic Classification Project (1982) National Cancer Institute sponsored study of classification of non-Hodgkin's lymphoma. *Cancer*, *47*, 2112-35
13. Beahrs, O. & Meyers, M. (eds) (1983) *Manual for Staging of Cancer*, 2nd edn, J.B. Lippincott, Philadelphia
14. Martin, D. (1981) The scientific basis for adjuvant chemotherapy. *Cancer Treat. Rev.*, *8*, 169-89
15. Simpson, E., *et al.* (1983) Absence of parathyroid messenger RNA in nonparathyroid tumours associated with hypercalcaemia. *NEJM*, *309*, 325-30
16. Editorial (1982) Oxford's tumour marker. *Lancet*, *ii*, 25-6
17. Gold, J. (1974) Cancer cachexia and gluconeogenesis. *Ann. N.Y. Acad. Sci.*, *230*, 103-10
18. Stein, T. (1978) Cachexia gluconeogenesis and progressive weight loss in cancer patients. *J. Theor. Biol.*, *73*, 51-9
19. Costa, G., *et al.* (1965) Changes in the composition of human muscle during the growth of malignant tumours. *Proc. Am. Assoc. Cancer Res.*, *6*, 12
20. Spector, A. (1975) Fatty acid metabolism in tumours. *Prog. Biochem. Pharmacol.*, *10*, 42-75
21. Carter, A., *et al.* (1975) Metabolic parameters in women with metastatic breast cancer. *J. Clin. Endocrinol.*, *40*, 260-4
22. Conyers, R., *et al.* (1979) Nutrition and cancer. *Br. Med. J.*, *i*, 1146
23. King, W. & Greene, G. (1984) Monoclonal antibodies localise oestrogen receptor in the nuclei of target cells. *Nature (Lond.)*, *307*, 745-7
24. Greene, G.L. & Jensen, E.V. (1982) Monoclonal antibodies as probes for estrogen receptor detection and characterization. *J. Steroid Biochem.*, *16*, 353-9
25. Downward, J., *et al.* (1984) Close similarity of epidermal growth factor receptor and v-erb B oncogene protein sequences. *Nature (Lond.)*, *307*, 521-7
26. Waterfield, M., *et al.* (1983) Platelet-derived growth factor is structurally related to the putative transforming protein $p28^{sis}$ of simian sarcoma virus. *Nature (Lond.)*, *304*, 35-9
27. Cooper, G. & Finkel, T. (1984) Detection of a molecular complex between ras proteins and transferrin receptor. *Cell*, *136*, 1115-21
28. Nicolson, G. (1984) Cell surface molecules and tumor metastasis. *Exp. Cell Res.*, *150*, 3-22
29. Miller, A., *et al.* (1982) In R. Flamant & C. Fohanno (eds) *Evaluation of Methods of Treatment and Diagnostic Procedures in Cancer*, U.I.C.C. Technical Report Series, vol. 70, Geneva, pp. 101-9
30. Hutton, R. (1981) Is cured early cancer truly cancer? *Cancer*, *47*, 1215-20

31. Arthur, M., *et al.* (1984) Hepatitis B, hepatocellular carcinoma, and strategies for prevention. *Lancet*, *i*, 607-18

Further Reading

General Oncology

Halnan, K. (ed.) (1982) *Treatment of Cancer*, Chapman and Hall, London

De Vita, V., Hellman, S. & Rosenberg, S. (eds) (1982) *Cancer. Principles and Practice of Oncology*, J.B. Lippincott, Philadelphia

U.I.C.C. (1978) *Clinical Oncology. A Manual for Students and Doctors*, Springer-Verlag, Berlin

Pathology

Walter, J. & Israel, M. (1979) *General Pathology*, Churchill Livingstone, Edinburgh

Robbins, S. & Cotran, S. (1984) *Pathologic Bases of Disease*, W.B. Saunders, Philadelphia

Histochemistry

Filipe, M. & Lake, B. (eds) (1983) *Histochemistry in Pathology*, Churchill Livingstone, Edinburgh

Bancroft, J. & Stevens, A. (eds) (1982) *Theory and Practice of Histological Techniques*, 2nd edn, Churchill Livingstone, Edinburgh

Metastasis

Liotta, L. & Hart, I. (eds) (1982) *Tumour Invasion and Metastasis*, Martinus Nijhoff, The Hague

Carter, R. (1982) Some aspects of the metastatic process. *J. Clin. Pathol.*, *35*, 1041-9

Staging

American Joint Committee on Cancer (1983) *Manual for Staging of Cancer*, 2nd edn, J.B. Lippincott, Philadelphia

Clonality

Woodruff, M. (1982) Interaction of cancer and host. *Br. J. Cancer*, *46*, 313-22

Kinetics

Steel, G. (1977) *Growth Kinetics of Tumours*, Clarendon Press, Oxford

Conceptual Problems of Diagnosis

Wallace Park, W. (1980) *The Histology of Borderline Cancer, with Notes on Prognosis*, Springer-Verlag, Berlin

Cancer Biology

Cameron, I. & Pool, T. (eds) (1981) *The Transformed Cell*, Academic Press, London

Currie, G. & Currie, A. (1982) *Cancer: the Biology of Malignant Disease*, Edward Arnold, London

Ruddon, R. (1981) *Cancer Biology*, Oxford University Press, New York

Cachexia

Editorial (1984) Cancer cachexia. *Lancet*, *i*, 833-4

Levin, L. & Gevers, W. (1981) Metabolic alterations in cancer. *S. Afr. Med. J.*, *59*, 518-21/533-56

Racker, E. & Spector, M. (1981) Warburg effect revisited: merger of biochemistry and molecular biology. *Science*, *213*, 303-7

Saudek, C. & Telig, P. (1976) The metabolic events of starvation. *Am. J. Med.*, *60*, 117-26

Symreng, T., *et al.* (1983) Nutritional assessment reflects muscle energy metabolism in gastric carcinoma. *Ann. Surg.*, *198*, 146-50

Ectopic Hormone Production

Ratcliffe, J. (1982) Ectopic production of hormones in malignant disease. In J. O'Riordan (ed.) *Recent Advances in Endocrinology and Metabolism*, vol. 2, Churchill Livingstone, Edinburgh, pp. 187-209

Hormone Therapy

Lippman, M. & Eil, C. (1982) Steroid therapy in cancer. In B. Chabner (ed.) *Pharmacologic Principles of Cancer Treatment*, W.B. Saunders, Philadelphia

Sem. Oncol., *10*, No. 4, Suppl. 4 (1983)

Stoll, B. (ed.) (1982) *New aspects of breast cancer*, vol. 5, *Endocrine Relationships in Breast Cancer*. William Heinemann Medical Books, London

Nuclear Magnetic Resonance

Gadian, D. (1982) *Nuclear Magnetic Resonance and its Applications to Living Systems*, Oxford University Press, New York

Flow Cytometry

Diamond, L., *et al.* (1981) Flow cytometry in the diagnosis and classification of malignant lymphoma and leukaemia. *Cancer*, *50*, 1122-35

Melamed, M., *et al.* (eds) (1979) *Flow Cytometry and Sorting*. John Wiley and Sons, New York

Monoclonal Antibodies

Greaves, M., *et al.* (1982) Analysis of leukaemic cells with monoclonal antibodies. In A. McMichael & J. Fabre (eds) *Monoclonal Antibodies in Clinical Medicine*, Academic Press, London

Targeting

Gregoriadis, G. (1981) Targeting of drugs: implications in medicine. *Lancet*, *ii*, 241-7

Biological Modification

Alexander, P. (1982) Need for new approaches to the treatment of patients in clinical remission, with special reference to acute myeloid leukaemia. *Br. J. Cancer*, *46*, 151-9
Editorial (1983) Reversal of cancer. *Lancet*, *i*, 799-800
Markert, C. (1968) Neoplasia: a disease of cell differentiation. *Cancer Res.*, *23*, 1908-14
Moore, M. (ed.) (1983) *Maturation Factors and Cancer: Progress in Cancer Research and Therapy*, vol. 23, Raven Press, New York
Van R. Potter, (1978) Phenotypic diversity in experimental hepatomas: the concept of partially blocked ontogeny. *Br. J. Cancer*, *38*, 1–23

Genetic Aspect

Bishop, J.M. (1983) Cellular oncogenes and retroviruses. *Ann. Rev. Biochem.*, *52*, 301-54
Editorial (1981) Gene therapy: how ripe the time? *Lancet*, *i*, 196-7
Hamlyn, P. & Sikora, K. (1983) Oncogenes. *Lancet*, *ii*, 326-9
Marx, J. (1984) What do oncogenes do? *Science*, *233*, 673-6
Marx, J. (1984) Oncogenes linked to growth factor receptor. *Science*, *233*, 806
Ponder, B. (1980) Genetics and cancer. *Biochim. Biophys. Acta, 605*, 369-410
Weinburg, R. (1983) A molecular basis of cancer. *Sci. Am.*, *249*, 102-16

History of Cancer

Rather, L. (1978) *The Genesis of Cancer. A Study in the History of Ideas*, Johns Hopkins University Press, London and Baltimore
Sem. Oncol., *6*, No. 4 (1979)

2 THE CELL-SURFACE MEMBRANE IN MALIGNANCY

J.T. Gallagher

Contents

2.1 Why Study Cell Surfaces in Malignant Disease?

The cellular micro-environment plays a central role in regulating the growth and development of normal cells. A cell will interact with adjacent cells and with structural components of the extracellular matrix via the external surface of its plasma membrane. Although we still know very little about the mechanisms of cellular interactions, it is well established that cell development requires that the surface membrane should be capable of receiving and transmitting regulatory signals from the micro-environment. Cancer is a disease in which abnormalities of both cell growth and cell development are found. Tumour cells are capable of indefinite proliferation. Some specialist non-malignant cells have similar proliferative potential but tumour cells are able to overcome some of the physiological and mechanical constraints that restrict the population growth and territorial expansion of normal cells. In addition, tumour cells are frequently immature and lack the functional attributes of normal cell counterparts. It is therefore possible that cell-surface abnormalities are important factors in determining the characteristic features of malignant disease. If these abnormalities can be identified and related to specific behavioural properties they would throw some light on the molecular mechanisms of cancer and perhaps also represent target sites for chemo- or even immuno-therapy.

2.2 Structure of Biological Membranes

2.2.1 General Features

Membranes are supramolecular structures which form a series of enclosed compartments in which particular biochemical activities are confined. Lipids and proteins are the major membrane constituents together with a small but significant amount of carbohydrate. Membranes are formed from phospholipid bilayers in which the hydrophobic fatty acyl chains of the amphipathic phospholipids (Figure 2.1) are oriented toward the interior of the membrane with the polar 'head' groups at the membrane surface in contact with water (Figure 2.2). The bilayer is stable in an

Figure 2.1 Structure of Phospholipids

Components of phospholipids:

Nitrogen base
|
Phosphate — Polar head
|
Glycerol
|
$(\text{Fatty acid})_2$ — Hydrophobic tail

Phospholipids are amphipathic since they combine polar (hydrophilic) and non-polar (hydrophobic) substituents in the same molecule

The nitrogen base may be choline $HOCH_2\text{-}CH_2N^+(CH_3)_3$, ethanolamine $HOCH_2CH_2NH_3^+$ or serine $HOCH_2\text{-}CH_2COO^-.NH_3^+$. They are bound in ester links through the free OH group to phosphate. The nitrogen base plus phosphate forms the polar head of the phospholipid. The phosphate group and two fatty acids are linked to glycerol. The fatty acids form the hydrophobic tail. Usually, one fatty acid is saturated and the other is unsaturated (i.e. contains one or more double bonds). A typical structure for the phospholipid, phosphatidylcholine, is illustrated below:

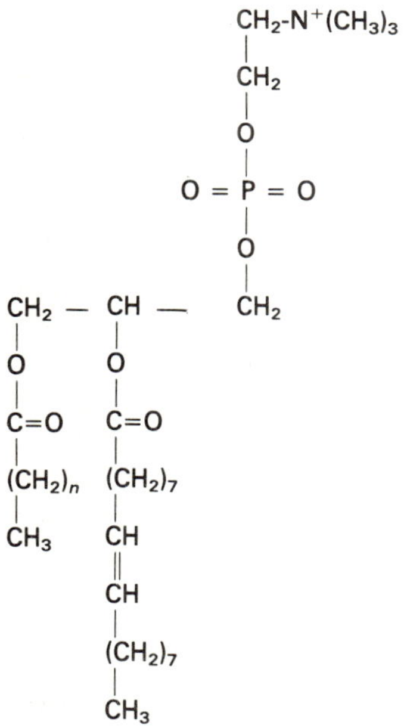

In the fatty acid chains the value of *n* can vary from 14 to 24. The unsaturated fatty acid shown is oleic acid with just one double bond. Unsaturated fatty acids can contain up to four double bonds

aqueous environment because the phospholipids are in their most favourable conformation and phospholipids in solution will form bilayers spontaneously. Despite their stability membranes are 'fluid', the hydrophobic regions having a viscosity comparable with a light oil. Fluidity can be correlated with the degree of unsaturation of the fatty acid chains and

Figure 2.2 Phospholipid Bilayer of Biological Membranes with Transmembrane Glycoproteins

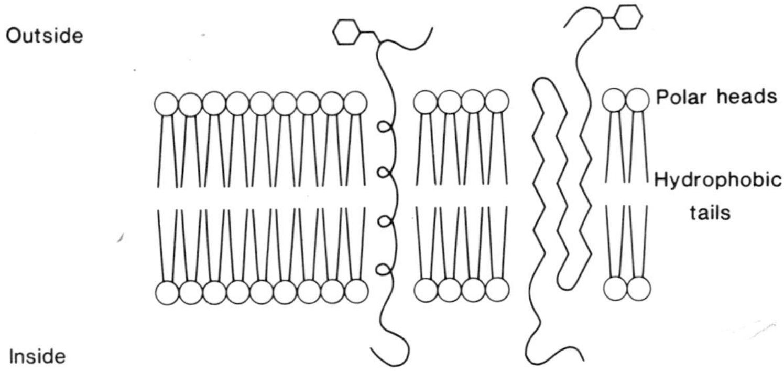

Proteins which span the phospholipid bilayer adopt regular conformations of either an α-helix (left) or β-sheet (right) within the hydrophobic domain. Both structures are stabilised in the bilayer of hydrogen bonding between the NH and CO groups of the amide bonds between the amino acids. The side-chain substituents of the amino acids in the bilayer spanning domain of the polypeptide are hydrophobic.

also with the cholesterol content which is high in cell-surface membranes. When the idea of membrane fluidity was first proposed by Singer and Nicolson in 1972[1] it had a major impact on our general concepts of membrane structure and function. Although the lipid bilayer had been recognised for many years, membranes were believed to be rigid structures with proteins restricted to the exterior or the bilayer. Together with the fluidity concept, Singer and Nicholson also introduced the notion of integral or transmembrane proteins. Such proteins can be inserted into and across the hydrophobic domain to provide a direct means of communication between the inner and outer region of membrane-enclosed compartments (Figure 2.2). The largest of these compartments is, of course, the cell itself and the limiting membrane, the plasma membrane, is a typical 'mosaic' of lipids and proteins which embodies both general features common to other membrane systems and also specific properties which reflect the highly specialised biological and biochemical activities that are restricted to the cell surface. Paramount among these are the transport of ions and nutrients, binding of extrinsic effectors such as hormones, growth factors and neurotransmitters and as already discussed the mediation of information exchange between cells and their micro-environment. Cell surfaces undergo considerable changes in composition and function during the growth and differentiation of normal cells and this is an important consideration when attempting to define surface abnormalities in malignancy (see section 2.4).

2.2.2 *Asymmetry of Plasma Membrane*

In cell-surface membranes the carbohydrate residues are located exclusively on the external (transcytoplasmic) surface. Most of the integral membrane proteins are in fact glycoproteins and contain a rich variety of covalently bound oligosaccharides which represent the major groups of membrane carbohydrates. The remaining carbohydrate is linked to fatty acyl-sphingosine (ceramide) to form membrane glycolipids (Figure 2.3).

Figure 2.3 Glycolipids

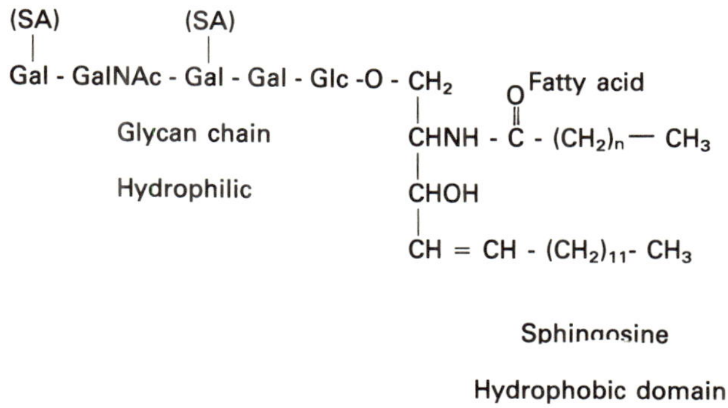

The structure formed by combining a fatty acid with sphingosine is called ceramide. Glycolipids are glycosylated derivatives of ceramide. Glycolipids are called gangliosides when they contain sialic acid. The sugar chains can vary enormously in length, composition and nature of the linkage between the sugars. Cell transformation is accompanied by a reduction in the concentration of the more complex gangliosides (i.e. more highly sialyated structures) and a selective increase in neutral glycolipids with short saccharide chains. There is an extensive body of literature on this subject and detailed reviews are available.[2,3]

2.2.3 *Plasma Membrane Glycoproteins*

An enormous range of structurally and functionally diverse glycoproteins are found in plasma membranes (Figure 2.4). Some are cell-type specific (e.g. immunoglobulins on B-lymphocytes), others, such as the Na^+, K^+-activated ATPase essential to the normal ion balance of the cell, are found in all plasma membranes. Membrane glycoproteins commonly occur as dimers or oligomers in which the subunits may be covalently linked by disulphide bonds or tightly associated by non-covalent interactions. The transferrin receptor is a disulphide-linked dimer made up of very similar glycoprotein subunits, whereas the HLA-DR antigens (also called class II or immune-response antigens) consist of two dissimilar non-covalently linked glycoproteins (HLA = histocompatibility antigen). Peripheral membrane proteins and glycoproteins are sometimes found in combination with integral constituents: notable examples are membrane immunoglobulin light chains disulphide-linked to integral

Figure 2.4 Glycoproteins of the Plasma Membrane: Macromolecular Associations and Interactions with the Lipid Bilayer

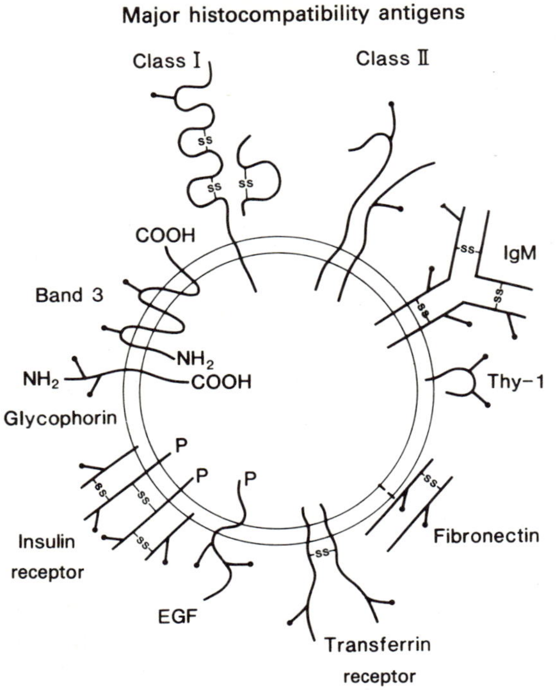

Plasma membrane glycoproteins are a heterogeneous group of macromolecules which vary in size, type and degree of glycosylation, subunit organisation and mechanism of interaction with the lipid bilayer. Transmembrane glycoproteins are usually orientated with the amino (NH_2) terminus on the outside of the cell and the carboxyl terminus on the inside. Glycoproteins normally span the bilayer once only but Band 3 glycoprotein of erythrocyte membranes traverses the bilayer several times with the carboxyl and amino ends of the polypeptide in reverse orientation to most other membrane glycoproteins. Inter- and intra-molecular disulphide bonds are essential to the structural integrity of many plasma-membrane glycoproteins. Fibronectin is a peripheral membrane glycoprotein that interacts with a specific, but unidentified, membrane receptor. Sugar residues are shown as —•

heavy chains and β_2-microglobulin which is non-covalently bound to the structurally polymorphic glycoproteins of the HLA, Band C histocompatibility complex (class I antigens; Figure 2.4). The Thy-1 glycoprotein antigen on T-cells is inserted only a small distance into the lipid bilayer and shows no apparent affiliations with other membrane glycoproteins. Membrane receptors for insulin and epidermal growth factor are glycoprotein kinases. Binding of the ligand stimulates kinase activity which resides in the cytoplasmic region of the receptor. Protein phosphorylation is the indicator of ligand binding and elicits further intracellular responses. These receptor molecules phosphorylate tyrosine residues in both themselves and other proteins. This is a particularly important area of contemporary cancer research since several viral oncogenes code for tyrosine kinases (see Chapter 4) which may have some structural homologies to membrane receptors.

In relation to what follows later on membrane-related changes in malignancy it must be emphasised that no causally related alterations in amino acid sequence have been found in tumour membrane glycoproteins and that the subunit interactions of these components seem, in general, to conform to the normal pattern. Changes that do occur with some consistency have been found only in the sugar residues but many of these are not associated exclusively with the malignant phenotype. Data interpretation requires careful consideration of the structure and synthesis of membrane carbohydrates of the cell systems under study, the validity of the normal cells chosen as controls and some understanding of the origin and progression of cancerous conditions.

2.3 Structure and Biosynthesis of Membrane Carbohydrates

2.3.1 Membrane Glycoprotein Oligosaccharides

To appreciate the nature and significance of surface oligosaccharide changes in tumour cells it is essential to discuss first some basic aspects of the formation of sugar chains and their intrinsic structural heterogeneity. In membrane glycoproteins three classes of carbohydrates can be recognised on the basis of the sugar–protein link, as follows:

(1) '*N*-linked' sequences in which an *N*-acetylglucosamine (GlcNAc) residue is linked to the side-chain amino (NH_2) group of asparagine. This is a common linkage in both membrane and serum glycoproteins.

(2) '*O*-linked' sequences in which *N*-acetylgalactosamine (GalNAc) is linked to the side-chain hydroxyl (OH) group of serine (Ser). This linkage is also found in the secretory mucous glycoproteins and is sometimes referred to as a mucin-type linkage.

(3) '*O*-linked' in which a xylose residue is bound to serine. Large polysaccharides such as chondroitin sulphate and heparan sulphate, collectively described as glycosaminoglycans (Gags), are joined to protein through this type of linkage (see later).

Sugar sequences found in membrane glycoproteins depend on the specificities and co-ordinated activities of glycosyltransferases which transfer single sugars from sugar nucleotide precursors to growing carbohydrate chains.[4] There is no known template for oligosaccharide synthesis comparable with mRNA for the synthesis of proteins. Sugar chains may be built up directly on the protein or initial assembly may occur on a lipid carrier.

2.3.1.1 N-*linked Saccharides* (Figure 2.5). These structures are first assembled on a lipid carrier, dolichol phosphate, as a branched dolichol oligomannose which contains two *N*-acetylglucosamine residues in the linkage region with dolichol and three glucose units in the periphery of one of the branches or 'antennae'. This 'high-mannose' oligosaccharide is then transferred to the apoprotein in the rough endoplasmic reticulum. Transfer is cotranslational (i.e. transfer before completion of protein synthesis) and so the glycan may influence the folding of the completed protein. Subsequently the glucose residues are removed and the carbohydrate chain may then follow one of three pathways: (a) remain unchanged; (b) lose the outer-chain mannose branches and gain new branches composed of *N*-acetylglucosamine, galactose (Gal) and sialic acid (SA) to form the complex *N*-linked oligosaccharides; (c) certain mannose units (Man) may become phosphorylated (this step is believed to be important in the transfer of glycoprotein enzymes to lysosomes).

These steps are summarised in Figure 2.5. In the processed *N*-linked sequences the level of complexity can vary enormously.[5] The simplest of the complex glycans is a 'biantennate' structure which may contain a fucose residue in the linkage region. Tri- and tetra-antennate structures may also be formed. These highly branched derivatives are found in relatively high concentrations on tumour cells (see section 2.7.3). Although each antenna is commonly composed of the sequence SA-Gal-GlcNAc-Man the length may be increased by repeat sequences of Gal-GlcNAc between the sialic acid and the mannose. Alternatively, the antennae may be incomplete and structures lacking the terminal sialic acid residues are relatively common. Although the mechanisms which determine the branching and overall size of *N*-linked sequences are unknown it is clear from many lines of evidence that the structural variability widely encountered in these molecules does not arise randomly but is the consequence of a tightly regulated assembly process. Presumably this so-called 'heterogeneity' in the structure of surface-membrane carbohydrates is essential to normal cellular functions.

2.3.1.2 O-*linked Oligosaccharides*. Tetrasaccharides of structure SA-Gal-GalNAc-Ser are the most common of the membrane glycans

```
        SA-Gal-GalNAc-Ser
                 |
                 SA
```

bound to protein through the GalNAc-Ser link. No lipid intermediates are involved in the biosynthesis of these structures, the sugars being added directly to the protein by sequential transfer from sugar nucleotide precursors. *O*-linked tetrasaccharides are arranged in 'clusters' along the protein core. In contrast *N*-linked sequences occur singly and with an overall frequency of around one to four chains per glycoprotein.

Figure 2.5 Biosynthesis of Biantennate Asparagine *N*-linked Oligosaccharides: (a) Biantennate *N*-linked Glycans; (b) Multibranched *N*-linked Oligosaccharides

a

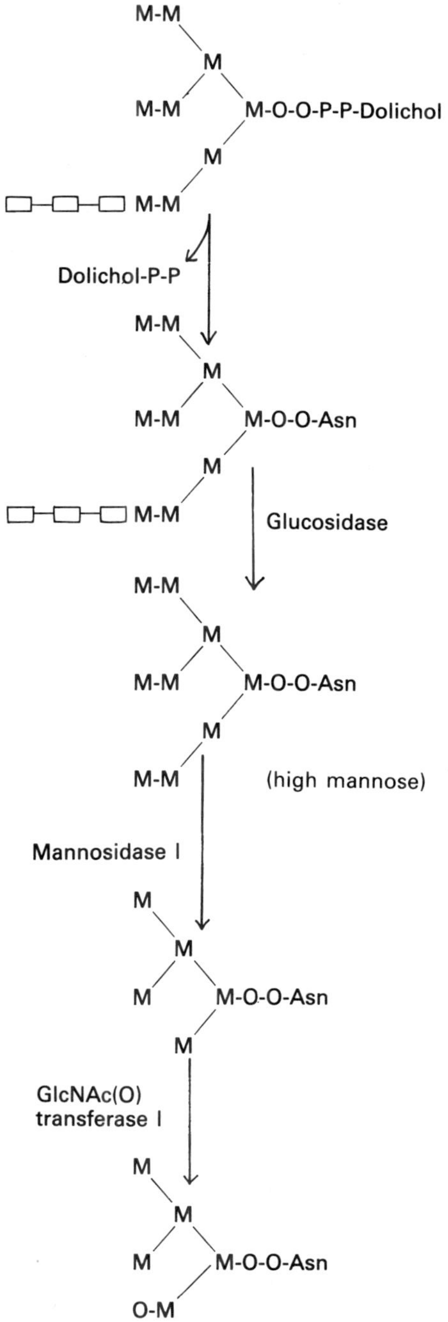

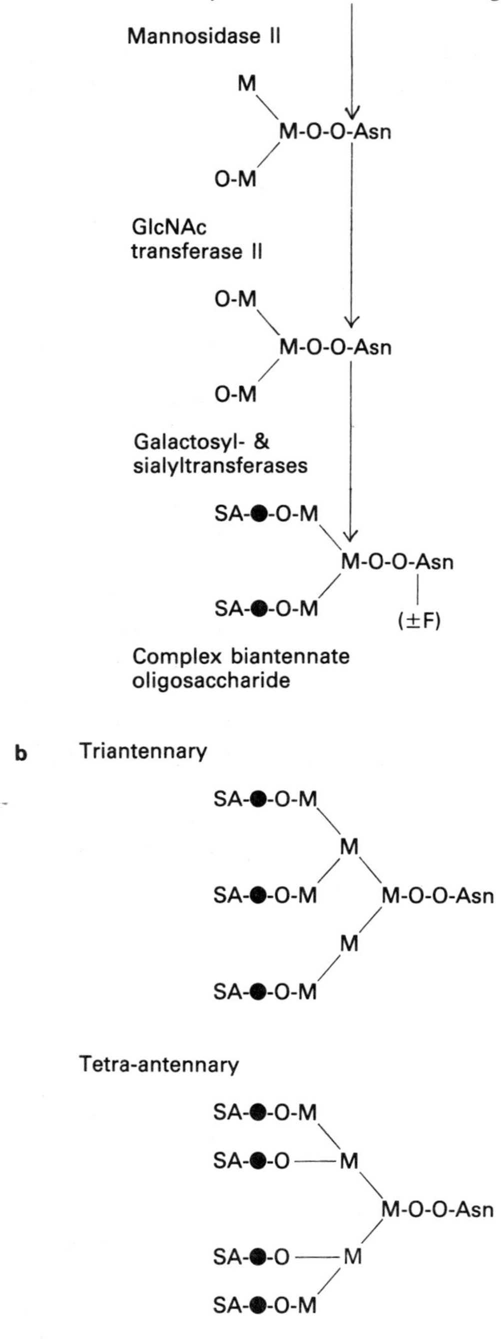

Key: M = mannose; F = fucose; SA = sialic acid; O = *N*-acetylglucosamine; ● = galactose; ▭= glucose.

In epithelial cell tumours disaccharides of sequence Gal-GalNAc-Ser are commonly found. This is the determinant of the 'T'-antigen and the corresponding antibodies are present in the serum of a large proportion of the human population. It is not clear whether the determinant is expressed in tumours as a result of incomplete synthesis or desialylation. Lectins from peanut and *Vicia graminae* can be used to identify the T-antigen and they could be useful for detecting premalignant conditions in individuals who may be at high risk of developing cancer.

2.3.2 Glycosaminoglycans

Saccharide chains linked to serine via xylose are very long linear structures often referred to as glycosaminoglycans (Gags).[6] The protein core plus attached Gags is called a proteoglycan to distinguish it from other glycoproteins which lack Gag chains. Gags are formed from repeat disaccharide sequences which usually consist of a uronic acid and an amino sugar. The uronic acids contain a carboxyl group at carbon 6 (C-6). There are two forms of the sugar acid, glucuronic acid (GlcUA) and iduronic acid (IdUA), which is an epimer of glucuronic acid with the carboxyl group in the opposite configurations. The amino sugar can be either *N*-acetyl-glucosamine or *N*-acetyl-galactosamine. On average there are 50–100 disaccharide repeat units in each Gag chain. Sulphate groups are found in all Gags except hyaluronic acid. As a result of these sulphate residues and the carboxyl groups the Gags are highly charged polyanions and they may have a profound influence on the cation-binding properties of cell surfaces.

In the linkage region a sugar sequence of GlcUA-Gal-Gal-Xyl-Ser connects the disaccharide repeats to the protein core. Thus the general structure of a Gag chain linked to protein is:

Serine–linkage sugars–disaccharide repeat sequences

Protein core — Gag region

Although data are incomplete, cell-surface proteoglycans seem to contain about three to six Gag chains per molecule.

There are several types of Gag. The disaccharide repeat unit in chondroitin sulphate consists of glucuronic acid and *N*-acetylgalactosamine, whereas dermatan sulphate is formed from iduronic acid and *N*-acetylgalactosamine (Figure 2.6). Some Gags contain both chondroitin sulphate and dermatan sulphate sequences in the same co-polymeric chain. During Gag biosynthesis dermatan sulphate is formed from chondroitin sulphate through the conversion, at the polymer level, of some or all the glucuronic acid residues into iduronic acid. Both Gags are sulphated at the 4 and/or 6 position of *N*-acetylgalactosamine and a 2-*O*-sulphate group is sometimes located on the iduronic acid residue.

Considerable differences in structure may thus be found along a typical co-polymeric chain.

Figure 2.6 Repeat Unit Sequences in Chondroitin and Dermatan Sulphates

```
Chondroitin sulphate         GalNAc-β(1→4)-GlcUA
                               |
                               S

Dermatan sulphate            GalNAc-β(1→4)-IdUA
                               |           |
                               S          (S)

Co-polymeric dermatan        GalNAc-β(1→4)-GlcUA-β(1→3)-GalNAc-β(1→4)-IdUA
  sulphate                     |                          |           |
                               S                          S          (S)
```

Each disaccharide in chondroitin and dermatan sulphate contains an average of one sulphate residue. The sulphate group (S) is an ester or *O*-sulphate which may be found in one of three positions: most commonly, *N*-acetylgalactosamine contains a sulphate group at either C-4 or C-6 and the iduronic acid residue in dermatan sulphate can be sulphated at C-2.

Structural variability is most pronounced in the heparan sulphate (HS) Gag chain. The primary product of heparan sulphate biosynthesis is a non-sulphated polymer composed of disaccharide repeats of glucuronic acid and *N*-acetylglucosamine [GlcUA-β(1→4)-GlcNAc]. A complex series of postpolymerisation modification reactions then occur to yield a polysaccharide with a highly irregular distribution of sulphate groups, in contrast to the more uniform distribution found in chondroitin sulphate and dermatan sulphate. The critical step in heparan sulphate synthesis is the de-acetylation of the *N*-acetylglucosamine residue to form glucosamine (Figure 2.7). The free amino group of glucosamine is then rapidly sulphated to yield *N*-sulphated glucosamine ($GlcNSO_3^-$) which is the characteristic sulphated monosaccharide constituent of heparan sulphate. The conversion from *N*-acetyl into *N*-sulphate groups only occurs in about 50 per cent of the *N*-acetylglucosamine residues. The conversions are localised to a significant degree so that domains of high *N*-sulphate concentrations are found in all HS-Gag chains together with other regions of the chain which remain predominantly *N*-acetylated. *N*-Sulphated disaccharides are mainly present as either repeat units of the type GlcUA-$GlcNSO_3^-$ or as alternating sequences with *N*-acetylated disaccharides, GlcUA-GlcNAc-GlcUA-$GlcNSO_3^-$. Additional polymer modifications then occur in the highly *N*-sulphated domains: the glucuronic acid residue can be converted into iduronic acid and additional sulphate groups (ester or *O*-sulphate residues) may be added to C-2 of iduronic

Figure 2.7 Modification Reactions in the Synthesis of Heparan Sulphate

GlcNAc-α(1→4)-GlcUA (1)

↓ deacetylation

GlcNH$_3^+$-GlcUA (2)

↓ *N*-sulphation

GlcNSO$_3^-$-GlcUA (3)

↓ epimerisation

GlcNSO$_3^-$-IdUA (4)

↓ *O*-sulphation (*O*-S)

GlcNSO$_3^-$-IdUA-2-*O*-S (5)

↓ *O*-sulphation

GlcNSO$_3^-$-IdUA-2-*O*-S (6)
|
6-*O*-S

Heparan sulphate is synthesised initially as a non-sulphated polymer of repeat sequences of GlcNAc and GlcUA. The extensive postpolymerisation modifications that occur can convert non-sulphated disaccharides (structure 1) into disaccharides containing three sulphate groups with the original GlcUA residue converted into IdUA (structure 6). The modification sequence does not go to completion in all cases and monosulphated (structures 3 & 4) and disulphated (structure 5) derivatives commonly occur.

acid and C-6 of *N*-sulphated glucosamine. *N*-Acetylglucosamine residues in the alternating regions may also acquire a sulphate group at C-6. Sulphation of this particular sugar residue is significantly reduced in transformed cells (section 2.7.4). These structural modifications are summarised in Figure 2.7. The modification reactions are often incomplete so that all the disaccharide intermediates shown in Figure 2.7 can be present in the same heparan sulphate chain. As a result of these polymer-level modifications, enormous structural diversity can be generated in heparan sulphate and it is the most complex of all mammalian polysaccharides.[6] Heparan sulphate is the most abundant Gag found at cell surfaces and it shows some variation in sulphate concentration from one cell type to another. It is tempting to speculate that the considerable potential for structural diversification in heparan sulphate implies an important role for this polysaccharide in specific cellular recognition processes. This point is taken up in more detail in section 2.7.4.

2.3.2.1 Keratan Sulphate. This is the only Gag that does not contain uronic acid. It consists of repeat disaccharides of Gal-GlcNAc in which one or both residues may be '*O*'-sulphated. This component has not been found on mammalian cell surfaces.

2.3.2.2 Hyaluronic Acid. This is the largest of the Gag chains. It may contain several thousand disaccharides of structure GlcUA-GlcNAc. The molecule is not sulphated and is not covalently linked to protein. It is found on cell surfaces bound to hyaluronic acid receptors of undefined structure. In the embryo an environment rich in hyaluronic acid is essential for the migration of immature cells to their site of differentiation and it is often claimed that a high concentration of hyaluronic acid in the vicinity of tumour cells promotes their aberrant migration. However, it has proved difficult to get firm evidence for this interesting proposal.

2.4 Cell Differentiation and Evolution of Malignant Clones[7]

In most organs and tissues a balance is maintained between cell renewal and cell loss. Mature cells have a finite life-span and they are replaced by new cells which originate from the so-called stem-cell compartment. Stem cells are very small in number but have the distinctive property of being able to self-renew and to maintain their population size. They may also differentiate to form progenitor cells that will ultimately mature and increase in number to yield functional end cells which are often incapable of further cell division (Figure 2.8). In the haemopoietic system stem cells are pluripotent as any individual stem cell can give rise to all the mature cells found in the peripheral blood. There are considerable differences in cell-surface properties between stem cells and functional end cells so the question arises of what cell population represents an appropriate control for malignant cells.

Carcinogens are mutagens and the initial event in carcinogenesis is an inheritable change in the genome. Many cells may be affected by carcinogenic agents but committed cells undergoing amplification division are transit cells and will ultimately be lost in the process of cell renewal. Stem cells, however, are fixed within the organism and as the time-lag between the initiation and expression of malignancy is usually very long it is most probable that stem cells, or their immediate progenitors, are the target cells in carcinogenesis. This will allow time for tumour promoters to entrench the initial genetic lesion within the genome and lead to the selection of an initiated stem-cell clone with a growth advantage over the normal stem-cell pool. A second mutation seems necessary for the full expression of malignancy. Clearly therefore an important step would be to compare the surface properties of normal and malignant stem cells but, for technical reasons (low cell numbers, problems of cell identification), this has not been possible so far.

Figure 2.8 Stem Cells in Normal Cell Development and Malignancy

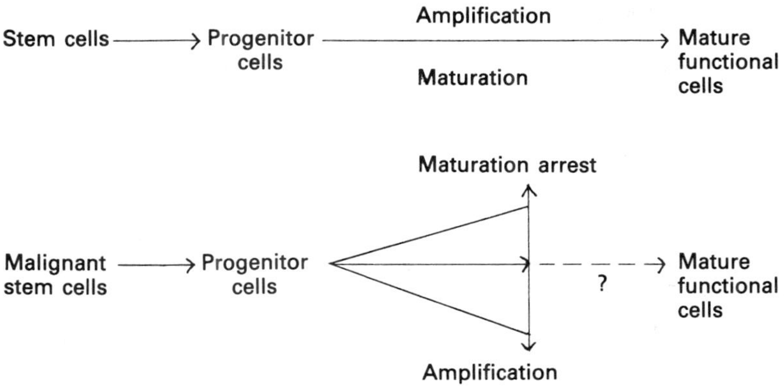

In normal organs and tissues cell renewal occurs through the commitment of stem cells to the differentiation pathway and the amplification and maturation of the progenitor cells to mature functional 'end cells'. Progenitor cells are the immediate products of stem-cell differentiation. They retain some self-renewal capacity but they are all destined to develop into mature cells. The progenitor cell compartment needs to be continuously replenished from the stem-cell pool. Malignant stem cells may give rise to apparently normal progenitors but cell development is arrested before full maturity is achieved. Continuous proliferation of the immature malignant cells exceeds cell loss and leads to the formation of a tumour.

The malignant stem cell and its progeny have common genetic abnormalities but differ in cell-surface phenotype. Malignant clones spawn large numbers of abnormal cells which are often unable to mature fully, a consequence of the uncoupling of cell differentiation and cell proliferation (Figure 2.8). This is an entirely different concept from the view of malignancy as an example of de-differentiation, an idea for which there is little firm evidence. The reasons for maturation arrest in tumours are unclear and will remain so until we gain more insight into normal cell development. To identify surface abnormalities in the immature cells which constitute the bulk of tumours it is essential to appreciate the phenotypic progression which occurs in the maturation pathways of relevant normal cell populations. In the majority of studies such objectives have not been achieved and all too often the immature phenotype of the tumour cells is compared with normal mature cellular phenotypes.

Some of these objections are not applicable to certain investigations carried out on haematological malignancies. Immature cells can be identified in haemopoetic tissues such as bone marrow and spleen and there is considerable knowledge of surface glycoprotein changes that occur during the development of different types of blood cells. In the leukaemias the malignant cells accumulate in large numbers in the peripheral blood and they may be easily isolated in viable form. In contrast, separation of malignant cells from solid tumours is often very difficult and preparations can be heavily contaminated with infiltrating normal cells such as lymphocytes, macrophages and fibroblasts.

2.5 Lymphocytic Leukaemias

2.5.1 Maturation Arrest

Leukaemias may arise in all cell lineages found in the peripheral blood but for simplicity only leukaemias of lymphoid cells will be considered here. Both acute and chronic forms of lymphocytic leukaemias [acute lymphoblastic leukaemia (ALL) and chronic lymphoblastic leukaemia (CLL) respectively] may occur in humans. ALL has a rapid progression unless treated, whereas CLL develops slowly and patients can survive for many years with little treatment. At the cell level the diseases are distinguished by major differences in morphology and differentiation. In ALL the cells are arrested early in the maturation pathway, whereas in CLL maturation arrest occurs only at a highly differentiated stage. Accordingly CLL cells have many surface properties of mature lymphocytes. Both forms of leukaemia can arise in the B-lymphocyte (antibody-producing) and the T-lymphocyte (thymically processed) populations but those in the B-cell axis are most common.

2.5.2 Surface Markers

There are four classes of ALL: ‘T’, ‘B’, ‘common’ and ‘null’.[8] Each class expresses a characteristic set of cell-surface markers. The majority of these markers are antigenic determinants on membrane glycoproteins and they are detected by the use of specific antibodies. T-ALL are recognised by using one of several so-called pan-T monoclonal antibodies which bind to immature and mature T-cells, whereas the characteristic feature of B-ALL is the presence of surface-membrane immunoglobulin (SmIg). Common-ALL, the most frequent form of leukaemia in children, expresses the c-ALL-antigen, a 100 000 mol.wt. glycoprotein which is developmentally regulated and cannot be detected on the mature cells which accumulate in CLL. Common-ALL originates from cells in the B-lineage since the c-ALL-antigen is sometimes found in association with cytoplasmic Ig, a marker of pre-B-cells. Null-ALL cells lack characteristic surface markers. They express HLA-DR glycoproteins but these are also found in common and B subtypes. The most immature leukaemic populations are found in null-ALL and the accumulated cells are similar to the earliest lymphoid progenitors. Four important points must be made about immunological surface markers used in the classification of ALL, as follows:

(1) They demonstrate clearly the cellular heterogeneity within a given group of closely related malignant neoplasms
(2) None of the markers is exclusive to leukaemic cells. They may be found on normal immature lymphoid cells in the bone marrow and thymus. Thus they are cell differentiation markers, not leukaemia markers, and they reflect the differentiation status of the leukaemic population
(3) The markers are expressed in a strict cell-lineage related fashion, i.e. T-cell markers are not found in association with B-cell markers and vice versa.[9] This is a significant point in the broader context of cellular phenotypes in malignant diseases. Gene expression, as reflected in cell-surface properties, shows no evidence of 'transdifferentiation' (the term here refers to a shift in surface phenotype from one cell type to another): i.e. the detectable cell-surface antigens are entirely legitimate to the cell lineage within which the malignant elements arise
(4) Although not specific for leukaemia these markers are extremely valuable in diagnosis and prognosis. Common-ALL respond well to therapy and have good long-term survival whereas T- and B-ALLs have a very poor prognosis and represent a group that requires an improvement in current treatment.

Despite the fact that the cell-surface glycoproteins in leukaemia can also be found in normal immature cells, the leukaemic surface may not be an exact replica of an immature phenotype. Some loss of synchrony in expression could occur so that normal surface glycoproteins may be found in abnormal combinations. Such combinations could have a significant impact on the growth properties of the leukaemic cells. It is also possible that some of these 'normal' glycoproteins on ALL-cell surfaces may have abnormal patterns of glycosylation (see section 2.7.3).

2.5.3 Electrophoretic Analysis of Membrane Glycoproteins[10]

Surface-membrane glycoproteins may be radiochemically labelled, *in situ*, at sialic acid and galactose residues. Sugar labelling requires the generation of aldehyde groups (CHO) and their reduction with tritiated sodium borohydride to a tritiated primary alcohol (CH_2OH). Aldehydes may be formed in the acyclic region of the sialic acid molecule with periodate and at C-6 of galactose by treatment with galactose oxidase. Under careful experimental conditions both methods radiolabel only cell-surface saccharides. Whole cells may then be lysed with detergents and the soluble glycoproteins separated by electrophoresis on polyacrylamide gels. Radioactive glycoproteins can be detected by radioautography as individual bands of different electrophoretic mobilities. Many bands have been identified in this way and, although the functions of the labelled glycoproteins are mostly unknown, the banding patterns represent a valuable fingerprint of surface glycoprotein composition. In general the results are in accord with the immunological surface-marker approach in that T- and B-cell leukaemias show distinctive and cell-lineage related patterns. So far the major impact of the electrophoretic data has been in identifying a subset of c-ALL, which normally has a favourable prognosis, that responds poorly to conventional treatments.

2.5.4 Lectin Binding

As an alternative strategy to the analysis of intact membrane glycoproteins by electrophoretic or immunological methods, the cell-surface carbohydrates of human leukaemic cells have been studied *in situ* by using lectins. Lectins are a distinctive class of carbohydrate-binding proteins which bind with high affinity to specific sugar sequences.[11,12] Two or more carbohydrate-binding sites are found on each lectin molecule and as a consequence of this bi- or multi-valency they rapidly agglutinate cells with the necessary surface carbohydrate acceptors by forming an organised network of intercellular connections. Historically this property of lectins was very important as transformed or tumour cells in culture were shown to be more readily agglutinated than normal cell counterparts. Although many exceptions to these observations have been noted, the results were important in drawing attention to cell-surface sugars as possible sites for structural change in malignant disease.

Lectins are distinguished on the basis of their sugar specificities and it is now well established that lectins can be extremely fastidious with respect to the oligosaccharides that they can recognise. They can be used to investigate the composition and branching patterns of membrane carbohydrates. Concanavalin A (Con A) for example, shows strong binding to high-mannose structures, whereas lentil lectin expresses high affinity for biantennate, complex glycans with a core fucose (see Figure 2.5). The mitogenic lectin from *Phaseolus vulgaris* (L-PHA) binds most strongly to the tetra-antennary oligosaccharides which are not recognised by Con A or lentil lectin. By the use of these three lectins a small but significant reduction in cell-surface density of high-mannose and complex *N*-linked oligosaccharides has been detected in ALL cells by comparison with normal mature lymphocytes. The CLL cells were abnormal only with respect to a low reactivity with Con A. CLL cells are quite mature and this finding indicates that normal lymphocytes develop a full complement of high-mannose structures only very late in development.

There is also an interesting group of lectins that bind to oligosaccharides which have a GalNAc residue exposed at the non-reducing terminus. One of these lectins, from the haemolymph of the edible snail, *Helix pomatia* (HP), has been particularly interesting in investigations of human leukaemia.[13] Normal and leukaemic human lymphoid cells must be treated first with neuraminidase before HP binding occurs because the

HP acceptor sugars are substituted with sialic acid in sequences of the type (SA-GalNAc-R) where 'R' represents additional sugars between the HP-reactive GalNAc and the protein core. These sequences are probably linked to serine through a second GalNAc residue (*O*-linked, see section 2.3.1) and not to asparagine. After enzyme treatment, normal lymphocytes bind HP to varying degrees with T-cells showing higher levels of binding than the majority of B-cells. In ALL cells there is a significant reduction in the density of HP acceptor sequences especially in the T phenotype ALLs. Several other GalNAc-binding lectins also interact poorly with ALL-cell surfaces but none show the striking decreases that are found with HP. In T-ALL HP binding is less than 20 per cent that of normal T-cells.

CLL cells express normal numbers of HP binding sites so a reduction in sugars complementary to HP is not an essential feature of leukaemic transformation in lymphoid cells. Since adequate models of lymphoid cell development are not available it is not known whether the reduced binding of HP in ALL cells reflects their relative immaturity or is a property associated specifically with the acute leukaemic condition. HP binding does not correlate with cell division since mitogen-stimulated normal T-cells show a slight increase in density of HP acceptors.

HP-binding sugars in lymphocyte surface membranes are found mainly in association with a 150 000 mol.wt. glycoprotein. Reduced HP binding in ALL could be due to a low level of synthesis of the protein core of this glycoprotein or to a decrease in the activities of the glycosyltransferases which synthesise the relevant sugar sequences.

HP binding is very low in acute leukaemias of cells in the macrophage/granulocytic lineage (the myeloid lineage), but normal peripheral blood granulocytes also have weak reactivity with HP. Thus, although lymphoid and myeloid acute leukaemia in man show very weak expression of HP-reactive sugars, this property only differentiates lymphoid leukaemia cells from their normal cell counterparts. This finding emphasises again the importance of correct identification of the cellular origin of malignant populations.

2.6 Membrane Glycoproteins in Solid Tumours

In plasma membranes of many solid human and animal tumours a glycoprotein of about 50 000 mol.wt. has been detected by isoelectric focusing.[14] The glycoprotein was strongly hydrophobic and the saccharides were bound to the protein by O-glycosidic links. Although absent from the surface membranes of non-malignant tissues a very similar glycoprotein, with perhaps a slightly reduced sialic acid content, was found in the nuclear membranes of normal cells. The insertion of this glycoprotein into the surface membrane of tumour cells could result from an abnormality in membrane flow, perhaps occasioned by changes in saccharide composition, so that the glycoprotein is transferred to the wrong subcellular destination. Incorrectly sited it may then perturb normal membrane functions.

2.7 Analysis of Malignancy *in vitro*

2.7.1 *Problems of Interpretation*

Because of the considerable difficulties in studying malignancy *in vivo*, simpler, more controllable systems *in vitro* are widely used. Many rodent cell lines have been established in culture that can be exposed to oncogenic agents (e.g. viruses, chemicals, radiation) leading to the development of transformed variants which show properties that may be associated with their potential for tumour formation. However, the transformed state is essentially defined by parameters *in vitro*. For example, unlike normal cells the motility of transformed cells is not inhibited by cell contact (contact inhibition) and they form multilayers rather than monolayers on the culture substratum, which is usually plastic. Transformed cells often have a reduced serum requirement for growth and will form colonies in soft agar whereas normal cells, with the notable exception of haemopoietic cells, require a solid surface for growth (i.e. their growth is anchorage-dependent, whereas transformed cells are anchorage-independent). Most cells that display these transformation properties will form tumours when injected into genetically compatible hosts. Unfortunately many so-called normal cell lines will also form tumours in animals but they are less tumorigenic, that is, larger numbers of cells must be inoculated to form tumours than transformed cell counterparts. Analysis of tumour formation is often carried out at the simplest level. For example, transformed cells may be injected subcutaneously and tumour formation identified as a lump at the site of the innoculum. The characteristic property of a malignant cell is its capacity to invade surrounding normal tissues and break through architectural barriers such as basement membranes. Invasive properties are rarely measured and often this deficiency is neatly side-stepped by defining malignancy as the ability of cells to grow continuously and kill the host. However, benign tumours (non-invasive) can kill if allowed to grow unchecked, although it is not normal clinical practice to let this occur! Invasion and, in particular, metastasis are the main problems in cancer treatment; cells transformed in culture rarely metastasise *in vivo*.

How seriously should we take data obtained with transformed cells? A reasonable view is to consider normal cell lines, even those of relatively low tumorigenicity, as broadly equivalent to cells initiated by a carcinogen *in vivo*. Transformed cells could then represent variants at a much later stage in tumour promotion. These cells have probably acquired further mutations but still await the final throw of the genetic switch which is essential for full expression of malignant behaviour. Clearly transformed cells can represent an important stage in the evolution of cancer but neither they nor the normal cultured cells with which they are compared stand at opposite points in the spectrum between normality and malignancy.

It is also noteworthy that several cell lines have been developed from malignant neoplasms in humans and often they show many of the properties of cells transformed in culture. However, human tumours display considerable clonal heterogeneity and the clones which grow in culture may not be those which sustain the tumour in the patient. Also it is often very difficult to identify and grow the normal cell of origin.

The following sections describe studies of normal, transformed and tumour-derived cells in culture.

2.7.2 Cell Hybridisation and Analysis of Transformed Cell Surfaces

Alterations in structure and composition of membrane glycoproteins and glycolipids are commonly found in transformed cells. It is always difficult, however, to determine whether the changes observed are essential to the continued expression of transformation characteristics. Cell hybridisation offers an ingenious means of tackling this problem. When transformed cells are fused with normal diploid cells in culture, the hybrids formed are usually normal by the criteria discussed above. With continued cultivation, however, chromosomes are lost from the hybrids with the result that genes for cell transformation may be segregated from the normal, presumably dominant, genes and transformed variants emerge from within the population. These segregant cells provide an exacting system for biochemical analysis since any property essential to their aberrant behaviour, and incompatible with normal cell characteristics, should always be expressed by the parental and the segregant tumour cells but must be absent from cells in which transformation is suppressed.

By using cell fusion systems a glycoprotein has been identified in the surface membranes of transformed cells that shows a change in its pattern of glycosylation.[15,16] The glycoprotein is a disulphide-linked dimer of mol.wt. around 200 000 that binds considerably more Con A but less of the lectin from wheatgerm (WGA) than a similar glycoprotein from normal cells.

The change in saccharide structure detected by these lectins probably represents an increase in high-mannose sequence at the expense of complex *N*-linked glycans (see Figure 2.5). There are a number of glycoprotein dimers of similar size to this 200 000 mol.wt. component, most notably glycoproteins involved in glucose transport and the transferrin receptor, both of which are essential for cell growth. The function of this transformation-associated glycoprotein has not been unequivocally established but the similarity with the glucose transporter prompted investigation into glucose uptake. The experiments showed that glucose-transporting systems in transformed cell hybrids showed an increase in affinity for glucose (decrease in K_m). This property could confer a growth advantage on tumour cells *in vivo* in an environment where blood supply may be restricted. It was further postulated that the change in carbohydrate composition might be responsible for enhancing the affinity of the transporter for glucose. This still remains to be established. In this connection, it is also worthwhile emphasising a point made earlier in this chapter, namely that carbohydrate residues may be altered in tumour or transformed cells while the protein components to which they are attached remain substantially unchanged.

A further development with the cell fusion system has been the detection of a high-molecular-weight cell-surface glycoprotein (350 000–390 000) which also consistently segregates with transformation. This component is heavily glycosylated (60 per cent by weight is carbohydrate) with short *O*-linked saccharides of the mucin type (see section 2.3.1.2). A specific monoclonal antibody has been used to detect this glycoprotein, termed the Ca-antigen, in a wide variety of malignant human tumours.[17] Although antibody binding was not unique to tumour tissues, no reactivity was observed against benign tumours, and against normal tissue it showed reactivity only with very specialised normal epithelium found in Fallopian tubes and the urinary tract. These findings are important because they establish that systems *in vitro* can be used to identify glycoproteins that are expressed with some selectivity on natural tumours. The mucin-like nature of the Ca-antigen suggests that it may have a protective function.

2.7.3 Glycopeptides of Transformed Cell Surfaces

Membrane glycoproteins may be radiolabelled biosynthetically in their carbohydrate chains by using radioactive sugar precursors such as glucosamine and fucose. Fucose is particularly useful since it is not metabolised to other sugars and in cell surfaces it is localised almost entirely in N-linked saccharide chains. Fucose-labelled glycopeptides have been isolated from the surfaces of normal and tumour cells by trypsin treatment and trypsin extracts were then incubated with a broad-spectrum proteinase such as pronase to remove most of the attached amino acids. When the glycopeptides were fractionated on the basis of their molecular size with a gel-filtration column (usually Sephadex G-50) the largest glycopeptide fractions were found in higher concentrations in extracts from tumour cells.[18,19] If the glycopeptides were then treated with neuraminidase to remove terminal sialic acid, the molecular size differences were eliminated. The precise structures of all these high-molecular-weight glycans have not been determined. It is evident, however, that they probably represent mixtures of sialylated tri- and tetra-antennary N-linked glycans that contain a fucose residue in the core region. The branching patterns of the oligosaccharides are similar on normal and transformed cells but the latter are able to enrich their cell surfaces with sequences containing a full complement of sialic acid terminal sugars. For example, fully sialylated tetra-antennary glycans of the structure shown in Figure 2.5 have been identified in trypsin extracts of transformed cells.[20] Complete sialylation is less common in homologous structures from normal cells.

In response to these interesting findings many investigations on sialyltransferase activities of transformed cells were carried out but the findings were inconsistent. In one study, however, this problem was examined by first taking the high-molecular-weight sialylated glycopeptides and removing the sialic acid residues with neuraminidase. When these desialylated derivatives were used as substrates a significant increase in sialyltransferase activity was found in transformed cells. This suggests that a specific sialyltransferase is involved in addition of sialic acid to multibranched oligosaccharides and that the concentration or activity of this enzyme is elevated after transformation.[19] Presumably sialyltransferases with different substrate specificities are responsible for the sialylation of bi-antennary structures.

Not all transformed cells exhibit an enrichment of high-molecular-weight glycopeptides, for example, chemically transformed mesenchymal cells derived from kidney tissues. Some normal cells in culture may also express these glycopeptides during rapid growth. In relation to carbohydrate function it is intriguing to note that the composition of many of the individual surface-membrane glycoproteins of transformed cells demonstrate a shift to heavily sialylated multibranched glycans. A normal oligosaccharide composition appears to be unnecessary for the association of these glycoproteins with the membrane surface. Most glycoproteins do, however, contain several oligosaccharide chains and one of these may be unaffected and act as the address mechanism for delivery to the plasma membrane.

Some recent interesting observations on the saccharide sequences of human granulocyte cell surfaces have suggested a possible relation between high-molecular-weight glycopeptides and cell migration.[21] Granulocytes develop in the bone marrow and when mature they migrate from the marrow spaces through capillary walls into the bloodstream. These mature migratory cells show a transient expression of the large glycopeptides found on many transformed cells in culture. On reaching the peripheral blood the granulocyte surface reverts to a predominant expression of lower-molecular-weight species. Immature granulocytes also have low concentrations of the larger glycopeptides and they are confined to the bone marrow. It is therefore most interesting that in acute leukaemia of the granulocytic lineage (the myeloid leukaemias) the neoplastic cells, which are immature by most cytological and immunological criteria, strongly express high-molecular-weight transformation-type glycopeptides and they migrate into the peripheral blood. It would seem that in these leukaemias such large glycopeptides are synthesised precociously. They may contribute to maturation arrest by encouraging premature cell movement away from essential differentiation signals originating from the bone marrow micro-environment.

2.7.4 Glycoaminoglycans and Proteoglycans

These groups of complex acidic polymers are found in highest concentrations in cartilage and other connective tissues. For many years they were looked on mainly as structural components with little capacity to directly influence cell behaviour. This view changed dramatically when the considerable structural variability of proteoglycans was recognised and when it became apparent that they were common constituents on the surfaces of the vast majority of animal cells. Many studies have attempted to define changes that occur in transformed cells and most investigators have managed to pick out some alteration of apparent significance. However, no consistent patterns emerged and it was not possible to specify a Gag or proteoglycan phenotype for tumour cells.[22]

As an alternative to studies on the general composition of Gags in transformed cells detailed investigations have now been carried out on the fine structure of the Gag chain and particular attention has been focused on heparan sulphate.[23] Results have shown that after cell transformation the *N*-sulphate content of heparan sulphate is unchanged and the close topographical relation between *N*- and *O*-sulphate groups is conserved (Figure 2.7). However, cell transformation is associated with a reduction in the overall concentration of *O*-sulphate groups. The decrease is found only in those regions of the polymer chain where alternating sequences of *N*-sulphated and *N*-acetylated disaccharides are found (section 2.3.2):

-GlcUA - GlcNAc - GlcUA - $GlcNSO_3^-$

(6-*O*-S) (6-*O*-S)

In these sequences the C-6 of GlcNAc or $GlcNSO_3^-$ may be substituted with *O*-sulphate residues and after transformation the number of *O*-sulphate groups is 30 per cent below that of normal heparan sulphate. Transformation does not affect *O*-sulphation in those regions of the Gag chain where repeat sequences of *N*-sulphated disaccharides are found (section 2.3.2). The specificity of this alteration in polymer sulphation strongly suggests that it might be an important factor in the progression toward malignancy. Structural changes in heparan sulphate could affect cellular interactions within tumour tissues. Heparan sulphate chains from normal cells have the intriguing property of self-association, whereas in heparan sulphate from transformed cells, the self-associative interactions are very weak. Heparan sulphate is found at both cell surfaces and in the extracellular matrix and its self-affinity properties, or lack of them, may have a major influence on the outcome of intimate molecular 'negotiations' within the pericellular domain. In this connection it is worthwhile pointing out that heparan sulphate binds specifically to structural glycoproteins of the extracellular matrix (e.g. laminin, fibronectin; see also section 2.8.2) and it may contribute to the formation of the organised molecular scaffolding that can be observed around normal cells. In transformed cells in culture, and frequently in naturally occurring tumours, the extracellular matrix appears as a disorganised structure which is deficient in the foregoing matrix constituents. This deficiency may be a significant factor in promoting tumour-cell invasion *in vivo*.

2.8 Cancer Metastasis[24]

2.8.1 Dissemination of Malignant Disease

The spread of tumour cells from the primary organ or tissue in which the neoplasm initially occurs to secondary sites is called metastasis. Metastasis usually forms part of the natural history of malignant tumours and is the major cause of death in cancer. The characteristic property of a malignant cell is the ability to migrate into or 'invade' surrounding normal tissues. Metastasis must be seen as an additional consequence of this abnormal migratory behaviour which is associated with the ability to break through limiting barriers such as basement membranes. As a result of their invasive activities tumour cells ultimately appear in the bloodstream and become transported to other sites in the body.

To sustain growth at a secondary site the metastatic cells must be derived from the stem-cell compartment of the tumour (see section 2.4). These cells represent only a small fraction of the malignant-cell population. Experimental models have confirmed that metastatic cells are found with low frequency as pre-existing variants within the primary tumour. They seem to arise spontaneously as a result of the genetic instability of malignant cells.

Multiple and intricate cell/cell and cell/matrix interactions occur during metastatic dissemination that reflect the adapatability of the surface membrane of tumour cells and their capacity for survival in potentially hostile environments. Certain critical steps can be identified in metastasis.

Primary tumour
↓
Invasion of normal tissues
↓
Breaching of tissue–blood barrier
↓
Transport in bloodstream
↓
Arrest in capillaries
↓
Breaching of blood–tissue barrier
↓
Colonisation of secondary site

To express invasive properties tumour cells must show a relative increase in adhesion to normal cells and their surrounding extracellular matrix in comparison with their adhesion to the main tumour-cell mass. In motile cells the adhesion sites must be dynamic with adhesive bonds constantly breaking and reforming. The majority of tumours originate in epithelial cells. Before they reach the bloodstream the malignant cells must breach two basement membranes, the first separating the epithelium from the underlying mesenchyme, and the second surrounding the translumenal surface of the endothelial cells which line blood vessels. After release into the bloodstream the tumour cells must survive the 'shear' forces imposed by the pressure of blood flow. Viability seems to be assisted by the development of small mixed aggregates of tumour cells and normal blood cells, in particular platelets and lymphocytes. These cell aggregates then lodge in small capillaries where they may become enveloped in a fibrin network. Inhibitors of fibrin formation decrease metastatic potential. Tumour cells then migrate from the capillary network into the adjacent tissues (extravasation) where they establish a secondary colony. A blood supply for the developing tumour is ensured by the release of angiogenesis factors which stimulate growth and directed migration of the vascular endothelium.

2.8.2 Site Selection of Secondary Tumours[25]

Several experimental models of metastasis have now been established. Transplantable tumours, i.e. tumours that can be maintained by growth and transplantation in animals, have been most widely used in these investigations. Metastatic spread is normally evaluated after the injection of tumour cells directly into the bloodstream although on some occasions they are injected subcutaneously or directly into muscle tissues. Cells have been selected with different metastatic properties. For example, from the mouse B-16 melanoma cell line, variants have been developed which show preferential growth in the lung, brain or other organs. The surface properties of these variant cell lines and the original 'parental' cell populations can then be compared to see if organ specificities or metastatic potential may be correlated with particular physical or chemical characteristics. Methods of analysis of cell surfaces have been similar to those described in section 2.5. Unfortunately, despite extensive studies, we still know very little of the molecular determinants of metastasis. This is perhaps not surprising, as the mechanisms of normal cell migration and interaction are poorly understood. It is thus very difficult to relate the presence or absence of cell-surface molecules to the aberrant behaviour of tumour cells. Also, much of the data that have been acquired are inconsistent and probably reflect the different cell systems used and different methods of cell selection and experimental design.

2.8.3 Cell-surface Modification[26]

There is little doubt that altering cell surfaces modifies metastatic behaviour. For example, treatment of cells with neuraminidase or trypsin can cause gross changes in the sites of tumour-cell arrest and a significant decrease in metastatic activity. Incubation of cells with tunicamycin, an inhibitor of *N*-linked protein glycosylation (see section 2.3.1), also causes a considerable reduction in metastasis. Although their modes of action are very different, one of the common changes induced by the foregoing methods was a considerable reduction in cell-surface sialic acid. In view of this it was proposed that high levels of sialic acid may be directly correlated with metastatic potential.[27] This view has not been substantiated, however, and the prevailing evidence suggests that there is no simple relationship between the surface-membrane concentration of sialic acid (or any other sugar) and metastatic properties. It should also be borne in mind that the cell membrane is a dynamic structure and surface changes induced by chemical or enzymic methods will be gradually reversed by the continuous turnover of membrane glycoproteins. In such experiments metastasis may be ultimately determined more by the rapidity with which specific surface components or organised membrane domains can be reconstituted rather than by the membrane properties existing at the time of inoculation.

An interesting feature of many cells grown in culture is that they shed into the growth medium small fragments of their plasma membrane in the form of closed vesicles. These vesicles can be harvested by centrifugation and under suitable experimental conditions they can be induced to fuse with the surface membrane of viable cells. When shed vesicles from a highly metastatic melanoma cell line, which showed strong preference for lung colonisation, were fused with cells of low metastatic activity the recipient cells displayed the potent metastatic behaviour and organ preference of the vesicle donor cells. It is not known which components in the shed vesicles influence cell behaviour (see section 2.8.5 for some interesting possibilities) but these experiments provide very strong evidence for the importance of surface membranes in the development of metastasis. The new material introduced into the cell membrane by vesicle fusion has a relatively short lifetime and, like the changes caused by enzyme treatments of the cell surface, the original surface-membrane composition is re-established by normal pathways of membrane turnover. This being so, an important general conclusion can be drawn from experiments on surface-membrane modification. Metastatic potential and organ specificities are determined mainly by early events following the appearance of tumour cells in the bloodstream. It follows that cellular interactions in the vascular endothelium are a particularly important, indeed determinative, step in metastasis. It is here rather than through interactions with the major cell populations in an organ or tissue that the site selection occurs. There must be distinguishing features of the vascular endothelium in different tissues that are recognised by tumour cells. Unfortunately, at the present time, their identity is unknown.

2.8.4 Interaction with Vascular Endothelium

The need to understand more about cellular recognition in the capillary wall has led to some fascinating new observations on the multiple influences that the tumour cell can exert on a previously normal endothelium. When tumour cells bind to endothelial cell surfaces, the confluent endothelial monolayer obligingly retracts and exposes the underlying basement membranes. Tumour cells then bind to the exposed regions from where they gain access to the neighbouring tissue. Fibronectin and laminin are two important structural components of basement membranes and other connective tissues.[28] They are both large glycoproteins that have specific binding sites for collagen and proteo-glycans and they are clearly essential to the functional integrity of the extracellular matrix. They are also cell-adhesion molecules and contain distinctive cell-binding regions which interact with specific receptors on cell-surface membranes. Malignant cells probably attach to laminin and fibronectin during extravasation; however, if transfer from blood to tissue is to proceed cell binding must be accompanied by cell migration. It has now been established that laminin and fibronectin have a chemotactic influence on metastatic cells, i.e. the cells will migrate along a concentration gradient of either component.[29] Cell migration through the basement membrane could be promoted by the release of proteinases, known to be common secretion products of tumours, at the leading edge of the motile metastastic cells. These proteinases could degrade laminin and fibronectin and so maintain a concentration gradient of chemotactic substances. Collectively, these adhesive, degradative and chemotactic properties must provide many of the essential requirements for successful colonisation of distant sites by blood-borne metastatic cells.

2.8.5 Monoclonal antibodies

Monoclonal antibodies have been developed that recognise antigenic determinants on the surfaces of metastatic cells.[30] Selection of antibodies was based on their ability to inhibit tumour-cell attachment to cultured endothelial cells, a system which, to some extent, reproduces the interaction of tumour cells with the capillary endothelium *in vivo*. If monoclonals selected in this way are incubated with tumour cells before inoculation into animals they significantly reduce, or even prevent, the development of metastases. These antibodies block metastases by binding to a distinctive group of adhesive antigenic glycoproteins associated with the surfaces of tumour cells. Similar glycoproteins are found on early embryonic cells but they have not so far been identified on normal adult cells. They probably play an important role in regulating cell/cell interactions in fetal development. Genes coding for these adhesive glycoproteins must be re-expressed during the malignant transformation of adult cells. The monoclonals described here were raised after immunisation with R-16 mouse melanoma cells but they have activity against other types of tumour cell and, significantly, they are not species-specific. For instance, the anti-(mouse) melanoma monoclonals are active against some human tumour cells. Monoclonal antibodies could open up new possibilities for immunotherapy in the treatment of malignant disease (see Chapter 10).

2.9 Tumour Antigens and Tumour Immunity

An enormous amount of time and energy has been spent in the quest for tumour-specific antigens. If they exist (see Introduction) and the immune system could be primed to recognise them, then potentially every malignant cell could be specifically eliminated without harm to normal cells. Immunotherapy of cancer has had its ups and downs and presently it is down, but not yet out (see section 2.8.5). There is, however, no convincing evidence for tumour-specific antigens in naturally occurring human malignancies. Invariably antigens claimed to be tumour-specific are subsequently identified in fetal tissues or immature cells in the adult (see sections 2.4 and 2.5). Although unsuccessful so far in the search for unique malignancy markers, data accumulated by immunological investigations have had an enormous impact on our fundamental concepts of malignant disease. As described in several sections of this chapter antibody probes have been vital to the identification of adhesion-, proliferation- and differentiation-related glycoproteins on tumour-cell surfaces. We now need to learn more about how the expression of the constituents is regulated and the roles they play, be they obligatory or synergistic, in the growth and spread of cancer.

2.10 General Functional Implications of Surface-membrane Changes in Malignancy

It is extremely difficult to rationalise all the changes that have been identified in cancer-cell surfaces in terms of a coherent and meaningful concept of how the molecular organisation of the malignant plasma membrane leads to aberrant patterns of cell growth and development. Evidence has already been discussed for the role of high-molecular-weight sialylated glycopeptides, present in abundance on tumour cells, in promoting premature migration of immature malignant variants. The alterations in heparan sulphate fine structure could act synergistically with the glycopeptide changes to encourage undesirable migratory and invasive activities. Normal immature cells are mobile cells in transit from their sites of formation to their final position within an organ or tissue. If malignancy is indeed an expression of maturation arrest, the constituent surface-membrane glycoproteins will be compatible with migratory properties. The Ca-antigen could then serve a useful protective function for the malignant cell as it wanders into alien territories. Misplaced glycoproteins (section 2.6) could antagonise normal surface-membrane functions as well as introducing new and 'sinister' biochemical activities. However, not all the changes that have been described occur in all tumours and it is probable that during the evolution of cancerous conditions the selection of different combinations structurally altered and/or asynchronously expressed surface-membrane glycoproteins could lead to a common malignant end-point. While plausible though unsubstantiated mechanisms can be proposed for abnormal cellular migration little information is available on the reasons for the progressive growth of tumours. Answers to this problem will only be found through a deeper understanding of normal cellular homeostasis and the roles played by specific growth-regulatory substances and their surface-membrane receptors.

Appendix: Structures of Monosaccharides Found in Membrane Glycoproteins and Proteoglycans

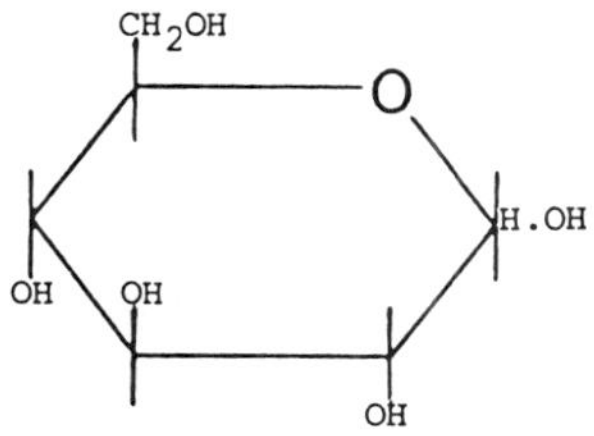

Glucose-Glu

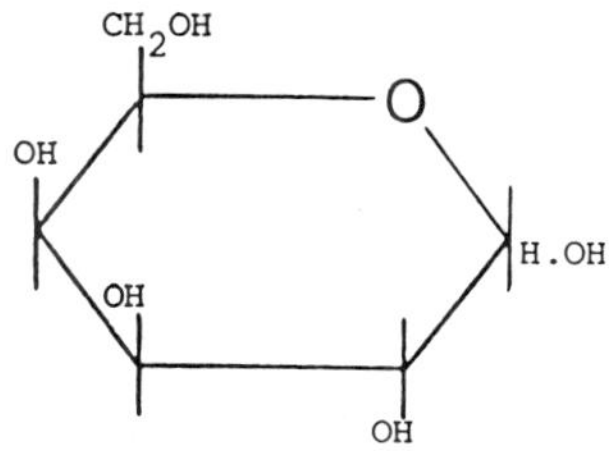

Galactose-Gal

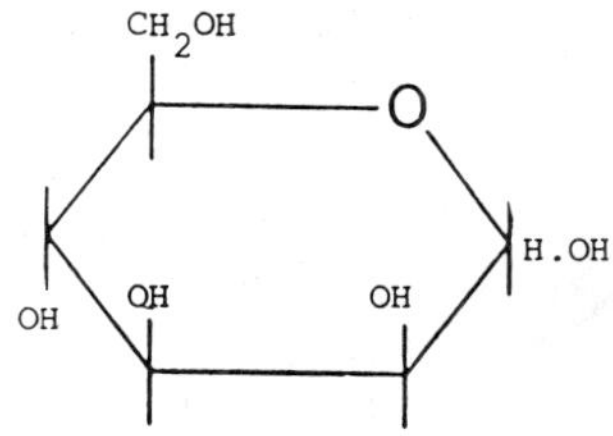

Mannose -Man

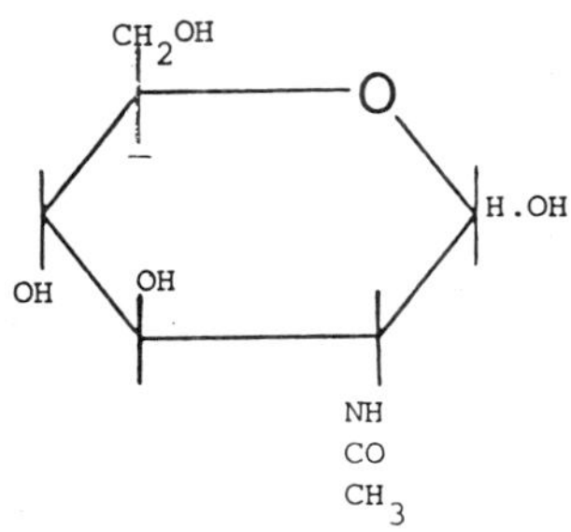

N-Acetylglucosamine -GlcNAc

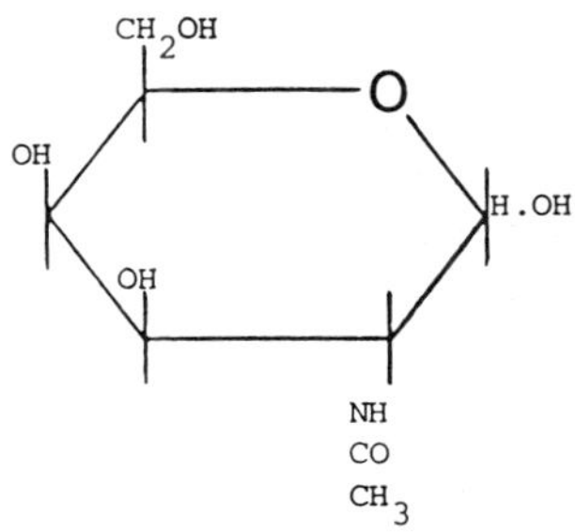

N-Acetylgalactosamine -GalNAc

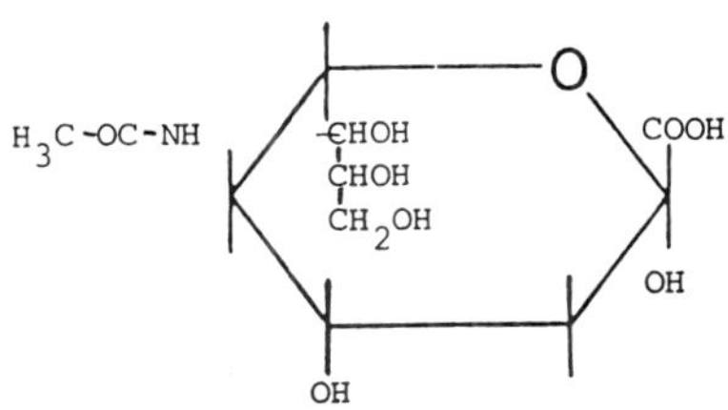

Sialic Acid (SA) or
N-Acetylneuraminic Acid

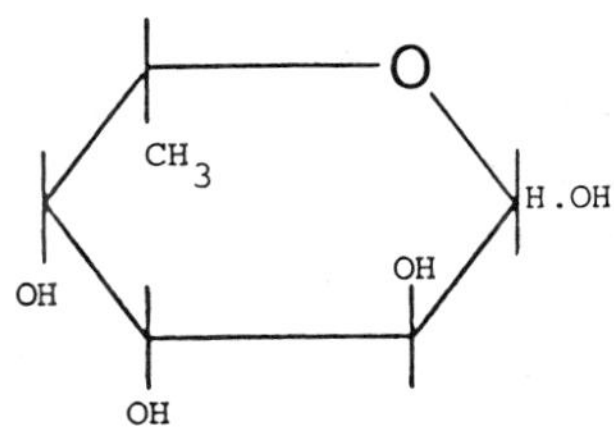

Fucose (F)

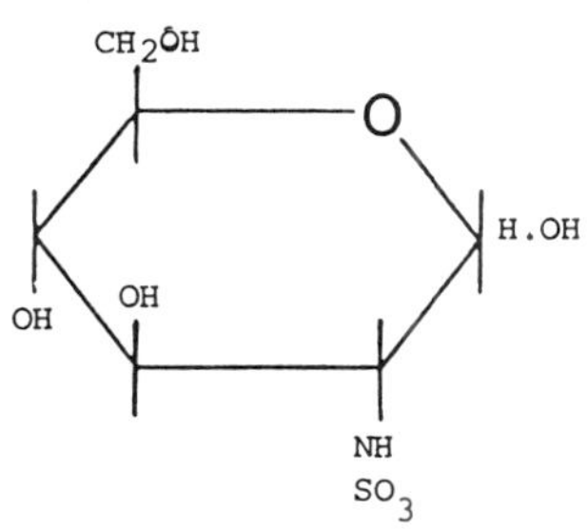

N-sulphated glucosamine ($GlcNSO_3$)

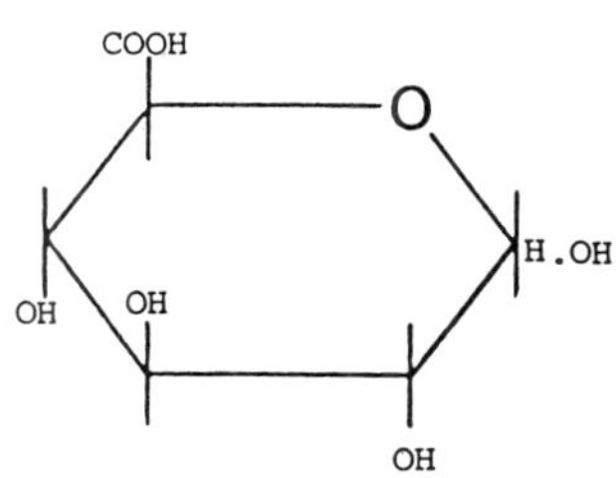

Glucuronic Acid (GlcUA)

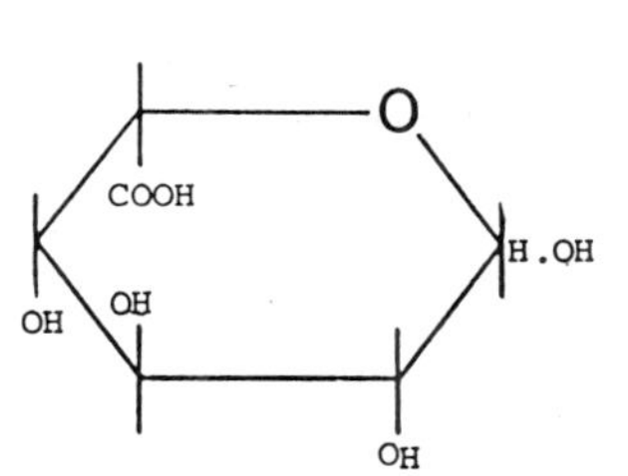

Iduronic Acid (IdUA)

References

1. Singer, S.J. & Nicolson, G.L. (1972) The fluid mosaic model of the structure of cell membranes. *Science*, *175*, 720-31
2. Critchley, D.R. (1979) Glycolipids as membrane receptors important in growth regulation. In R.O. Hynes (ed.) *Surfaces of Normal and Malignant Cells*, John Wiley and Sons, Chichester and New York, pp. 63-101
3. Hakomori, S. (1975) Structures and organisation of cell surface glycolipids dependency on cell growth and malignant transformation. *Biochim. Biophys. Acta*, *417*, 55-89
4. Schachter, H. & Roseman, S. (1981) Mammalian glycosyltransferases: their role in the synthesis and function of complex carbohydrates and glycolipids. In W.J. Lennarz (ed.) *Biochemistry of Glycoproteins and Proteoglycans*, Plenum Press, London and New York, pp. 85-147
5. Montreuil, J. (1980) Primary structure of glycoprotein glycans. Basis for the molecular biology of glycoproteins. *Adv. Carbohyd. Chem. Biochem.*, *37*, 157-233
6. Lindahl, U. Höök, M. (1978) Glycosaminoglycans and their binding to biological macromolecules. *Ann. Rev. Biochem.*, *47*, 385-417
7. Fialkow, P.J. (1976) Clonal origins of tumours. *Biochim. Biophys. Acta*, *458*, 283-321
8. Greaves, M.F. (1979) Cell surface characterisics of human leukaemic cells. In P.N. Campbell & R.D. Marshall (eds) *Essays in Biochemistry*, vol. 15, Academic Press, London, pp. 78-124
9. Greaves, M.F. (1982) Target cells, cellular phenotypes and lineage fidelity in human leukaemia. *J. Cell Physiol.*, Suppl. 1, 113-26
10. Gahmberg, C.G. & Anderson, L.C. (1982) Surface glycoproteins of malignant human leukocytes. *Biochim. Biophys. Acta*, *651*, 65-83
11. Goldstein, I. & Hayes, C. (1978) The lectins: carbohydrate binding proteins of plants and animals. *Adv. Carbohydr. Chem. Biochem.*, *35*, 128-340
12. Lis, H. & Sharon, N. (1977) In M. Sela (ed.) *The Antigens*, vol. IV, Academic Press, London and New York, pp. 429-529
13. Harding, M., Crowther, D. & Gallagher, J.T. (1983) Lectin binding patterns of normal and leukaemic lymphoid and myeloid cells. *Lectins*, III, 225-33
14. Price, M.R. & Stoddart, R.W. (1976) A characteristic protein of the surfaces of neoplastic cells. *Biochem. Soc. Trans.*, *4*, 673-5
15. Bramwell, M.E. & Harris, H. (1978) An abnormal membrane glycoprotein associated with malignancy in a wide range of different tumours. *Proc. R. Soc. Lond. B Biol. Sci.*, *201*, 87-106
16. Bramwell, M.E. & Harris, H. (1978) Some further information about the abnormal membrane glycoprotein associated with malignancy. *Proc. R. Soc. Lond. B Biol. Sci.*, *203*, 93-9
17. Bramwell, M.E., Bhavanandan, V.P., Wiseman, G. & Harris, H. (1983) Structure and function of the Ca antigen. *Br. J. Cancer*, *48*, 177-83
18. Glick, M.C. (1979) Membrane glycopeptides from virus-transformed hamster fibroblasts and the normal counterpart. *Biochemistry*, *18*, 2525-32
19. Warren, L., Buck, C.A. & Tuszynski, G.P. (1978) Glycopeptide changes and malignant transformation. A possible role for carbohydrate in malignant behaviour. *Biochim. Biophys. Acta*, *516*, 97-127
20. Takasaki, S., Ikehira, H. & Kobata, A. (1980) Increase of asparagine-linked oligosaccharides with branched outer chains caused by cell transformation. *Biochem. Biophys. Res. Commun.*, *92*, 735-42
21. Van Beek, W., Tulp, A., Bolscher, J., Blanken, G., Roozendahl, R. & Egbers, M. (1984) Transient versus permanent expression of cancer-related glycopeptides on normal versus leukaemic myeloid cells coinciding with marrow egress. *Blood*, *63*, 170-6
22. Kraemer, P.M. (1979) Mucopolysaccharides: cell biology and malignancy. In R.O. Hynes (ed.) *Surfaces of Normal and Malignant Cells*, John Wiley and Sons, Chichester and New York, pp. 149-198
23. Winterbourne, D.J. & Mora, P.T. (1981) Cells selected for high tumorigenicity or transformed by Simian virus 40 synthesise heparan sulphate with reduced degree of sulphation. *J. Biol. Chem.*, *256*, 4310-20

24. Fidler, I.J. & Hart, I.R. (1982) Biological diversity in metastatic neoplasms. *Science*, *217*, 998-1003
25. Nicolson, G.L. (1979) Cancer metastasis. *Sci. Am.*, *240*, (3), 50-60
26. Nicolson, G.L. (1982) Cancer metastasis. Organ colonisation and the cell surface properties of malignant cells. *Biochim. Biophys. Acta*, *695*, 113-76
27. Yogeeswaran, G. & Salk, P.L. (1981) Metastatic potential is positively correlated with cell surface sialylation of cultured murine tumour cell lines. *Science*, *212*, 1514-16
28. Yamada, K.M. (1983) Cell surface interactions with extracellular materials. *Ann. Rev. Biochem.*, *52*, 761-800
29. McCarthy, J.B. & Furcht, L.T. (1984) Laminin and fibronectin promote the haptotactic migration of B16 mouse melanoma cells *in vitro*. *J. Cell Biol.*, *98*, 1474-80
30. Vollmers, H.P. & Birchmeier, W. (1983) Cell adhesion and metastasis: a monoclonal antibody approach. *Trends Biochem. Sci.*, *8*, 452-5

Further Reading

Cameron, I.L. & Pool, T.B. (eds) (1981) *The Transformed Cell*, Academic Press, New York

Friedman, S.J. & Skehan, P. (1981) Malignancy and the cell surface. In I.L. Cameron & T.B. Pool (eds) *The Transformed Cell*, Academic Press, New York, pp. 67-133

Hay, E.D. (1981) Extracellular matrix. *J. Cell Biol.*, *91*, 205s-23s

Hynes, R.O. (ed.) (1979) *Surfaces of Normal and Malignant Cells*, John Wiley and Sons, Chichester and New York

Lennarz, W.J. (ed.) (1980) *The Biochemistry of Glycoproteins and Proteoglycans*, Plenum Press, London and New York

Olden, K., Parent, J.B. & White, S.L. (1982) Carbohydrate moieties of glycoproteins. A re-evaluation of their function. *Biochim. Biophys. Acta*, *650*, 209-32

Rudden, R.W. (1981) *Cancer Biology*, Oxford University Press, Oxford and New York

Stoddart, R.W. (1984) *The Biosynthesis of Polysaccharides*, Croom Helm, London and Sydney

3 RADIATION CARCINOGENESIS

J.E. Coggle

Contents

3.1 Introduction

The biological effects of radiations are caused by the absorption of the radiation energy in cells and tissues and by the distribution of that energy. These interactions can arise either from external radiation sources outside the tissues or from internal contamination by radioactive substances. If radiations pass through tissues without leaving any energy behind they have no biological effect. The absorption of thermal radiation (heat) energy can cause a detectable rise in body temperature whereas a dose of ionising radiation sufficient to kill a human cannot be detected by the body because it will not increase its temperature by more than 0.001°C. Ionising radiations are composed of photons (X- and γ-rays) or particles (electrons, neutrons, protons, etc.) with sufficient energy to cause ionisation, that is, to remove orbital electrons from the atoms of the material through which the radiation is travelling. The absorption of radiation energy is a random process and so in a complex system such as a mammalian tissue, which is 70–90 per cent water, most of the energy will be involved in ionisation and excitation of water molecules. This radiation splitting (radiolysis) of water produces chemical species known as free radicals and it is these which are primarily responsible, via a series of interactions, for the biochemical and, ultimately, the biological harmfulness of ionising radiation.

At high doses (tens of gray[a]) radiation can cause the cessation of metabolism and cellular disintegration and necrosis, a type of death known as interphase or non-mitotic death. However, much lower doses (1–2 gray) will kill some two-thirds of a population of cells by inhibiting their proliferative ability (see Chapter 9). There is much evidence to link such reproductive or 'mitotic death' with DNA damage and in particular with a loss of chromosomal material. So the response of a tissue to radiation is the summation of the sublethal and the lethal damage its cells sustain coupled with their capacity to repair such damage. The extent of

---

[a] Grays are the unit of absorbed dose: the amount of energy imparted to unit mass of matter such as a tissue. gray (Gy) = 1 joule per kilogram: 1 mGy = 1/1000 Gy; 1 cGy = 1/100 Gy.

the damage suffered is a complex function of physical and biological factors. The physical factors include the size of the radiation dose, the so-called 'quality' of the radiation that is related to density of its ionisations[b] and whether the exposure is a single, a fractionated or a protracted dose.

Radiation may be divided into somatic and hereditary effects. Somatic effects occur in the ordinary cells of the body, whereas the hereditary effects occur in the cells of the reproductive organs and they may be passed on to the offspring of the irradiated individuals.

Somatic effects may occur within hours, days or weeks of exposure and there is general agreement that such 'acute effects' are due to a depletion in the cellularity of the tissues as a result of the delay or prevention of mitosis and to the mitotic death of cells. Radiation also produces late biological effects that do not manifest themselves for months, years or even decades after. One school of thought suggests that late effects are due primarily to vascular damage that leads to the slow and insidious degeneration of the parenchyma of a tissue and to a late fibrosis of its connective tissues. Another school of thought suggests that the wide diversity of late radiation damage, from fibrosis to the induction of cataracts, is best explained in terms of the gradual mitotic death of cells that will inevitably occur at a very slow rate in a tissue with a very slow rate of cell turnover.

We have already noted that radiation-induced cell lethality is probably due to severe DNA damage such as rearrangement and loss of chromosomal material. Double-strand breaks in DNA have often been proposed as the critical molecular lesion but there are data against this hypothesis and it remains a moot point whether lethality is due to a class of unrepaired DNA strand breaks. Radiation is able to induce sublethal gene and chromosomal mutations and the increase in such damage is the basis of its hereditary effects. Similar 'mutations' induced in somatic cells probably form the basis of radiation carcinogenesis.

3.2 Somatic Mutation Theory and Induction of Cancer by Radiation

The somatic mutation theory of cancer suggests that the DNA of a cell becomes altered or mutated so that its information content is changed. The mutation might be an invisible gene mutation (i.e. a point mutation or a frameshift mutation) or an actual breakage or loss of a chromosome might occur. If such a defective cell is capable of indefinite division

---

[b] The quality of radiation or its linear energy transfer (LET) is the rate of loss of energy along the track of an ionising particle and is expressed in keV per micrometre of track. High LET radiations such as neutrons and α-particles generally have denser tracks than low LET (X- and γ-ray) radiations.

independently of the control mechanisms of the body, it is tantamount to a cancerous cell. So the mutation theory amounts to little more than a re-description of the term 'cancer', e.g. a cancer cell is a cell with a mutation giving it cancerous potential. However, the theory is attractive because it is fairly consistent with the experimental evidence both in radiation and chemical carcinogenesis. In particular, recent advances in molecular biology involving cellular transforming genes (cancer genes or oncogenes; see Chapter 4),[1] transfection experiments[2] and work with cell hybrids[3] and cybrids[4] have enormously strengthened the theory that most cancers have a mutational origin. But carcinogenesis is undoubtedly more than mutagenesis: it is a multistage progressive process. At its simplest it is probably a two-stage process consisting of an initiation and a promotion step (see Chapter 5) and a recent hypothesis has suggested that mutagenesis is responsible for the initiation phase and an epigenetic mechanism is the basis of the reversible promotion step.[5] Error-prone repair or replication of DNA damage may be the molecular basis of mutagenesis, whereas promotion is the process that allows such initiated (precancerous?) cells to escape the anti-mitotic control mechanisms of normal cells. Promotion by epigenetic factors is defined as those changes in a cell that modify the expression of the cancer genes either by their redepression or by post-translational modification of their genetic information.

Information on radiation-induced mutations in somatic cells is rudimentary. Most of the quantitative information is from studies of just a few loci, such as mutations at the hypoxanthine-guanine phosphoribosyltransferase (HG-PRT) locus and mutations to ouabain and to methotrexate resistance. There is good evidence that a significant proportion of such mutations induced by radiation are associated with chromosomal deletions and gene rearrangements but the details of the initial molecular mutational lesions have not been identified. There is even less knowledge of the molecular lesions induced by radiation that transform a normal cell into a cancer cell; suffice to say that radiation seems to act as a complete carcinogen, i.e. it is both an initiator and a promoter. And as we shall see the complexity of the dose–response relations for radiation carcinogenesis wholly support the multistage nature of carcinogenesis.

One of the first observations of cancer after radiation was the appearance of skin tumours on the hands of early X-ray workers, and up to 100 such cancers were reported by 1911, i.e. within 15 years of Roentgen's discovery of X-rays. Since then many systematic studies on animals have shown that radiation causes an increase in the incidence of almost all types of naturally ('spontaneously') occurring cancers. The malignant tumours do not appear until long after the exposure, the delay may be as long as 30–40 years in the case of some human cancers.

Between the exposure and the appearance of the tumour there may be no observable defect in the tissues that eventually become cancerous.

Radiation is capable of inducing tumours in almost all of the tissues of the body, although tissues vary in their susceptibility to radiation-induced cancer (see Table 3.1).

Table 3.1 Relative Sensitivity of Some Human Tissues to Radiation-induced Cancer

High sensitivity	Moderate sensitivity	Low sensitivity
Bone marrow	Breast	Skin
(acute leukaemias	Lung	Bone
myeloid leukaemia)	Salivary glands	Stomach
Thyroid		Pancreas

The fact that radiation may induce tumours in nearly all tissues distinguishes it from most chemicals and viruses that can cause tumours only in a few specific tissues.

As is the case for many other effects of radiation, a high dose rate is generally more effective in causing tumours than a low one, and high linear energy transfer (LET) radiation is generally more effective than sparsely ionising radiation in causing tumours.

Tumours most commonly appear in the tissues that have been directly exposed to radiation, but there are some cases where cancer induction is an indirect process and in which systemic effects play an important role. For example, thymic-lymphoma induction in mice seems to require a complex interaction between the killing of bone marrow and lymphoid tissue, the release of virus and the incorporation of the virus into the unirradiated thymus cells.[6] Or again, ovarian-tumour induction seems to require a sufficient level of oocyte-killing such that the ovary–pituitary axis then upsets the gonadotropin levels and this in turn triggers the development of ovarian tumours.[7]

Cancers induced by radiation are indistinguishable from naturally occurring cancers and so the quantitation of the risks of cancer development has to be made on a statistical basis and can never be absolute. It is the object of most of the experimental and epidemiological work with radiation carcinogenesis to reduce this inherent statistical uncertainty in the risk estimates for radiogenic cancer.

We shall divide the remaining discussion into three parts, essentially the three sources of information that are available for estimation of cancer-induction rates. These are (1) aspects of the fundamental radiobiology of dose–response curves, (2) experimental animal data and (3) human epidemiological data.

3.3 Qualitative Aspects of Dose–response Curves for Cancer Induction

In this section we shall consider the question 'what is the shape of the curve relating the incidence of cancer with the radiation dose?'

It should be stated at the outset that this question will not be satisfactorily answered until there is a greater understanding of the mechanism of carcinogenesis. Nevertheless, certain assumptions about the shape of the dose–response curve can be made by using basic radiobiological concepts.[8] These theoretical curves can then be compared with the experimental data from animal studies and with human epidemiological data to see which, if any, of the general shapes is consistent with the empirical evidence.[9,10]

Figure 3.1 shows a variety of theoretical dose–response curves for radiation-induced cancer.

Figure 3.1 Theoretical Dose–response Curves for Radiation Carcinogenesis

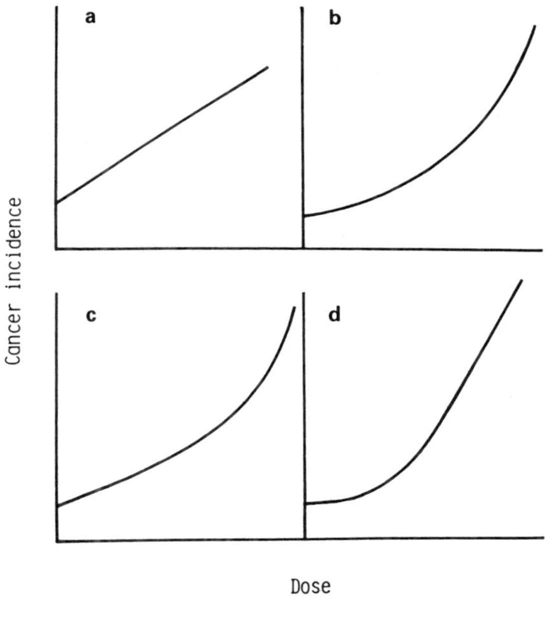

a, Linear; **b**, quadratic; **c**, linear quadratic; **d**, threshold and curvilinear curve.

The linear relation between dose and incidence of induced cancer was first adopted some 25 years ago as the simplest working hypothesis for radiation protection purposes. It is still held by many authorities to be consistent with much of the experimental and human data. The linear function is given by

$$I = c + aD \tag{1}$$

where I is the cancer incidence after dose D, c is the cancer incidence of the unirradiated control population and a is the coefficient that determines the slope of the linear curve in Figure 3.1**a**.

If two independent initiating events are required to transform a normal cell in to a cancer cell then the form of the dose–response curve would be a quadratic given by the function

$$I = c + bD^2 \tag{2}$$

in which the cancer incidence (I) varies with the (dose)^2 (Figure 3.1**b**).

A somewhat intermediate dose–response relation is the linear-quadratic form expressed by the function

$$I = c + aD + bD^2 \tag{3}$$

which combines equations (1) and (2). This linear-quadratic model has received much attention in the last few years because certain of the principles of microdosimetry, which are outside the scope of this chapter, would favour such a hypothesis. The important point about the linear-quadratic model is that the initial part of the dose–response curve, i.e. the low-dose region, will show no threshold and will be linear. At higher doses the cancer incidence will increase more steeply than predicted by the linear relation (Figure 3.1**c**).

Finally, Figure 3.1**d** shows a theoretical curvilinear relation that has a threshold of dose, i.e. a dose at which there is no risk of cancer induction. We need not concern ourselves with this relation since it is not possible to prove experimentally the existence of a threshold at low doses.

Both human and experimental cancer-induction curves often tend to plateau or even show a peak incidence at a certain 'turnover dose point' after which the cancer incidence falls with increasing dose. Figure 3.2 shows a theoretical relation of this type. The seemingly complex curve is explained by assuming a competition between two processes, the induction by radiation of a malignant transformation of a normal cell and the possibility that such a transformed cancer cell will be killed or at least lose the ability to divide.

The general relation between dose and cell-killing can be represented by the equation

$$f = \exp\left[-\left(\alpha D + \beta D^2\right)\right] \tag{4}$$

If we combine equations (3) and (4) we get the most general function for the dose–response relation

Figure 3.2

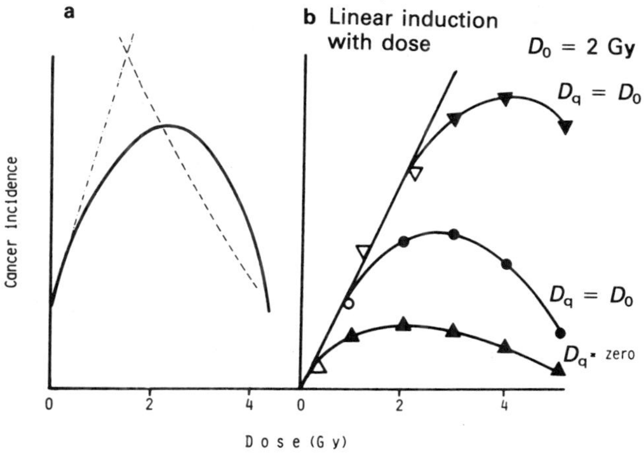

a, Observed cancer incidence (———) as a function of the competition between induction (–·–·–) and cell death (- - - - -).
b, Diagrammatic representation of the way in which the shape of the cell-killing response modifies the observed cancer yield.

Induction is linearly proportional to dose. Open symbols, points unaffected by cell-killing; closed symbols, points where the observed yield is reduced in proportion to cell sterilisation. D_q, the quasi threshold dose, is a measure of the size of the shoulder on survival curves; D_o is the dose required to reduce survival by a factor of e^{-1} (0.37) along the exponential part of a survival curve.
Source: Mole, R.H. (1975) *Br. J. Radiol.*, **48**, 157; courtesy the author and publisher.

$$I = (c + aD + bD^2) \exp [- (\alpha D + \beta D^2)] \tag{5}$$

which with suitable coefficients will fit the shape of the curve in Figure 3.2**a**.

Figure 3.2**b** illustrates the effect changing the parameters of the cell-inactivation term in equation (5) would have on the final cancer-incidence curve. For example, if there is no shoulder to the killing curve (D_q = zero), the initial linear part of the induction curve rapidly plateaus and tends to decrease as the doses increases (Figure 3.2**b**).

As we shall see in the next sections both the human and the experimental cancer-induction curves show a wide variety of shapes. Also the individual data points that make up a cancer-induction curve usually have rather large confidence intervals. In fact such is the variability of the data that statistical curve-fitting is often inadequate to distinguish between the different models of dose–response relations shown in Figure 3.1. This is especially true at the low-dose region of the curves, i.e. the region of greatest interest for cancer-risk assessment.[11] Despite the relatively large body of data on human populations there is still great uncertainty, and therefore scope for controversy, about the cancer risks associated with low doses and low dose rates. All but one of the major national and international committees that considered cancer-risk estimation in the last decade favoured a linear no-threshold relation for human radiogenic cancer. The committees all accepted that a number of non-linear curves could be drawn through the limits of error of the data. They were simply affirming that the linear fit was not only the simplest and administratively convenient but was also 'consistent with the data'. The linear hypothesis also has the merit that it is conservative. For low LET radiation it may even overestimate the real cancer risk at low doses. Curve B in Figure 3.3 is an idealised curvilinear cancer-induction curve for low LET radiation from animal and human data. The broken line drawn through the curve to the origin is the linear extrapolation that would be used to estimate the cancer incidence at the low-dose region where there are no dose points. One can see this linear extrapolation will overestimate the risk in the low-dose region. In contrast, for high LET radiation, since in some experiments the data points turn over at relatively low doses (curve A) the broken line which is a linear extrapolation consistent with the points might underestimate the low-dose cancer risk.

Figure 3.3 Diagrammatic Curves of Cancer Incidence vs Dose After a High (curve A) and Low (curve B) LET Radiation

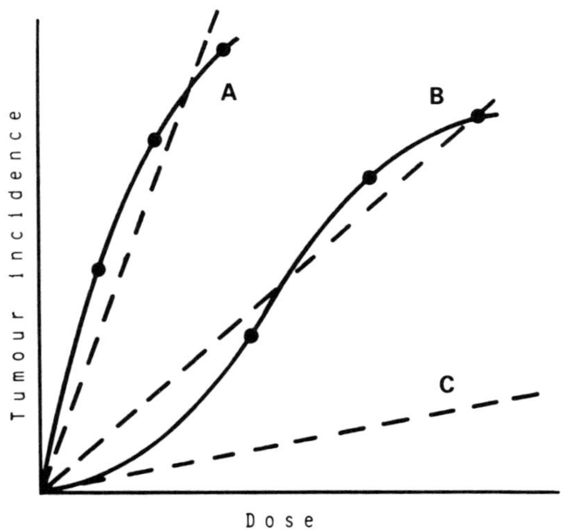

Curve C shows the likely reduced effectiveness of low dose rate, low LET radiation.

Source: Fry, R.J.M. (1981) *Radiat. Res.*, **87**, 224; courtesy the author and publishers.

Figure 3.3 also illustrates another important point: the influence radiation quality has on the cancer-incidence curve. High LET radiation induces more cancer per unit of dose than low LET radiation and, because of the shape of the curve, this relative carcinogenic efficiency is likely to rise sharply at low doses.

Another factor that often modifies the efficiency with which radiation induces cancer is dose rate (curve C, Figure 3.3). This will be dealt with in detail in the next section.

3.4 Experimental Radiation Carcinogenesis

The induction of fatal cancer is regarded as the most significant risk of radiation to man. Consequently much effort has been put in to experiments to understand the mechanisms of cancer induction and to provide answers to such important questions as What is the shape of the dose–response curve at low doses? Does the shape vary from tissue to tissue? What effect do radiation quality and radiation dose rate have on the dose–response curve for cancer? and, finally, To what extent are experimental dose–effect relations, obtained by using animals, relevant to human radiogenic cancer risk rates?

Most experiments have been carried out with small rodents and for acute high dose-rate exposures the animals are irradiated for perhaps a few seconds or minutes, whereas for chronic low dose-rate exposures the animals may be continuously irradiated for their entire life-span. Both control and irradiated mice are housed under identical conditions and the tumour incidences at various times postirradiation and at death can be compared. Figure 3.4 shows a small selection of the dose–response curves that have been obtained in mice and rats for a variety of tumours.[12-19] The immediate impression is the wide range of shapes, from linear to highly curved, some with plateaus, some with turnover points (Figure 3.2**a**) and some with thresholds.This variability is not surprising when one considers that different cancers may arise by a different sequence of events. For example, in most instances, radiation really induces tumours, whereas in other cases it merely accelerates the time of appearance of a spontaneous incidence of cancer without increasing the overall lifetime risk. It is likely that the mechanism(s) involved in these two distinctly different processes is different and the shapes of the incidence curves are also likely to differ. Similarly, when one considers the multiplicity of factors that may intervene between the initiating event(s) [somatic mutation(s)?] and the final expression of the cancer it is not surprising that dose–response curves are highly variable. The final expression of transformed 'precancerous cells' can be modulated by immunological, hormonal, vascular and nutritional factors and together with cell-

Figure 3.4

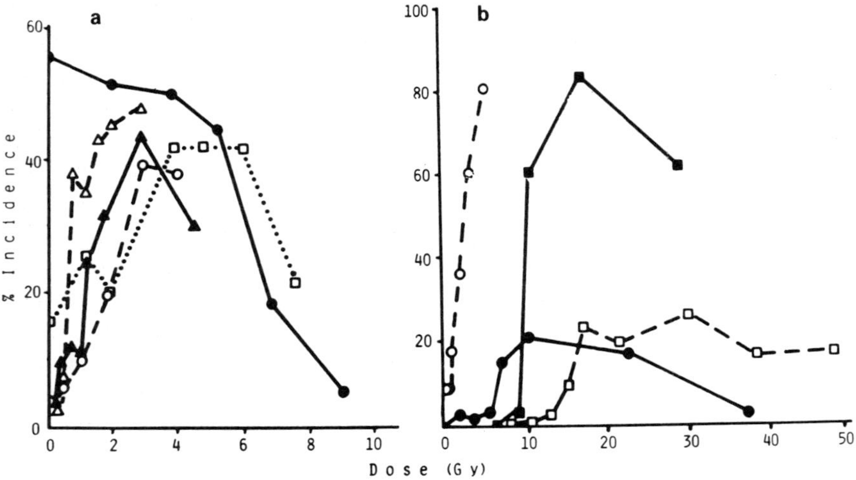

a, Dose–effect relation in mice for the induction of myeloid leukaemia[12] (▲), thymic lymphoma[13] (○), reticular[14] (●), ovarian[15] (△) and lung[16] (□) tumours.
b, Dose–effect relation in rats for the induction of mammary (○), kidney[18] (■) and skin[19] (□, ●) tumours.

Table 3.2 Factors that Affect Dose–response Curves for Radiation Carcinogenesis

Physical	Biological
Whole body, partial body; external, internal radiation; uniform or point source radiation; single, fractionated, protracted doses; dose rate, radiation quality	Age at exposure; tissue at risk; sex, hormonal, immunological status; genetic traits; 'spontaneous' incidence; repair capacity; interaction with other chemical, physical or biological (viral, etc.) carcinogenic agents

proliferation kinetics they may all influence the final cancer incidence (Table 3.2). We saw in Figure 3.2**b** how small alterations in the cell-killing parameters D_o (the sensitivity of the cells) and D_q (the repair capacity of the cells) drastically alter the final shape of the induction curve.

One of the important lessons from experimental radiation carcinogenesis is that the variety of dose–response curves probably implies different mechanisms for different types of tumours.

The variables above contribute to the large standard deviations that are characteristic of the individual points on many incidence curves (see Figures 3.7–3.9). Such statistical variations usually make it impossible to define accurately the exact shape of the cancer incidence at low doses (<< 0.5 Gy). This is true even for the largest and most rigorously controlled experiments. The statistical problems of the low-dose region also mean that experimental carcinogenic data cannot be used, with any confidence, to provide numerically useful data for human-risk estimation. The uncertainty of extrapolating mouse data to man outweighs the uncertainties and flaws that characterise the human data themselves.

It is probably true to say that the experimental search for the true shape of the dose–response curve as opposed to theoretical considerations (see section 3.3) is as doomed as the search for the holy grail. Given the multiplicity of factors the choice of a linear dose–response curve is probably scientifically unsound, even if it is utterly sensible from an administrative radiation-protection point of view.

In an attempt to avoid the biological complexity of carcinogenesis experiments *in vivo*, workers have turned to systems *in vitro*. Such assay systems have been developed recently in which it is possible to quantify the number of normal cells that have been transformed by radiation into malignant cells. At present the technique only works with a very few cell types such as Syrian hamster embryo cells, mouse $C_3H/10T\frac{1}{2}$ and 3T3 cells and human skin fibroblasts.[20] This last involves irradiating a known number of cells and observing the subsequent colony growth. The colonies produced by transformed cells are very distinctive and the cells of such colonies can be tested *in vivo* for their malignancy; less than one cell in a million will spontaneously transform into a cancer cell. By using this system it is possible to obtain rather precise dose–response relations over a wide range of doses, starting at dose levels that are impractical in animal experiments, e.g. $\leqslant$ 0.01 Gy. These transformation-incidence curves are proving to have unexpectedly complex shapes. Figure 3.5 shows such a curve for single doses of X-rays. It consists of three parts: at doses above 1 Gy the data are consistent with a quadratic dependence on dose; below 0.3 Gy the data show that the number of transformed cells is directly proportional with the dose, i.e. a linear relation; while between 0.3 and 1 Gy the frequency of transformed cells does not vary with dose. The data as a whole cannot be fitted by a single straight line and a linear regression line (the broken line, Figure 3.5) would underestimate the incidence of transformed cells at low doses. These curves *in vitro* remain unexplicable. One might have expected a somewhat simpler dose–response relation *in vitro* since the assay is of cancer initiation and does not involve any of the complicating factors *in vivo* noted earlier.

Figure 3.5 Dose–response relation for Transformation *in vitro* of Normal Mouse Cells ($C_3H/10T^1{}_2$ cells) to Malignant Cells After Exposure to Single Doses of X-Rays

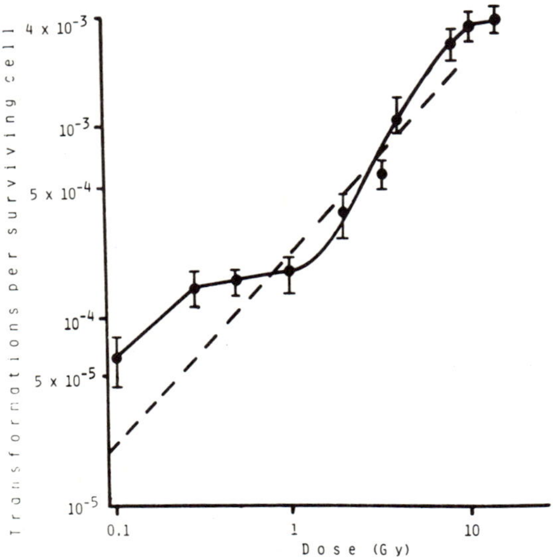

The broken line is a linear regression line fitted to the points.

Source: Hall, E.J. & Miller, R.C. (1981) *Radiat. Res.*, **87**, 208; courtesy the authors and publishers.

It is firmly established that protracted or fractionated doses of low LET radiation are less biologically effective for the same total dose than single acute doses. This is also true for cancer induction and discussion of what factor ought to be introduced to allow for the possible reduced effectiveness of low dose rate in carcinogenesis is always a topic of lively debate in radiation-protection circles. Since there are so few useful human data regard has to be paid to animal evidence. Figure 3.3, curve C, shows the theoretical cancer incidence expected for low LET radiation at a low dose rate. It is derived by linearly extrapolating the low-dose region of curve B in Figure 3.3. Figure 3.6 shows some experimental animal data for cancer induction for a range of doses given at either a high or low dose rate. A 'dose-rate effectiveness factor' (DREF) can be derived from such curves by taking the ratio of the slopes of the high and the low dose-rate curves. Values between 1 and 10 have been obtained for such DREFs for different tumours, i.e. for some tissues changes in dose rate do not alter the number of tumours induced, while for other tissues radiation at a high dose rate is ten times as effective as

Figure 3.6 Influence of Dose Rate for a Variety of Experimental Mouse Tumours

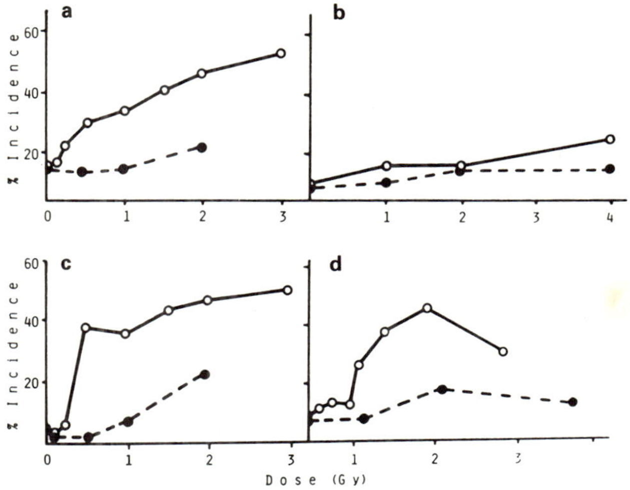

a, Thymic lymphoma at either 0.45 (○) or 0.08 (●) Gy/min; **b**, mammary tumours at either 0.45 (○) or 0.08 (●) Gy/min; **c**, ovarian tumours at either 0.45 (○) or 0.08 (●) Gy/min; **d**, myeloid leukaemia at either 0.8 Gy/min (○) or 0.75 (●) Gy/day.

Source: **a-c** Ullrich, R.L. and Storer, J.B. and **d** Upton, A.C. (1980) National Council on Radiation Protection and Measurements Report, No. 64; courtesy the authors and publisher.

low dose-rate radiation. Recognition of such dose-rate effects in radiation protection is important, since human occupational exposure is usually at very low dose rates, while cancer-risk estimates are generally derived from acute high dose-rate exposure data (see later). Intriguingly, there are some animal data to indicate that for high LET radiation (e.g. α-particles) protracted and fractionated doses are more effective at inducing tumours than acute doses. At present there is no entirely satisfactory explanation for such findings.

Finally, there is considerable experimental evidence that high LET radiation is more effective at inducing cancer than low LET radiation. A wide variety of so-called 'relative biological effective' (RBE) values can be found in the research literature. (Relative biological effectiveness is the ratio of doses of two types of radiation to produce the same effect.) Such RBE values at doses above 1 Gy are often close to unity. However, because of the relative shapes of the high and low LET dose–response curves (Figure 3.3), RBE values tend to increase sharply at low doses and may be >> 10 for low doses (around 0.01 Gy) and low dose–rate high LET radiation.

We have just touched on some of the variables known to affect the dose–response curve for radiation carcinogenesis in experimental animals and Table 3.2 lists some of the other factors which, if they were fully understood, would probably give us a fairly definitive understanding of radiation carcinogenesis.

3.5 Human Data

Radiogenic cancer in man has been studied since 1902 when a skin cancer was reported on the hand of a radiation worker. This was some seven years after Roentgen's discovery of X-rays. There can of course be no cancer-induction experiments in man comparable with those in animals. Nevertheless, observations on people deliberately or accidentally exposed to radiation have shown that radiation doses too small to cause macroscopic tissue damage, carry an increased risk of cancer. It is widely accepted that this increased risk is approximately proportional to dose (i.e. approximately linear with dose) down to doses of the order of 0.01 Gy.[21] The risk also applies to most of the organs or tissues of the body.

There are many problems associated with risk estimation:

(1) The latent period between exposure and the appearance of a tumour may be decades, which necessitates long follow-up periods
(2) Dose assessment is often imprecise. For the A-bomb survivors the dosimetry is all retrospective and is currently being revised, yet again (see later). For internal radiation there are problems of dose distribution and the choice of the relevant dose especially where the radioactive substance has a long biological half-life. There is also often a chemical toxicity associated with the isotope that has to be taken into account: for example, 'thorotrast' (thorium oxide) in the liver causes a great deal of cell death from chemical toxicity and separating the chemical and radiation effects in assessing the risk of induction of liver cancer is impossible
(3) The dose levels may not be suitable for risk estimation. They may cover such a narrow range that linear extrapolation to different doses may be invalid (see Figure 3.3). The dose distribution may also be important in risk assessment. It may not be justifiable to assume that a high dose to a small fraction of the body carries the same carcinogenic risk as the same total dose given uniformly to the whole body
(4) Then there is the important problem of what 'controls' to compare with the exposed population. This is particularly a problem with radiotherapy patients and studies may lead to false conclusions unless the correct control population is considered. For instance, it was thought that treatment with radioactive iodine for hyperthyroidism increased the

incidence of leukaemia, but when an equivalent group of people with hyperthyroidism who had not been treated with radiation was studied it was found that these people had a high natural incidence of leukaemia. The selection of a control population to compare with the survivors of Hiroshima and Nagasaki is also difficult. A comparison with Japan as a whole, a largely rural population, is invalid because disease incidences are different in urban and rural communities, so a suitable urban Japanese population must be found. The control group most often used is the population in the two cities who received less than 0.1 Gy. But even then the exposed group is an unrepresentative urban population as there were no men of military age in the cities at the time of the bombings. Also, there was a great deal of immediate mortality that must have been to some extent selective and which may account for the generally lower risk estimates obtained from the A-bomb survivors compared with other irradiated groups

(5) Finally, there are statistical problems associated with the need to observe large populations for long periods to detect the small increases in tumour incidence. These are often insurmountable and lead to wide confidence limits on risk estimates.

Despite the difficulties, there are more than a dozen different organs of the body for which there is good retrospective epidemiological data. These involve populations of known size whose doses are well documented and who have been followed for prolonged periods and with an adequate control population. These populations are listed in Table 3.3. Each population gives us data on different modes of exposure whether single, acute or whole-body doses of external radiation (A-bomb survivors),[22] partial-body irradiation (radiotherapy exposures)[23] or protracted internal exposures (dial painters,[24] uranium miners[25] and thorotrast patients[26]). The populations include the victims of initial and radioactive fallout from nuclear weapons, various categories of patients including those with ankylosing spondylitis (see section 3.5.1) and those given a radioactive contrast medium 'thorotrast' (colloidal [^{232}Th]thorium oxide), and various groups who have received significant occupational exposure including uranium miners and dial painters. This last group received internal contamination with long-lived radium isotopes which used to be present in luminous paint.

Thyroid, breast, lung and bone cancer and leukaemia are the best documented of human radiogenic cancers. Evidence of these five cancers comes from more than one source. In Table 3.4 each + represents an independent epidemiological survery. Such independent risk estimates are of great corroborative value. Risk estimates for other cancers are less reliable but we shall return to the specific values of risk estimates later.

Until 1980, the most useful source of information on the carcinogenic

Table 3.3 Populations that are Used for Cancer-risk Estimates

Nuclear weapons	
	A-bomb survivors (Hiroshima and Nagasaki)
	Tests (Marshall Islanders)
	Other fallout
Medical radiotherapy	
	Spondylitics
	Other non-malignant conditions, e.g. pelvis, scalp
Medical diagnosis	
	Fluoroscopy (TB)
	Thorotrast
	Pelvimetry (fetus)
Occupational	
	Uranium and other miners
	Dial painters
	Radiologists
	Nuclear industry
Natural environment	
	Areas of high natural background

Table 3.4 Sources of Risk Estimates for Different Radiogenic Cancers

	Leukaemia	Thyroid	Breast	Lung	Bone
Japanese survivors	+	+	+	+	−
Marshall Islanders	−	+	−	−	−
Radiotherapy patients	+++	++	+	+	+
Medical diagnosis	−	−	+	−	+
Uranium miners	−	−	−	+	−
Dial painters	−	−	−	−	+

hazard of both high LET (neutrons) and low LET (γ-rays) radiation came from the cancer incidences in the Japanese survivors of the atomic bombings. A joint Japanese–USA team has followed the medical history of some 100 000 A-bomb survivors for over 30 years. However, a re-evaluation in 1980–2 of the dosimetry of the A-bombs showed significant differences from the previously computed dose values in 1965. A great deal of work still needs to be done before the 'new' A-bomb dosimetry can provide a firm basis for a revised set of radiation-risk standards.[27] Figure 3.7 shows a preliminary assessment of the difference in the dose–response curves for Hiroshima and Nagasaki by using the 1965 dose estimates (**a**) and the 1980–1 dose estimates (**b**). The difference between the cities in **a** was attributed to the high neutron component at

Hiroshima. The 1980–2 re-evaluation of the A-bomb dosimetry suggests that there were virtually no neutrons at either city and that the γ-irradiation dose at Hiroshima was underestimated and at Nagasaki it was overestimated. These adjustments effectively abolish the differences between the cities and both sets of data fit the same linear quadratic curve at least up to about 2 Gy (Figure 3.7**b**).

Figure 3.7 Dose–response Curves for Leukaemia for Hiroshima (●) and Nagasaki (○)

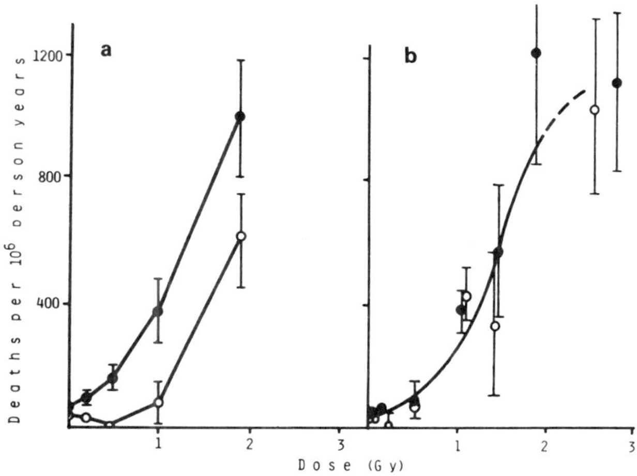

a, By using the 1965 dosimetry level; **b**, by using the 1980–1 dosimetry level.

Source: **a**, Rossi, H.H. & Mays, C.W. (1978) *Health Phys.*, **34**, 353; courtesy the authors and Health Physics Society.
b, Straume, T. & Dobson, R.L. (1981) Lawrence Livermore Laboratory Reprint, UCRL, 85446; courtesy the authors.

Figure 3.7**b** also illustrates two important points made earlier in this chapter. First, the wide confidence limits on the individual leukaemia-incidence points make any risk estimates statistically uncertain. Secondly, the linear-quadratic dose–response relation means that the use of a linear extrapolation would tend to overestimate the real leukaemia risk at least for this set of data (see also Figure 3.3).

In contrast to the linear-quadratic in Figure 3.7**b**, the bone-cancer incidence in Figure 3.8 for α-particle irradiation shows a negatively (upwardly) sloping relation with dose. The data come from persons exposed to the bone-seeking radionuclide radium-226 between 1915 and 1935. The group includes dial painters and some patients given radium therapeutically. A linear extrapolation would underestimate the real

Figure 3.8 Dose–response Data for Bone Cancer in a Group of Radium-226-exposed Dial Painters and Radiotherapeutically Exposed Patients

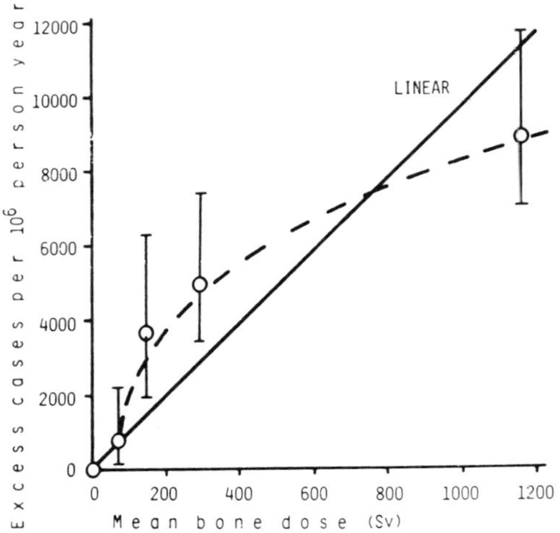

Source: National Research Council, Advisory Committee on the Biological Effects of Ionizing Radiation, 1972. BEIR I Report; courtesy the National Academy of Sciences, USA.

bone-cancer risk for such a curve. These data and those from other bone-cancer studies show a tendency to plateau at high doses.

The effect of dose and cancer induction was discussed earlier and it was noted that, in general, it is expected that for low LET radiation the risk per unit dose will decrease as the dose rate decreases. However, this is not always the case and for cancer of the breast there is no significant difference in the risk rate between women receiving several small doses in multiple fluoroscopic examinations carried out over several years, when compared with the risk of large doses from high dose-rate radiotherapy. And for high LET radiation there are even data to suggest that protracted doses are more effective than acute doses. For example, Figure 3.9 shows the incidence of bone cancers (osteosarcomas) in juveniles and adults after intravenous injection of a colloidal preparation of the α-particle-emitting isotope radium-224 called 'peteosthor'. As the dose is spread out in time the carcinogenic effect per unit dose increases from 40 to 200 per 10^4 per Gy of α-particle irradiation. Such data are very relevant to any assessment of α-particle effectiveness when received over long periods as in occupational exposure.

From the numerous surveys that have been conducted into carcinogenesis in man we can obtain both qualitative information and

Figure 3.9 Dose–response Data for Bone-sarcoma Risk vs Protracted Doses of Radium-224

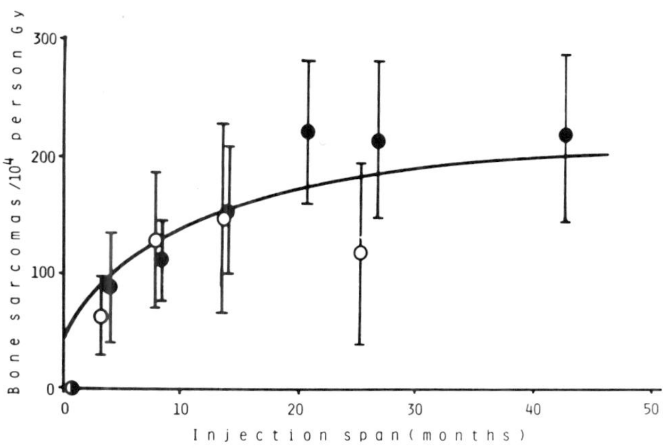

Protraction in children (●) and adults (○).

Source: Speiss, H. & Mays, C.W. (1973) In *Radionuclide Carcinogenesis*, US Atomic Energy Agency Symposium Series, vol. 29, p. 442; courtesy the authors.

quantitative estimates of human radiogenic cancer risks. Before giving these risk estimates it will be instructive to give details of two such surveys.

3.5.1 Ankylosing Spondylitics

In the 1950s there was concern in the UK that the radiotherapy given to patients with ankylosing spondylitis (arthritis of the spine leading to complete stiffness of the back) might have very serious long-term effects. Consequently, several epidemiological surveys were carried out and to date 14 111 patients with spondylitis treated with X-rays between 1935 and 1954 have been followed and compared with a control group of 1021 patients with spondylitis not given X-ray treatment.[23] Significantly more cancer deaths were recorded in all sites that were heavily irradiated, e.g. the oesophagus, stomach, lungs, spinal cord and nerves. No real excess in cancer was recorded in sites that received minimal doses from the scatter of the X-ray beam that was primarily directed at the spinal region.

The excess mortality from leukaemia occurred between three and eight years after X-ray treatment, with no excess recorded at more than 20 years after treatment. In contrast, excess mortality from other cancers was maximal between nine and 20 years with a few cancers appearing at latent periods greater than 20 years. The relation between leukaemia incidence and dose was linear for mean spinal bone-marrow doses between less than 1 Gy and 6 Gy. The data indicate that a total of ten fatal leukaemias would be expected per cGy per 10^6 persons over a 15-year period (upper limit of 13 cases per 10^6 per cGy).

3.5.2 Childhood Cancers from Irradiation in utero

A survey of cancer in children in the UK was started in 1954 to discover why there had been a progressive increase in leukaemia deaths in young children in the interwar years. The study covering the period 1953–1967 involved a retrospective survey by medical officers interviewing mothers of 8513 children who had died, mostly under the age of 10 years, from a malignant disease. An equal number of mothers whose children had not died were also interviewed. All mothers were asked about their radiological history during pregnancy and their answers checked with hospital records. The outcome of this UK survey and another in the USA is that it is now firmly established that obstetric diagnostic radiography causes an increased risk of malignant cancer in childhood. The overall estimate of this risk is that there will be some 200–250 cases of fatal malignant disease in children under 10 years old as a result of the antenatal exposure of 10^6 human embryos and fetuses to 10 mGy. Of the 20–25 malignant cancers per year half will be leukaemia and half will be solid tumours. Furthermore the excess fatal cancer risk increases linearly with the number of X-ray exposures, i.e. from doses between 2 and 200 mGy. The relative overall risk of such radiography was about 1.47, i.e. 47 per cent more fatal cancers will occur in the children of mothers who received radiography than those that did not. In the UK study the excess risk was found to be five times greater for radiation given in the first three months of pregnancy than for the second and third trimesters. So a dose as low as 10 mGy given in the first few weeks of pregnancy might cause cancer in 1 in 1000 children. The natural prevalence of fatal cancer in children less than 10 years old is 1 child in 2000. These findings were intially greeted with much scepticism but they are now widely accepted as part of the increasing body of evidence that low doses can cause cancer.[28]

3.6 Quantitation of Cancer Risk

Epidemiological data such as that just mentioned, is too imprecise to determine the exact shape of the dose–response relation for human radiogenic cancer. Also it is usually necessary to estimate the cancer risk at low doses ($\leqslant 0.01$ Gy) by extrapolation of observations made at high doses (> 1 Gy). Such extrapolation over one or two orders of magnitude add to the uncertainty of our risk estimates.

Table 3.5 ICRP Risk Factors for Induction of Cancer (Death Rates per Sievert)[a]

Leukaemia	2.0×10^{-3}
Breast	2.5×10^{-3}
Lung	2.0×10^{-3}
Bone	0.5×10^{-3}
Thyroid	0.5×10^{-3}
All other tissues	5×10^{-3}
Total risk	1.25×10^{-2}

[a] For low LET radiation: dose (Sv) ≡ dose (Gy)

Nevertheless quantitative risk assessments have to be made and have been made by a number of national and international committees including the International Commission on Radiological Protection (ICRP), the United Nations Scientific Committee on the Effects of Atomic Radiation (UNSCEAR) and by the US Academy of Sciences in their Biological Effects of Ionizing Radiations (BEIR Reports). Table 3.5 gives the risk rates for fatal cancer that are used by the ICRP for low LET radiation.[29] These rates are a measure of the probability of fatal cancer induction per unit of dose. For low LET radiation the dose units, sieverts,[c] used in the Table are equivalent to a dose in grays. These factors are intended to apply irrespective of age and sex and since men are not susceptible to breast cancer the factor for females is twice that given in the Table. The total mortality risk for cancer according to ICRP is 1.25×10^{-2} (1 in 80). This means that for every 80 people exposed to 1 Gy of low LET X-ray or γ-rays, one would subsequently die from radiation-induced cancer. The latent period before the appearance of the cancer will vary with the type of cancer and may be as short as two to three years for leukaemia or might be more than 20 years for skin and lung cancer. The total cancer rise rate (fatal and non-fatal cancers) is probably two to three times the fatal risk rate. Despite the apparent precision of these internationally agreed risk factors they are only estimates and should be seen in conjunction with other risk statistics. Table 3.6 shows the wide range of values given by some other committees; those of the ICRP are at the lower end of the range. The final estimate, 'without the bomb data', is one that excludes the Japanese data. It is reassuring that if one excludes the bomb data, because of the

[c] Sievert (Sv) is the unit for the dose equivalent, the quantity obtained by multiplying absorbed dose by a factor to allow for the different biological effectiveness of ionising radiation in causing harm.

Table 3.6 Average Cancer Death Risk per 10^4 Persons per Sievert of Low Dose Rate: Low LET Radiation

Committee	Range of estimate
BEIR II 1972	115–620
UNSCEAR 1977	75–175
ICRP 1977	125 (no range given)
BEIR III 1980	158–501
UNSCEAR (without the bomb data)	100–440

Table 3.7 Annual Fatal Accident Rate in Some UK Industries and Estimated Fatal Cancer Risk in Radiation Workers: Deaths Per Million Workers per year

Industry	Rate
Deep sea fishing	2500
Coal mining	250
Construction	200
Average radiation worker	50
Textiles	25
Clothing and foot wear	3
All employment	50

current uncertainty over the dosimetry, the ICRP risk values are not increased by more than a factor of two.[30]

Finally, for comparison, Table 3.7 lists the annual fatal accident rates in some UK industries and the estimated cancer risk among radiation workers.[31] The latter is based on an average effective dose received by UK workers of 4 mSv per year and an overall cancer risk of 1.25×10^{-2} per sievert. This implies an overall fatal cancer risk for the average UK radiation worker of 5×10^{-5} per year or 50 deaths per million workers per year. This risk of death is comparable with that of other 'safe' occupations. Some workers receive doses well above average which takes their risks out of the safest category.

Despite intensive research over many decades there remain important gaps in both our theoretical and practical understanding of radiation carcinogenesis. Our ignorance is highlighted by the fact that we do not have answers to the following important questions. Is there a threshold dose for cancer in some tissues and not others? How do dose rate and radiation quality affect the process of radiation carcinogenesis? How applicable to man are the dose rate and other studies in animals? To what extent do non-carcinogenic secondary factors play a role in radiation carcinogenesis? While such questions go unanswered there will be both doubt surrounding the cancer-risk estimates and an urgent need for more research.

3.7 Summary

Cancer induction is the most intensively studied and most significant late effect of radiation. The most favoured hypothesis is that radiation causes cancer by inducing somatic mutations. It is capable of inducing tumours in almost all the tissues of the body, although the tissues vary greatly in their susceptibility. Great empirical and theoretical efforts have been made to understand the fine detail of the relation between cancer incidence and dose. But, despite this, the simplest relation, i.e. linearity, is still held to be consistent with most of the experimental and human data at least at low doses. At higher doses, many experimental animal-cancer incidences plateau or even show peak values that suggest a competitive effect between cancer induction and cell-killing.

There have been numerous epidemiological studies of human radiogenic cancer and from these, internationally agreed risk estimates have been obtained. Nevertheless, until we know more about cancer in general and radiation cancer in particular, such risk estimates will remain tentative and subject to change.

References

1. Cooper, G.M. (1982) Cellular transforming genes. *Science*, *218*, 801-6
2. Newbold, R.F. & Overell, R.W. (1983) Fibroblasts immortality is a prerequisite for transformation by EJc-Ha-ras oncogene. *Nature (Lond.)*, *304*, 648-51
3. Harris, H. (1979) Human genetics. *CIBA Foundation Symp.*, *66*, 311
4. Shay, J.W., Lorkowski, G. & Clark, M.A. (1981) Suppression of tumorigenicity in cybrids. *J. Supramol. Struct. Cell Biochem.*, *16*, 75-82
5. Trosko, J.E. & Chang, C.C. (1981) The role of radiation and chemicals in the induction of mutations and epigenetic changes during carcinogenesis. *Adv. Radiat. Biol.*, *9*, 2-35
6. Kaplan, H.S. (1967) On the natural history of murine leukaemias. Presidential address. *Cancer Res.*, *27*, 1325-40
7. Fould, S.L. (1975) *Neoplastic Developments*, vol. 2, Academic Press, New York
8. Goodhead, D.T. (1982) An assessment of the role of microdosimetry in radiobiology. *Radiat. Res.*, *91*, 45-76
9. Fabrikant, J.I. (1980) The BEIR III controversy. *Radiat. Res.*, *84*, 361-8
10. Radford, E.P. (1980) Human health effects of low doses of ionizing radiation: the BEIR III controversy. *Radiat. Res.*, *84*, 369-94
11. Goodhead, D.T. (1980) Models of radiation inactivation and mutagenesis. In R.E. Meyn & H.R. Withers (eds) *Radiation Biology in Cancer Research*, Raven Press, New York, pp. 231-47
12. Upton, A.C., Randolph, M.L. & Conklin, J.W. (1970) Late effects of fast neutrons and gamma rays in mice as influenced by the dose rate of irradiation: induction of neoplasia. *Radiat. Res.*, *41*, 467-91

13. Clapp, N.K., Darden, E.B. & Jernigan, M.C. (1974) Relative effects of whole body sublethal doses of 60 MeV protons and 300 kVp X-rays on disease incidences in RF mice. *Radiat. Res.*, *57*, 158-86
14. Metalli, P., Covelli, V., DiPaola, M. & Silini, G. (1974) Dose incidence data for mouse reticulum cell sarcoma. *Radiat. Res.*, *59*, 21
15. Darden, E.B., Cosgrove, G.E., Upton, A.C., Christenberry, K.W., Conklin, J.W. & Davies, M.L. (1967) Late somatic effects in female RF/Un mice irradiated with single doses of 14 MeV fast neutrons. *Int. J. Radiat. Biol.*, *12*, 435-52
16. Coggle, J.E. & Peel, D.M. (1978) Relative effects of uniform and non-uniform external radiation on the induction of lung tumours in mice. In *Late Biological Effects of Ionizing Radiation II*, IAEA, Vienna, pp. 83-94
17. Shellabarger, C.J., Bond, V.P., Cronkite, E.P. & Aponte, G.E. (1969) Relationship of dose of total body^{60} Co radiation to incidence of mammary neoplasia in female rats. In *Radiation Induced Cancer*, IAEA, Vienna, pp. 161-72
18. Maldague, P. (1969) Comparative study of experimentally induced cancer of the kidney in mice and rats with X-rays. In *Radiation Induced Cancer*, IAEA, Vienna, pp. 439-58
19. Burns, F.J., Albert, R.E. & Heimbach, R.D. (1968) RBE for skin tumours and hair follicle damage in the rat following irradiation with alpha particles and electrons. *Radiat. Res.*, *36*, 225-41
20. Borek, C. (1982) Radiation oncogenesis in culture. *Adv. Cancer Res.*, *37*, 159-232
21. Doll, R. (1981) Radiation hazards: 25 years of collaborative research. *Br. J. Radiol.*, *54*, 179-86
22. Kato, H. & Schull, W.J. (1982) Studies on the mortality of A bomb survivors 7. Mortality 1950-1978, part I: cancer mortality. *Radiat. Res.*, *90*, 395-432
23. Smith, P.G. & Doll, R. (1982) Mortality among patients with ankylosing spondylitis after a single treatment course with X-rays. *Br. Med. J.*, *284*, 447-60
24. Rowland, R.E., Stehney, A.F. & Lucas, H.F. (1983) Dose-response relationships for radium induced bond sarcomas. *Health Phys.*, *44*, 15-31
25. Waxweiler, R.J., Roscoe, R.J., Archer, V.E., Thun, M.J., Wagoner, J.K. & Lundin, F.E., Jr (1981) Mortality follow-up through 1977 of the white underground uranium miners cohort examined by the United States Public Health Service. In *Radiation Hazards in Mining*, pp. 823-30, Society of Mining Engineers, New York
26. Van Kaick, G. *et al.* (1983) Recent results of the German Thorotrast Study. *Health Phys.*, *44*, 299-306
27. Loewe, W.E. & Mendelsohn, E. (1982) Neutron and gamma ray doses at Hiroshima and Nagasaki. *Nucl. Sci. Eng.*, *81*, 325-50
28. Totter, J.R. & MacPherson, H.G. (1981) Do childhood cancers result from prenatal X rays? *Health Phys.*, *40*, 511-24
29. International Commission on Radiological Protection (1977) Publication 26. Annals of the International Commission on Radiological Protection, vol. 1(3), Pergamon Press, Oxford
30. Charles, M.W. & Lindop, P.J. (1981) Risk assessment without the bombs. *J. Soc. Radiol. Protect.*, *1*(3), 15-19
31. National Radiological Protection Board (1981) *Living with Radiation*, HMSO, London (ISBN- 0 85951-145-6)

Further Reading

Becker, F.F. (ed.) (1982) *Cancer: A Comprehensive Treatise*, vol. 1, Plenum Press, London
Chapter 15. Physical carcinogenesis: radiation history and sources, A.C. Upton, pp. 551-67
Chapter 18. Radiation carcinogenesis, J.B. Storer, pp. 629-59
BEIR III (1980) National Research Council Advisory Committee on the

Biological Effects of Ionizing Radiation *The Effects on Populations of Exposure to Low Levels of Ionizing Radiation*, National Academy of Sciences, Washington, DC

Boice, J.D. & Fraumeni, J.F. (eds) (1984) *Radiation Carcinogenesis: Epidemiology and Biological Significance*, Raven Press, New York (ISBN 0-89004-907-6)

Cohen, B.L. (1980) The cancer risk from low-level radiation. *Health Phys.*, *39*(4), 649-78

Pizzarello, D.J. (ed.) (1982) *Radiation Biology*, CRC Press, Boca Raton, FL

Chapter 5. Radiation carcinogenesis, R.L. Ullrich, pp. 111-27

UNSCEAR (1977) United Nations Scientific Committee on the Effects of Atomic Radiation. Report to the General Assembly *Sources and Effects of Ionizing Radiation*. UN. E.77.IX.1, United Nations, New York

Annex G. Radiation carcinogenesis in man, pp. 361-423

Annex I. Experimental radiation carcinogenesis, pp. 565-654

4 VIRUSES AND CANCER

G. Peters

Contents

4.1 General Concepts in Tumour Virology

4.1.1 Introduction

It is generally believed that, whatever the cause, tumorigenesis usually involves specific genetic alterations in an individual cell which then proliferates to form the tumour mass. Since such alterations may be multiple, and each one as inconspicuous as substituting a single nucleotide base within a genome composed of over 10^9 base pairs, the task of detecting, let alone circumventing, these events at the molecular level is a daunting one. Considerable research efforts have therefore focused not on the cancer cell itself, but on particular viruses which have been shown to induce these genetic alterations in cells and to cause tumours in animals. Such infectious agents offer two obvious opportunities to the cancer researcher. The first is that if viral infections do indeed cause or predispose individuals to the risk of cancer, then prophylactic measures might prevent these effects. The second is that animal-virus model systems provide considerable scope for experimental manipulation. Most of the viruses discussed in this chapter have genomes about a millionfold less complex than those of the cells they infect. Not only are they simpler to analyse, but the ease with which purified virus particles can be isolated often facilitates the preparation of specific nucleic acid and immunological reagents. We hope that by understanding how these relatively simple biological agents influence the behaviour of cells, we might gain fundamental insights into the molecular basis of neoplasia. In view of the enormous impact that virus research continues to have in unravelling eukaryotic gene expression, and the recent developments in the study of viral and cellular oncogenes, this hope seems more than justified.

This chapter considers all aspects of tumour-virus research, from epidemiology to the latest ideas on the molecular basis of neoplasia. However, the subject is enormously diverse and to try to provide a comprehensive review would be futile (witness not the number but the sheer bulk of the texts suggested as further reading). The aim is therefore to highlight key developments and give prominence to generalisations and unifying theories, with the hope that this blend of factual information and conceptual overview will serve as a basic framework into which the reader can fit future developments as they arise.

4.1.2 Origins of Tumour-virus Research

The association between viruses and cancer was first recognised in the early 1900s when several laboratories, notably that of Peyton Rous, showed that tumours could be induced in chickens by transmission of an infectious filtrable agent. In the ensuing years virus-induced neoplasms have been described in members of all vertebrate classes. However, both the histology of the tumours and taxonomy of the viruses involved are quite diverse (see Tables 4.1 and 4.3) and there are no obvious links between the two.

Most of these early observations were based on tumours occurring in domestic or laboratory animals, which were amenable to both epidemiological and experimental studies. The tumours were principally, though not invariably, confined to connective and haemopoietic tissues and were prevalent in young or early adult animals. This is in sharp contrast to the situation encountered in humans, where a high proportion of cancers occur in epithelial tissues and a majority arise late in life. There is, in fact, a logarithmic relation between human-cancer incidence and age, suggesting that not one but several random events must be required for the development of a tumour. Consequently, it is clear that most human malignancies cannot be attributed to a single infectious agent as is the case with some animal models. However, rather than negating the possibility of virus involvement in human cancer, such considerations force us to re-evaluate how tumour viruses should be defined and to search for alternative ways of assaying for their effects.

Since virus infection may represent only one of the multiple steps involved, and could occur long before any overt signs of neoplasia, direct assays for tumorigenicity *in vivo* may be both inappropriate and not feasible. The realisation that cancer is a disease at the cellular level therefore represented one of the most significant advances in tumour virology since it focused attention on the influence that viruses might have on individual cells as opposed to the whole animal. Three major landmarks can be recognised in the formulation of this concept. The first, the development of inbred laboratory animals, not only underlined some direct associations between viruses and cancer, but opened up the possibility of performing transplantation studies, and led to the subsequent demonstration that a single aberrant cell is sufficient to give rise to a tumour. The second was the introduction of tissue-culture systems that support the growth of uniform eukaryotic cell types *in vitro*. The third, and perhaps the most significant of all, was the recognition that when such cultured cells are infected by tumour viruses, they frequently undergo dramatic changes in appearance and growth properties.[1] When these so-called 'transformed' cells are transplanted back into animals, they usually prove to be tumorigenic, suggesting that the events occurring in transformed cells *in vitro* might mimic those required for neoplasia *in vivo*. Transformation of cultured cells therefore represents both a convenient assay and an alternative criterion for judging the oncogenic potential of a virus.

The phenotypic changes associated with transformation are manifold and may vary between different cell types, so that the term describes a number of distinct phenomena. Most cultured cells attach firmly to the substrate provided by the culture vessel and grow into relatively organised arrays in a monolayer. In contrast, many transformed cells are not only altered morphologically but have lost their dependence on the substrate and grow into groups of disorganised overgrowing cells which are easily recognised against the background of normal cells. Each group or 'focus' of cells represents the progeny of a single transformant so that focus formation is both a qualitative and quantitative assay for transformation. In addition, eukaryotic cells survive for only limited periods in tissue culture, except for occasional spontaneous variants which have acquired the ability to grow into permanent cell lines. Infection by certain tumour viruses can extend the survival time of primary cells in culture, allowing them to grow into detectable colonies or in some cases into immortal lines. Thus, two distinct modes of transformation, focus formation and immortalisation, may represent discrete steps in the progression of a cell toward malignancy.

4.1.3 Taxonomy of Tumour Viruses

As judged by tumour formation in the animal or cell transformation in tissue culture, tumour viruses come from a surprising number of taxonomic groups (Tables 4.1 and 4.3 list examples). Virus replication is totally dependent on infection of an appropriate host cell and viruses of differing types and complexities require different combinations of cell components in the course of their life-cycle. However, an obvious division exists between viruses which carry RNA or DNA as their genetic materials.

Most viruses with RNA genomes replicate in the cytoplasm and are not associated with cancer. The exceptions are the retroviruses that have the unique property of replicating via the formation of a nuclear DNA intermediate and therefore being more akin to the viruses with DNA genomes. Thus, all tumour viruses enter the nucleus and introduce new DNA, a portion of which may become covalently attached or integrated into the chromosomal DNA of the cell. The amount of DNA carried into the cell and its organisation in the viral genome varies. For example, the genomes of the hepadna- and papova-virus families are circular and contain around 3000–8000 base pairs as compared with the linear genome of about 140 000–180 000 base pairs of the herpesviruses. The number of proteins which these genomes can encode is correspondingly variable and unfortunately there appear to be no common threads, such as the same gene or same enzyme activity, which are shared by all tumour viruses.

A number of the DNA viruses listed in Table 4.3 are extremely common infectious agents in the human population in many parts of the world. Even if virus infection is only one of several contributing factors, it is clear that cancer can be only a rare consequence. Unlike retroviruses, most DNA viruses kill the cells in which they replicate, so that their oncogenicity only becomes apparent in situations where they are no longer cytotoxic, for example, when some viral function is impaired or when the virus infects cell types which do not permit full replication. Thus many DNA tumour viruses are only tumorigenic in species other than their natural hosts. With human viruses tumorigenicity is often assessed in rodents or by transformation of rodent or primate cells. While this might appear to be an inappropriate assay for a biological agent, it should be remembered that most physical or chemical carcinogens are assessed in precisely this way and, by these criteria, some human viruses rank among the most potent carcinogens known.

Viruses that are only tumorigenic under non-permissive conditions become stranded in the tumour cell and therefore cannot be transmitted from cell to cell. In these situations, further investigations rely heavily on detecting residual evidence of viral DNA, RNA or proteins in the tumour cells, and much of the epidemiology of human tumour viruses is based on such a strategy. However, in some cases, most notably the acutely oncogenic retroviruses (see section 4.2) a defective component can be rescued from the tumour cell by the simultaneous presence of a non-defective 'helper' virus which can replicate normally and complement the missing function. Infectious virus particles can thus spread to others in the population, making it simple to correlate virus and tumour incidence. Here, as in most common virus infections, transmission can be 'horizontal', involving the transfer of an infectious particle (or possibly an infected cell) from individual to individual. An alternative mode of transmission is described as being 'vertical' since it involves transfer of the virus exclusively from parent to offspring. This can occur in either of two ways (i) by congenital infection of the egg or fetus in a viraemic female or (ii) by the genetic transmission of an endogenous virus as a Mendelian locus. This latter route must, by definition, be confined to viruses whose genomes integrate into chromosomal DNA and is probably unique to retroviruses which on rare occasions establish infection of the germ line (see section 4.2.2).

Attempts to relate virus infections and tumorigenicity can be complicated by two further considerations. The first is that the response of different individuals can be extremely variable, depending on factors such as age, sex, genetic background and, most importantly, immunological status. Thus, some viruses are only tumorigenic when injected into newborn animals, since adults mount an effective immune response to the infection. The second is that some viral infections early in life can lead to persistence of the virus in a latent state, throughout the life-span of the animal. Such infections may go unnoticed, since the symptoms are mild or inapparent, but may result in activation of the virus later in life or when the immune system is impaired. For example, members of the herpesvirus family are known to exist in a latent state in a high proportion of the human population and can give rise to conditions such as cold sores or shingles when activated by appropriate stimuli.

4.1.4 Tumour-virus Genetics

The obvious goal of tumour-virus research is to determine which of the relatively few functions that a virus can encode might be responsible for oncogenicity. A virus might introduce into the cell a single new gene capable of inducing all facets of transformation, as with many of the acutely oncogenic retroviruses (section 4.2.3), but in other examples the interplay of several viral genes may be required. The existence and identity of such 'oncogenes' can be investigated in two ways. The first involves the isolation of viral mutants impaired in their capacity to transform cells. Mutations can be either conditional, as in temperature-sensitive lesions, where the phenotype of the infected cell can be varied by altering the growth conditions, or non-conditional, where the gene has been irreversibly altered or deleted. This approach identifies specific virally encoded gene products, whose continued expression is required for the maintenance of transformation, and was successfully applied in identifying the genes encoding the SV40 and polyoma virus tumour antigens[2] and the *src* gene of Rous sarcoma virus (see sections 4.2.3 and 4.3.3).[3]

The alternative approach, that greatly benefited from the advent of recombinant DNA techniques, is to assess the transforming abilities of specific subregions of the viral genome, when introduced into cultured cells by DNA transfection. A notable example of this was the demonstration that only the left-most portion of the adenovirus genome is sufficient for full transformation and tumorigenicity.[4] For larger viruses, with relatively complex genomes, this is the method of choice as specific mutants are harder to isolate and analyse. A useful corollary is not to introduce fragments of DNA into cells, but to examine the tumour cells for residual vestiges of the viral genome. The hope here is to find a consistent pattern in all transformed cells that would indicate which viral functions might be involved. While successfully applied in the case of adeno- and papova-viruses, these strategies are proving much harder to interpret with the herpesviruses, where the genome is larger and detection correspondingly more difficult.

The foregoing considerations have assumed that viral DNA directly induces neoplastic changes in cells. This is probably an accurate reflection of what is most readily observed experimentally, but it need not apply to all situations and alternative mechanisms ought not to be discounted. For example, if the virus were to act indirectly, or as a type of 'hit and run' agent, no traces or vestiges of viral infection might be detected in the tumour cells. Indirect effects that might be envisaged include suppression of the immune system, leading to release of pre-existing tumour cells from immunological control, and the proliferation of specific cell types as a result of mitogenic stimulation by virally infected cells or as a regenerative response to non-specific tissue damage. With these latter mechanisms, the virus need never come into contact with the tumour cell itself but simply establishes a situation in which some cells are at greater risk from other carcinogenic stimuli. Although supporting evidence is inherently elusive, such notions illustrate the potential difficulties and uncertainties which can be encountered in establishing links between virus infections and cancer.

With these theoretical considerations as a background, the following sections illustrate how they might apply to specific examples of RNA (section 4.2)- and DNA (section 4.3)-containing viruses.

4.2 RNA Tumour Viruses

4.2.1 Introduction

Retroviruses were the first viruses to be recognised as tumorigenic and infect and induce tumours in a wide range of animals. Many encode specific oncogenes, directly responsible for transformation, that were derived from precursors, or proto-oncogenes, pre-existing in the DNA of normal somatic cells (see section 4.2.4). Thus, neoplasia may result not from expression of a truly viral function, but from the aberrant expression of a cellular gene. The recent demonstration that such oncogenes may be implicated in specific human cancers, with no known viral aetiology, lends support to this notion and has amply vindicated the study of these tumour viruses in animal model systems. Before reviewing the properties of viral and cellular oncogenes and the mechanisms involved in their activation, it is perhaps pertinent to consider the general structural and replicative features of retroviruses and how they may influence tumorigenicity.

4.2.2 Structure and Replication of Retrovirus Genome

Retroviruses, isolated from an enormous diversity of species, including fish, birds, rodents, domestic mammals, primates and even humans (Table 4.1), and associated with a wide spectrum of pathological

Table 4.1 Examples of RNA Tumour Viruses

Type of virus	Abbreviation	Specific examples or substrains
Avian sarcoma virus	ASV	Rous (various strains), Y73, ESH, URII, Fujinami, PRCII
Avian myeloblastosis virus	AMV	AMV, E26
Avian myelocytomatosis virus		MC29, CMII, OK10, MH2
Avian erythroblastosis virus	AEV	AEV-ES4, AEV-R, AEV-H
Avian lymphoid leukosis virus	ALV	Various
Avian reticuloendotheliosis virus	REV	REV-T
Mouse mammary tumour virus	MMTV	Various
Murine sarcoma virus	MuSV	Moloney, Harvey, Kirsten
Murine osteosarcoma virus	MuSV	FBJ, FBR
Murine leukaemia virus	MuLV	Gross, Moloney, Abelson, Graffi
Murine erythroleukaemia virus	MuLV	Friend, Kirsten, Rauscher
Feline leukaemia virus	FeLV	Various
Feline sarcoma virus	FeSV	Gardner-Arnstein, Snyder-Theilen, McDonough, Gardner-Rasheed
Bovine leukaemia virus	BLV	
Simian sarcoma virus	SiSV	
Gibbon ape leukaemia virus	GaLV	
Human T-cell leukaemia virus	HTLV/ATLV	Types I and II

consequences, are unified by the nature of their genomes and mode of replication. They contain RNA as their genetic material, but replicate by formation of a DNA intermediate. The viral genome, of around 6000–10 000 nucleotides, comprises only three genes, one of which encodes the RNA-dependent DNA polymerase (reverse transcriptase) that performs the key step of transcribing the single-stranded viral RNA to produce a double-stranded DNA intermediate. On entering the nucleus, this DNA is inserted, or integrated, into the chromosomal DNA of the host cell as a 'provirus' and is stably inherited during subsequent cell divisions. In contrast to the situation with DNA tumour viruses (see section 4.3), integration of the proviral DNA is an essential step in the replicative cycle of a retrovirus.

Being inherited during cell division, retroviral DNA can, in some circumstances, become established as a genetic trait. This occurs in the rare instances when the virus infects cells in the germ line, giving rise to a provirus which is then present in the DNA of all cells in the organism. Such proviruses are referred to as 'endogenous' and are common features in, for example, domestic cats and chickens, and in inbred laboratory mice. Endogenous proviruses may or may not be relevant to disease, but they result in a mode of transmission, vertically through the germ line, which is probably unique to retroviruses.

The retroviral provirus contains all the information present in the viral RNA, and in the same order. However, it is slightly longer than the RNA because of the duplication of short segments from each end of the genome, the net result being a structure which is flanked by identical sequences, known as long terminal repeats (LTRs). The LTR elements provide the promoter and control the expression of the viral genes, but the actual transcription of the provirus back into RNA is carried out by the host-cell RNA polymerase. Similarly, the resultant RNA is processed and translated into the viral structural proteins as if it were a typical cellular RNA. As well as encoding the viral proteins, a proportion of this RNA is packaged to form the genomes of progeny virus particles. These normally bud from the plasma membrane without causing cell lysis, so that, in virological terms, retrovirus infections are generally benign and result in little, if any, change in the phenotype of the cell.

4.2.3 Oncogenicity of Retroviruses

Retroviruses can be divided into two broad categories based on their tumorigenic properties. The first of these is described as non-acutely or weakly oncogenic since, although infected animals usually develop tumours of a specific type, there is typically a long delay between virus infection and detectable neoplasia. As far as is known, these viruses do not contain a specific gene whose expression is required for tumour induction, and they are unable to transform cells in tissue culture. Even in the animal, the vast majority of infected cells remain normal and conversion of a particular individual cell into a malignant one is rare. Although in many cases the molecular basis for this event remains unclear, recent evidence with avian leukosis virus (ALV), murine leukaemia virus (MuLV), and mouse mammary-tumour virus (MMTV), has linked tumorigenesis to the site in the cellular DNA at which the provirus integrates. The implication is that the provirus somehow perturbs either the structure or expression of cellular genes in the vicinity of the integration site (for review see Ref. 5). However, since the provirus can insert apparently anywhere in the cellular DNA, it can presumably influence any gene, and only under rare circumstances would this confer a selective growth advantage or neoplastic potential on the cell.

Retroviruses in the second category induce tumours in animals very rapidly, typically in a matter of days or weeks, and are consequently described as being acutely oncogenic. They also transform cultured cells, usually fibroblasts or cells of haemopoeitic origin, and can be readily assayed in this way. This has greatly facilitated the analysis of these agents, and one of the major advances was the demonstration that conditional and non-conditional mutants could be isolated which were impaired in their ability to transform cells. The first observations, made with Rous sarcoma virus (RSV) indicated that the viral genome contained a gene, or at least genetic information (given the acronym *src*), which was directly responsible for the ability of the virus to transform cells.[3] In the ensuing years, other viral oncogenes were found and we now recognise around 20 distinct examples, each with its respective three-letter code (see Table 4.2A).

RSV is now regarded as a special case because it contains the *src* gene as well as the three replicative genes normally present in the retroviral genome. In contrast, all other isolates of acutely oncogenic retroviruses are defective, in that the oncogene has replaced some of the viral replicative functions. They are therefore unable to replicate autonomously and can only be propagated as infectious particles in the continued presence of a competent helper virus which will complement the missing functions. In other words, the defective genome RNA of the acutely oncogenic agent can be packaged into a transmissable particle by the structural proteins and RNA-dependent DNA polymerase supplied by a non-defective virus. Usually, one of the weakly oncogenic retroviruses described previously provides the helper functions, and it is generally accepted that the defective viruses were derived from their helpers by recombination. Thus ALVs are presumed to be the progenitors of the avian sarcoma and acute-leukaemia viruses and likewise for the equivalents in mice and cats (see Table 4.1). Moreover, many acutely oncogenic viruses only arose under laboratory conditions during passage of weakly oncogenic retroviruses in animals. In epidemiological terms therefore they are insignificant as vectors for the transmission of malignant disease in animal populations, but as research tools their value and impact is irrefutable.

4.2.4 Origins of Viral Oncogenes

Since acutely transforming retroviruses have acquired specific oncogenic sequences at the expense of some of their replicative functions, the obvious questions that arise are Where did this information come from? and How did it get there? The answers lie at the heart of current research into the molecular basis of cancer because these viral oncogenes are not really viral genes at all, but are cellular genes in another guise. The pioneering work in this regard again centred on the *src* gene of RSV and the observation that sequences related to *src* were already present in the DNA of normal chicken cells.[6] A similar conclusion was soon reached for all viral oncogenes subsequently discovered, so that for each oncogene listed in Table 4.2A, it is necessary to distinguish between the viral (v-*onc*) and cellular (c-*onc*) forms of the gene. All the retroviral oncogenes known at the time of writing are included in the Table, but new examples are continually being uncovered. Nevertheless, it is encouraging to note that the list is likely to be limited in number, since in several cases the

Table 4.2 Origins and Properties of Oncogenes

Acronym	Original source	Species of isolation	Properties of gene product Cellular location/function
A. *Transforming genes in retroviruses*[a]			
src	Rous sarcoma virus	Chicken	Plasma membrane/tyrosine kinase
fps [b]	Fujinami sarcoma virus	Chicken	Tyrosine kinase
fes	Snyder-Theilen feline sarcoma virus	Cat	
yes	Y73 sarcoma virus	Chicken	Tyrosine kinase
abl	Abelson murine leukaemia virus	Mouse	Plasma membrane/tyrosine kinase
ros	URII Avian sarcoma virus	Chicken	Cytoplasmic membranes/tyrosine kinase
fgr	Gardner–Rasheed feline sarcoma virus	Cat	Tyrosine kinase
fms	McDonough feline sarcoma virus	Cat	Cytoplasmic membranes
Ha-ras 1	Harvey murine sarcoma virus	Rat	Plasma membrane/GDP or GTP binding
Ki-ras 2	Kirsten murine sarcoma virus	Rat	Plasma membrane
myc	MC29 avian myelocytomatosis virus	Chicken	Nuclear matrix/DNA binding
myb	Avian myeloblastosis virus	Chicken	Nuclear matrix
fos	FBJ murine osteosarcoma virus	Mouse	Nucleus
sis	Simian sarcoma virus	Woolly monkey	Cytoplasm/platelet-derived growth factor
erb B	Avian erythroblastosis virus	Chicken	Plasma membrane/EGF receptor
ski	Avian SKV770 virus	Chicken	
rel	Reticuloendotheliosis virus	Turkey	
mos	Moloney murine sarcoma virus	Mouse	
raf [b]	3611 murine sarcoma virus	Mouse	
mil	MH2 avian myelocytomatosis virus	Chicken	
ets[c]	E26 avian myeloblastosis virus	Chicken	
erb A[c]	Avian erythroblastosis virus	Chicken	

B. *Cellular sequences perturbed by retrovirus integration*[c]		
int-1	Mouse mammary-tumour virus integration region	Mouse
int-2	Mouse mammary-tumour virus integration region	Mouse
MLVI-1	Murine leukaemia-virus integration region	Rat
MLVI-2	Murine leukaemia-virus integration region	Rat
RMO Int-1	Murine leukaemia-virus integration region	Rat
Pim-1	Murine leukaemia-virus integration region	Mouse
C. *Oncogenes detected by DNA transfection*[d]		
Ha-ras 1	Bladder carcinoma	Human
Ki-ras 2	Lung carcinoma	Human
N-ras	Neuroblastomas/leukaemias/sarcomas	Human
B-lym	B-Cell lymphomas	Chicken/human
mam	Mammary carcinomas	Mice/human
neu	Neuroblastomas/glioblastomas	Rat

[a] In many instances these genes have been identified in several, independent, virus isolates, but only the prototype is listed

[b] These pairs represent the same gene from different species

[c] Transforming capabilities of these genes or sequences have not been directly demonstrated but they are implicated by virtue of transduction or insertional mutagenesis by oncogenic viruses

[d] *ras* and *B-lym* gene families have been cloned and characterised but the status and appropriate acronym for *mam* and *neu* remain uncertain

same or closely related oncogenes have been identified in a number of quite independent virus isolates. For example, parts of the *myc* gene are to be found in at least four different isolates of avian virus and *fps* and *fes* represent the same gene acquired by viruses of avian or feline origin respectively. Moreover, some of these genes can also be implicated in tumorigenesis by weakly oncogenic retroviruses as in the case of ALV-induced lymphomas and erythroblastosis. Here, the insertion of a provirus adjacent to either the c-*myc* or c-*erbB* gene, respectively, appears to be one of the critical events in the development of neoplasia.[5] By analogy therefore other specific integration regions recognised in tumours induced by MMTV and MuLV (Table 4.2B) may represent additional examples of cellular oncogenes.

Oncogenes, or proto-oncogenes as the cellular progenitors are sometimes referred to, are present in the DNA of all somatic cells so that it seems conceivable that carcinogenic agents other than viruses may also influence their behaviour. The advent of DNA-mediated gene transfer, introducing naked DNA into recipient cells in culture, showed that it was possible to induce a low but measurable frequency of morphological transformation, provided that the DNA was obtained from a tumour cell.[7] At first these findings were treated with some scepticism since only one particular line of mouse fibroblasts, called NIH 3T3 cells, could function as suitable indicator cells in this assay. Nevertheless, it soon became apparent that the dominant transforming genes detected in this way were in some instances closely related to the *ras* family of viral oncogenes characteristic of the Harvey and Kirsten murine sarcoma viruses. This confirmed the efficacy of the NIH 3T3-DNA transfection assay and provided direct evidence that a gene first identified in a virus model system could be involved in a human cancer with no known viral aetiology. Only a limited number of oncogenes have been identified in these DNA transfection studies (Table 4.2C), but this approach has also revealed examples of transforming genes that had not been previously recognised from studies on oncogenic retroviruses.

One of the striking conclusions drawn from these studies on viral and cellular oncogenes is that they have been highly conserved during vertebrate evolution. Thus sequences related to, for example, the *src*, *ras*, *myc* and *abl* genes have been detected in species as distant as humans and Drosophila, and in some cases even yeast. The implication is that in their normal cellular contexts, these genes may serve important, conserved, functions common to many cell types and species.

4.2.5 Activation of Oncogenes

Since cellular oncogenes exist in the DNA of normal cells, they cannot be oncogenic in their normal contexts, and some stimulus or perturbation must be necessary to activate their latent properties. This issue is the focus of much of the current research, the prevalent notion being that these genes may become oncogenic only when expressed in some structurally altered form or at inappropriately high levels, or perhaps a combination of both. The ways in which this might be achieved are manifold and may or may not involve viral infections, but some examples for which there is now good evidence are indicated below.

4.2.5.1 Transduction by a Retrovirus. The acutely oncogenic retroviruses represent the most obvious situation in which both structural alteration and enhanced expression of an oncogene might prevail. In the first place, many viral oncogenes are in fact fusions between the acquired cellular sequences and remnants of the viral structural genes, so that the product represents a hybrid protein containing both viral and cellular determinants. The presence of additional amino acids at either the amino or carboxy terminus could dramatically alter the structure, activity, target specificity or subcellular location of the oncogene product. Alternatively, more subtle differences between the v-*onc* and c-*onc* products might be generated by point mutations or changes in the coding frame as a consequence of the recombinational events required for transduction of the gene. Whether or not structural alterations are incurred, any sequences acquired by the viral genome are likely to be expressed at relatively high levels, since they will be brought under the control of the strong transcriptional promoter within the viral LTR. The oncogene product may therefore be synthesised in amounts far in excess of the basal levels required for normal cell metabolism or may be introduced into inappropriate cells in which the proto-oncogene is normally silent. An example of the latter would be the expression of a fetal gene in adult tissues. Some direct tests of these proposals have been performed in which the normal cellular counterpart of genes such as *mos* and *ras* have been linked to a retroviral LTR, by genetic manipulation, and shown to be capable of transforming cells. However, this is not a general phenomenon and other proto-oncogenes that have been tested do not appear to induce transformation under analogous circumstances.

4.2.5.2 Activation by Provirus Integration. Insertion of a provirus into the cellular DNA adjacent to or even within a latent oncogene may be sufficient to alter its properties or modulate its expression. For example, the majority of ALV-induced lymphomas in chickens show evidence of proviral integration next to the c-*myc* locus, such that the *myc* gene is expressed into mRNA which initiates at a viral transcriptional promoter. This mechanism, known as 'promoter insertion', is conceptually simple but alternative less-direct modes of activation clearly exist since in some of the lymphomas the provirus is in the wrong transcriptional orientation or is located downstream of the *myc* gene.[5] A similar situation has recently been found in mammary tumours induced by MMTV where the proviruses are integrated upstream or downstream of either of two specific cellular genes (Table 4.2B). Neither of these genes has yet been identified as an oncogene by other criteria so that tumours induced by weakly oncogenic retroviruses may provide further examples of cellular oncogenes which may not be amenable to transduction or assayable by DNA transfection. Moreover, the striking similarity betweeen proviruses and transposable genetic elements raises the possibility that the latter may also play some role in oncogene activation.

4.2.5.3 Chromosomal Abnormalities and Oncogenes. Tumour cells frequently show gross abnormalities in chromatin structure, such as the presence of an additional copy of particular chromosomes, loss or deletion of chromosomes, translocation of material from one chromosome to another, or amplification of specific segments of chromatin. Significantly, some tumours are reproducibly associated with a specific abnormality, as in the reciprocal translocation between chromosomes 9 and 22 in human chronic myelogenous leukaemia (the shortened form of chromosome 22 is termed the 'Philadelphia chromosome') and between chromosomes 8 and 14 in many cases of Burkitt's lymphoma. These examples have recently suggested further links between tumour-virus oncogenes and human cancer in that the break-point in chromosome 9 at which translocation occurs is very close to the human homologue of the *abl* oncogene and the break-point on chromosome 8 is adjacent to the *myc* gene. In this latter case, the translocation results in the apposition of the *myc* gene and the immunoglobulin heavy-chain locus on chromosome 14.[8] Interestingly, some rare cases of Burkitt's lymphoma show different translocations involving chromosome 8 and either 2 or 22 resulting in apposition of *myc* and either the immunoglobulin κ- or λ-type light-chain loci respectively. Not surprisingly, the effect of these and other translocations on the structure, expression and potential tumorigenicity of the oncogenes involved is currently the subject of intense interest and vigorous debate. In addition, the recent demonstration that small segments of chromatin which have undergone amplification in some tumour cell lines may carry multiple copies of oncogenes, such as *myc*, *myb* and *ras*, presents yet another link between oncogene expression and malignancy in non-viral contexts.[9]

4.2.5.4 Mutagenesis of Oncogenes. The realisation that cancer is a genetic disease at the cellular level suggested that mutagenic agents capable of altering the structure of DNA might also be carcinogenic, with the obvious corollary that cancer genes are mutated forms of normal genes. The viral and cellular oncogenes and the dominant transforming genes detected by DNA transfection provide a battery of genes with which to examine this issue. One striking proof of the relevance of mutation in neoplasia was the demonstration that the *ras* genes detectable by transformation assays with human tumour-cell DNA, differed from their counterparts in normal tissues at a single nucleotide, resulting in substitution of the twelfth amino acid in the protein product.[10] Similarly, the oncogenic variants of the *ras* genes carried by murine sarcoma viruses and the activated *ras* gene from chemically transformed cells are altered at the same position. The sophistication of recombinant DNA and sequencing techniques is such that not only are additional examples of point mutations being uncovered, but the effects of altering specific amino acid residues at any position in the protein can now be tested directly.

These potential mechanisms for oncogene activation are not mutually exclusive and can, and probably do, act in conjunction with one another. Thus, translocated c-*myc* may be both structurally altered and expressed at elevated levels, whereas the *ras* and *src* genes transduced by acutely oncogenic retroviruses are mutated forms of the cellular genes. The future therefore holds great promise for elucidating which parts of each oncogene may be relevant in tumorigenesis and how these may be affected by viral or environmental carcinogens.

4.2.6 Functions of Oncogenes

While a great deal is known about the structure of viral and cellular oncogenes, the complete nucleotide sequence and consequently the predicted amino acid sequence of the gene product usually sheds little light on its function. Some advances have been made, however, based largely on intuition or serendipity. For example, a number of oncogene products may have evolved from common ancestors as they are all phosphoproteins and have a demonstrable protein kinase activity (Table 4.2). Protein phosphorylation and dephosphorylation are well-known mechanisms of metabolic regulation and perturbation of such processes could lead to the multiple phenotypic differences between normal and tumour cells. Again the earliest observations centred on the product of the *src* gene of RSV which was found to be capable of transferring a phosphate moeity from ATP onto other protein substrates *in vitro*. Curiously, the amino acid target for the kinase activity of *src* and the other oncogene products of this type is tyrosine, rather than serine and threonine, the more common sites of protein phosphorylation.[11] However, despite considerable efforts to identify the intracellular substrates for these kinases and to assess the influence of tyrosine phosphorylation on cellular physiology, no clear picture has yet emerged to rationalise the various observations.

Further clues as to possible functions can be gleaned from the intracellular location of the oncogene products. For example, the majority of the protein kinase family appear to be associated with the inner surface of the plasma membrane or intracellular membranes, whereas others, like the *erb B* product, are glycoproteins that extend to the external surface of the cell. In contrast, the products of the *myc*, *myb* and *fos* genes are found almost exclusively in the cell nucleus and may therefore interact in some way with chromatin. DNA binding is obviously an attractive possibility to account for the pleiotropic effects of transformation.

Finally, there are now two examples of oncogenes for which a biological function has been established on the basis of direct structural analysis. Quite independent lines of investigation into the primary structures of the *sis* gene of simian sarcoma virus and human platelet-derived growth factor (PDGF) reveal irrefutable structural homology. PDGF is a peptide factor released from blood platelets at wound sites and has been shown to have very potent mitogenic effects on cultured fibroblasts. Along similar lines, the *erb B* gene product is closely related to the cell-surface receptor for epidermal growth factor and the viral oncogene may represent a truncated or mutated form of growth-factor receptor.[12] It is therefore feasible that overproduction of normal or aberrant growth factors and their receptors, or conceivably self-stimulation of factor-producing cells, might account for some aspects of malignancy.

4.2.7 Human Retroviruses

The few reported isolates of retroviruses from human tissues have mostly proved to be of primate origin, are not reproducibly associated with clinical symptoms and can in several cases be attributed to laboratory contamination. The exceptions are a group of retroviruses, known variously as the human T-cell leukaemia viruses (HTLV) or adult T-cell leukaemia viruses (ATLV), that have been isolated from the cultured lymphocytes of patients with certain forms of T-cell lymphoma or leukaemia.[13] The tumours occur in adults and are usually aggressive leukaemias of relatively mature T-cells. Both the virus and disease are very rare in the general population but an epidemiological link is suggested by the clustering of both in specific geographic regions, notably the south western islands of Japan and parts of the Carribean. However, unlike retroviruses in animal model systems, it is by no means clear what bearing HTLV has on the disease. In the endemic regions, up to 10–15 per cent of the population may show evidence of viral infection, but only a small proportion of virus-positive individuals develop tumours, suggesting that other environmental factors must be involved. Although there is a strong case for horizontal transmission, probably by intimate contacts such as the transfer of blood or infected cells, some primary tumours show little if any signs of virus expression until grown in culture, and virus harvested from such cultures does not infect other cells.

Nevertheless, when normal T-lymphocytes are co-cultivated with these leukaemic T-cells they do become infected with HTLV and their growth in culture becomes independent of T-cell growth factor, a component normally required for propagation of T-lymphocytes *in vitro*. Thus HTLV can be looked on as a transforming virus by these criteria, but how such properties can be correlated with the long delay between exposure to virus and the onset of neoplasia, and the influence of other environmental factors, awaits further investigation.

From studies at the DNA level, HTLV appears unrelated to other known retroviruses, apart from some weak homologies to bovine leukaemia virus, but isolates from different geographic loations seem to be very similar. The sequence of the provirus suggests that it can encode the three structural genes normally required for replication but also has the potential for a fourth gene, specifying a protein of mol. wt. about 27 000. Although the role of these extra sequences is intriguing, it seems unlikely that they constitute a classical oncogene. In the first place, they are not related to any normal cellular sequences and, secondly, the tumours are clonal as expected for a weakly oncogenic retrovirus. However, while it would appear that the HTLV provirus must contribute to tumorigenesis, since it is not found in normal tissues of ATL patients, there is as yet no evidence for regionally specific integration or insertional mutagenesis of a cellular oncogene.

4.3 DNA Tumour Viruses

4.3.1 Introduction

DNA-containing tumour viruses occur in at least four distinct families: hepadna-, papova-, adeno- and herpes-viruses (see Table 4.3). They therefore vary markedly in the size and organisation of their genomes, but in every case except the hepadnaviruses, the genomes can be divided into those regions or genes which are active early in the infectious cycle, before the onset of DNA synthesis, and the late functions whose expression is delayed until after DNA synthesis has begun. As a general rule, early functions are involved with genome replication and control, whereas late functions usually comprise the structural components of the viral particle or activities required for cell lysis. In contrast to retroviruses, replication of the genome does not require integration into the host-cell DNA. In the rare instances where integration is observed, as in transformed or tumour cells, it appears to be random with respect to how much and which part of the viral genome becomes linked to cellular DNA.[14] Compared with acutely transforming retroviruses, transformation by DNA tumour viruses is relatively inefficient and occurs at very low frequency, since the normal consequence of DNA virus replication is

Table 4.3 Examples of DNA Tumour Viruses

Virus family	Example	Natural host	Associated neoplasms
Hepadnavirus	Hepatitis B	Humans	Hepatocarcinoma
	Animal hepatitis viruses	Woodchuck	Hepatocarcinoma
		Ground squirrel	—
		Duck	—
Papovaviruses	Polyoma	Mouse	
	SV40	Old-world monkeys	None in species of origin
	JC	Humans	
	BK	Humans	
	Papilloma viruses	Humans	Warts, skin cancer, cervical carcinoma
		Cattle	Alimentary carcinoma
		Cottontail rabbit	Papilloma/carcinoma
Adenoviruses	Adenoviruses (1–31)	Humans	
	Animal adenoviruses	Monkeys	None in species of origin
		Cattle	
		Birds	
Herpesviruses	Epstein-Barr virus	Humans	Burkitt's lymphoma (BL), nasopharyngeal carcinoma (NPC)
	Herpes simplex virus 1	Humans	
	Herpes simplex virus 2	Humans	Cervical carcinoma?
	Cytomegalovirus	Humans	Kaposi sarcoma?
	Marek's disease virus	Chickens	Neurolymphomatosis
	Herpes virus ateles/saimiri	Monkeys	Lymphoma, lymphoblastic leukaemia

cell death. Thus, cells of the natural host, that are permissive for viral replication, can only be transformed by defective viruses, whereas non-permissive cells can be transformed by both defective and wild-type viruses. These considerations make it difficult to establish causal links between virus infection and naturally occurring tumours, particularly as the DNA tumour viruses tend to be ubiquitous and often non-pathogenic in their natural hosts. Much of the data therefore pertain either to tumorigenesis or cell transformation under laboratory conditions or to epidemiological association between virus and tumour incidence in natural populations.

4.3.2 Hepadnaviruses

This family includes human hepatitis-B virus (HBV) and the hepatitis viruses from woodchuck, ground squirrel and duck.[15] Their DNA comprises a long complete strand of around 3200 bases and a shorter complimentary strand of variable length, which are held together as an open circle. Viral DNA replication requires the synthesis and then reverse transcription of an RNA intermediate, rather analogous to but the converse of the retroviruses. The genome has the capacity to encode four polypeptides, by using overlapping sequences in different reading frames, and two of these presumably represent the viral surface antigen (HBsAg or 'Australia' antigen) and core antigen (HBcAg) which are diagnostic of HBV infections.

Primary infections by HBV result in liver damage but are normally self-limited and are resolved within a matter of months. However, in about 5–10 per cent of cases, and an even higher percentage of infected newborns, the disease fails to resolve and can result in persistence of viral antigens in the blood for years. It is estimated that approximately 10 per cent of the population in parts of Africa and Asia are chronic carriers of HBsAg compared with around 0.1 per cent in the USA and northern Europe. It was this differential geographical distribution of primary infection and chronic antigen expression that first suggested an aetiologic link between HBV and hepatocellular carcinoma, which is prevalent only in the endemic areas. This association has now been reinforced by further epidemiological surveys. For instance, long-standing carriers of HBV, and particularly those that develop cirrhosis, show a 200-fold greater incidence of hepatocellular carcinoma than control subjects that are negative for HBsAg.[16] Moreover, it is estimated that 43 per cent of male carriers above the age of 40 years will eventually die from liver cancer. Since the tumour is thought to account for over 500 000 deaths each year, it is clearly a major world problem and probably represents one of the commonest fatal neoplasms in humans.

Despite this relatively convincing epidemiological association, the role of HBV in tumorigenesis is still uncertain. Not only do many carriers escape the disease, but analogous, albeit rare, tumours arise in uninfected individuals. Considerable efforts have therefore been directed toward finding potential cofactors, the most tenable suggestion being dietary aflatoxins, though alcohol and even cigarette smoking have been weakly implicated. Research at the molecular level has been hampered by the inability to propagate the virus in culture, but this has been to some extent alleviated by the advent of recombinant DNA techniques and the realisation that a relevant model system (woodchuck hepatitis virus) exists in animals.[15] Recent indications are that viral DNA can be found integrated into the chromosomal DNA of the host but that integration may not be a prerequisite for tumorigenesis and can occur in normal cells of chronic virus carriers. Nevertheless, the detection of viral DNA in tumours casts some doubt on the intriguing suggestion that the influence of HBV in hepatocellular carcinoma is indirect, and that malignant clones arise within the rapidly dividing population of cells regenerating after virus-inflicted liver damage.

4.3.3 Papovaviruses

The name of this virus family derives from the initial letters of three characteristic members: *pa*pilloma virus and the slightly smaller and therefore distinct subgroup of *po*lyoma virus and simian *va*cuolating agent (SV40).

4.3.3.1 Polyomaviruses. This group includes the monkey virus SV40, the human viruses JC and BK and the mouse virus polyoma. They are relatively simple and a great deal is known about the organisation, expression and function of their genetic information, including their complete nucleotide sequences. They have therefore had a considerable impact on our understanding of viral tumorigenesis and of eukaryotic gene expression in general.

Although polyoma virus induces many types of tumours (hence its name) when injected into newborn mice, it is ubiquitous among wild and laboratory mice and is not usually associated with overt pathological consequences. Mouse cells are permissive for polyoma virus replication but other rodent cells are non-permissive and can be morphologically transformed by polyoma virus infection *in vitro*. Similarly, SV40 virus is normally lytic in monkey cells, its natural host, but transforms mouse cells in culture and induces tumours in newborn mice. Stable transformation by these viruses is very infrequent, yet a single virus particle or viral DNA molecule is sufficient to induce all aspects of the transformed phenotype. Consequently, mutant viruses that have lost this ability have been the source of much of the information regarding the viral functions required for transformation.

The polyomaviruses contain double-stranded DNA as their genetic material, organised as a covalently closed circular molecule of around 5000 base pairs, and encoding five or six distinct proteins. The viral genome is temporally and physically divided into early and late functions and analysis of mutants assigns transforming capacity to the early region. In the normal lytic cycle, viral DNA replication takes place autonomously, without association or integration into the chromosomal DNA of the host cell. In contrast, all transformed cells contain variable amounts of randomly integrated viral DNA.[14] In agreement with the data from viral mutants, all cells transformed by wild-type virus contain and express an intact early region of the viral genome.

The early regions of SV40 and polyoma virus encode a series of proteins known collectively as the tumour antigens, or T-antigens, since they were initially detected by using antisera from tumour-bearing animals. SV40 specifies two such T-antigens, referred to as large (T) and small (t). Although the coding sequences for these proteins overlap within the same region of viral DNA, it is possible to generate mutants which specify functional forms of only one or other of the two T-antigens. Analysis of such mutants indicates that large T alone is necessary and sufficient for full transformation, leaving the function of small t in some doubt. Large T-antigen from SV40 can therefore be regarded as the product of a viral oncogene but, in contrast to the retroviral oncogenes discussed in section 4.2, large T has no cellular counterpart. In terms of function, the properties of large T remain as elusive as other oncogene products, complicated by the potential differences between activities under permissive and non-permissive conditions. In permissive cells T-antigen occurs predominantly in the nucleus, plays an important role in the initiation of viral DNA synthesis and apparently stimulates host DNA synthesis; *in vitro* it binds to specific regions of viral DNA and has a readily assayable ATPase activity; in transformed cells it forms a tight and specific complex with a 53 000 mol. wt. protein encoded by the host. Which, if any, of these activities is implicated in the oncogenic properties of this protein remains to be seen, but is clearly a topic of intense interest.

The early region of polyoma virus can encode not two but three T-antigens, referred to as large (LT), middle (MT) and small (ST). As with SV40, these discrete products are encoded by the same DNA and are generated by differential RNA splicing. Although the situation is inherently more complex, analysis of viral mutants and the construction of genetically engineered DNAs, which express only one of the three possible products, have indicated that none of these T-antigens is individually capable of inducing full transformation of primary cells. As discussed in more detail in section 4.4, tumorigenesis by polyoma virus may therefore be an example of where two oncogene products are required for full expression of malignancy. The indications are that polyoma LT and MT are both involved and that they may display functionally separable properties.

Two human polyomaviruses called JC and BK, are tumorigenic in animals and transform non-permissive cells in culture in the same way as SV40 and polyoma. However, these human viruses, and indeed all members of the polyomavirus group, were isolated from non-neoplastic lesions and any suggestion of involvement in human cancers remains highly controversial and rather unlikely. This point is further emphasised by recalling that live SV40 virus, which is definitely known to transform human cells in culture, was unwittingly injected into a large number of human subjects as a contaminant of the early polio vaccines, yet in the ensuing 20 or more years, there have been no indications that these individuals have an increased cancer risk. Thus, despite their undisputed contribution to the molecular biology of neoplasia, the relevance of polyomaviruses in naturally occurring malignancies remains obscure.

4.3.3.2 Papillomaviruses. Papillomaviruses cause benign epithelial proliferations (i.e. papillomas or warts) in a wide range of animals, but have been most rigorously studied in humans, cattle and in the cottontail rabbit, where the prototype and incidentally the first example of a DNA tumour virus was originally recognised by Shope in 1933. Papillomaviruses have a closed circular genome of around 8000 base pairs and, in the absence of suitable culture systems in which to propagate these viruses, their analysis at the molecular level has depended largely on recombinant DNA techniques. Thus, instead of a serological classification, as commonly applied to other viruses, the papillomaviruses are classified on the basis of genome structure and DNA sequence homology relative to other isolates.

It is now possible to distinguish at least 24 different human papillomaviruses (HPV) and six bovine papilloma viruses (BPV), although it should be noted that some of these have only been detected at the DNA level and not as mature particles. Significantly, many of the different subtypes appear to be reproducibly associated with specific clinical conditions. In humans these include skin warts, laryngeal papillomas, anogenital warts, flat condylomas of the cervix and vagina, and the rare skin condition epidermaldysplasia verruciformis (EV). All of these lesions represent benign proliferations of epithelial cells, but there are indications that some may undergo conversion into malignant carcinomas in a low but significant percentage of cases. One of the most striking examples is EV, where approx. 30 per cent of persistent cases undergo conversion into squamous-cell carcinoma later in life.[17] It is far from clear what role the virus plays in this process, but in animals papillomavirus infections appear to act synergistically with other carcinogens or cofactors, the most notable example being the induction of carcinomas by the combined influences of cottontail rabbit papillomavirus and chemical carcinogens. Similarly, a high proportion of upland cattle in Scotland suffer from alimentary papillomas, as a result of infection by BPV-4, and around 5 per cent of these develop carcinomas. Here the cofactor has been identified as a chemical in the bracken fern on which these cattle graze. With EV in humans, it is suspected that ultraviolet light may serve as cofactor since malignant conversion normally occurs only on exposed areas of skin. However, these fairly definitive correlations between papilloma virus-induced lesions and neoplasia must be tempered by the observation that the tumour tissues do not always show signs of viral infection or viral DNA, and never appear to contain mature viral particles.

It is now clear that some, though not all, papillomaviruses are capable of inducing transformation of cultured mouse cell lines. In marked contrast to the polyomaviruses, the transformed cells contain viral DNA in an episomal non-integrated state, maintained by a trans-acting viral function. Experiments with subfragments of the viral genome have indicated that the transforming region of DNA may be confined to as little as 30 per cent of the genome, but it is still not known what functions it might encode.

4.3.4 Adenoviruses

The adenovirus genome is a linear DNA molecule of around 30 000–40 000 base pairs, six to eight times larger than polyoma, and coding for about 20–30 viral polypeptides. The viruses are ubiquitous in their species of origin and most humans have been exposed to several subtypes by their mid-teens. At least 31 distinct serotypes of human adenovirus have now been recognised and, depending on their association with particular symptoms (mainly acute respiratory and alimentary tract infections), these have been assigned into four separate groups. The demonstration that a human adenovirus (Ad12) could induce tumours in newborn hamsters was the first demonstration of the oncogenicity of a human virus and caused major repercussions, particularly the dropping of all plans to make an attenuated adenovirus vaccine for use in humans. This potential is not limited to Ad12, nor to human adenoviruses, but although some subtypes are highly oncogenic (e.g. Ad12, 18 and 31), some show only limited oncogenic capacity and others are evidently non-oncogenic. Nevertheless, all serotypes tested have the ability to transform rodent cells in culture.

Adenoviruses are lytic in cells of their natural host and only transform under non-permissive circumstances. The transformed cells contain variable amounts of integrated viral DNA, often multiple copies, but invariably contain and express early gene functions encoded within the left-most 10–15 per cent of the linear DNA genome. Similarly, transfection of cells with specific fragments of viral DNA confirmed that only this lefthand end, comprising the early regions E1a and E1b, was sufficient for transformation.[4] In a series of elegant experiments in which adenoviruses were constructed which were impaired in either or both of the E1a and E1b functions, it has been established that both are required for full expression of the transformed phenotype, again indicative of co-operation between oncogene products (see section 4.4).

4.3.5 Herpesviruses

Members of the herpesvirus family are extremely common agents in the human population, being the causes of conditions such as cold sores, chickenpox and glandular fever, and in many cases establishing infections without any appreciable symptoms. Once established, these viruses can persist in the body in a latent state throughout the lifespan of the individual. Later in life, virus expression may occasionally be reactivated as a result of some additional stimulus such as stress or immune suppression. This persistence and ubiquity of the herpesviruses therefore complicate any attempts to trace epidemiological links with human neoplasms. Nevertheless, they have long been considered as potential human tumour viruses for three major reasons. In the first place, some animal herpesviruses cause neoplastic disease in natural settings; secondly, inactivated herpesviruses can transform cells in culture, although much less efficiently than other DNA tumour viruses; thirdly, some of the epidemiological evidence that particular herpesviruses may represent contributory factors in human malignancies is quite compelling.

Research into the oncogenic properties of herpesviruses has been strongly influenced by the parallel studies on the polyoma and adenoviruses, the assumption being that the viral DNA must encode specific transforming functions, analogous to the T-antigens, and therefore that specific regions of the viral genome should be detectable in all transformed cells. Unfortunately, there is no consistent or convincing evidence that either of these situations prevail. One of the problems in interpreting the data is that the herpesviruses are very much more complex than the other tumour viruses, with a linear DNA genome of around 150 000–180 000 base pairs. This means that by comparison with, say, the early regions of polyoma and SV40, the transforming function could be contained within as little as 1–2 per cent of the total genome, making its detection within transformed cells or tumour tissue much more ambiguous. The situation is further complicated by the complex physical organisation of genomic DNA. The following discussion of individual members of the herpesvirus family will therefore dwell less on possible molecular mechanisms than on the arguments for or against their involvement in human cancers.

4.3.5.1 Epstein–Barr Virus. EBV has a high incidence world-wide; most people are affected very early in life, with no obvious clinical symptoms, and the virus establishes a persistent latent infection in B-lymphocytes, resulting in only sporadic activation and virus production. In contrast, when exposure is delayed until adolescence or early adulthood, the virus causes infectious mononucleosis (glandular fever). The reasons behind this variable response to infection are unknown, but the causal relation between EBV and infectious mononucleosis is unequivocal. Unfortunately, the same cannot be said about the association between EBV infection and cancer where the evidence, though strong, is at best circumstantial. Most data relate to two specific malignancies: Burkitt's lymphoma (BL) and nasopharyngeal carcinoma (NPC).[18]

Although the world-wide incidence of BL is low, it occurs at a relatively high frequency (three to four cases per 100 000 per year) in children in specific geographical locations. In view of the ubiquity of EBV, it is assumed that this geographical clustering of BL reflects the distribution of an additional contributing factor, the most likely candidate being endemic malaria. The evidence supporting EBV involvement comes from several types of observation. In the first place, BL patients have significantly higher titres of antibodies to EBV and retrospective studies have indicated that the patterns of EBV antigen expression may have prognostic relevance. Secondly, the tumour cells contain multiple copies of EBV DNA, usually episomal, although rare chromosomal integration cannot be excluded. Thirdly, the tumour cells express a nuclear antigen (EBNA) associated with EBV infection. Finally, by the criterion of immortalisation (see section 4.1.2), EBV can transform cultured B-lymphocytes.

In Africa, 98 per cent of BL cases show these positive indications of EBV involvement. However, the remaining 2 per cent are negative for any signs of the virus, even though the individuals have been exposed to both EBV and malaria. Moreover, BL occurs elsewhere in the world, at about the same frequency as these 2 per cent of African cases, and in other locations only a small proportion (15–20 per cent) of the tumours are EBV positive. Significantly, all cases of BL show evidence of chromosomal abnormalities in the tumour cell. These usually take the form of reciprocal translocations between chromosomes 8 and 14, although in a minority (approx. 10 per cent) the chromosomes involved are 8 and 2, or 8 and 22. As discussed previously (section 4.2.5), these events bring the cellular oncogene *c-myc* on chromosome 8 into juxtaposition with one of the immunoglobulin gene loci.[8] Although analogous translocations have been detected in other haemopoeitic neoplasms, they do not occur in B-lymphocytes immortalised by EBV infection.

Current theories as to the role of EBV in BL therefore centre on the notion that EBV infection somehow alters the growth properties of B-lymphocytes, particularly in individuals suffering immunosuppression as a result of malaria, and that these stimulated cells are more susceptible to other changes, including chromosome rearrangements, which lead eventually to the selection of malignant cell clones. Thus, while EBV infection might predispose individuals to BL in the endemic areas, the virus is not essential and the alterations required can be achieved by other routes.

The second malignancy that has been linked to EBV infection is undifferentiated nasopharyngeal carcinoma (NPC). This tumour is particularly common in southern China, among ethnic southern Chinese communities elsewhere in south-east Asia, and in parts of northern and eastern Africa. In the endemic areas of China, NPC represents the commonest tumour, accounting for some 30 cases per 100 000 per year, with a peak incidence in the 45–55 year age-group. As with BL, the incidence of NPC in other parts of the world is low and sporadic, so that the geographic clustering suggests the involvement of an additional factor. The evidence for EBV involvement in NPC again centres on the presence of increased antibody titres and the detection of EBV DNA and EBNA-antigen in the tumour cells. This association is true of all undifferentiated NPC irrespective of ethnic origin. Although EBV-negative cases also occur, these usually show a different histology, comprising more differentiated cell types. However, despite these positive correlations, some uncertainties still remain, such as why a lymphotropic virus like EBV should be implicated in an epithelial tumour and why there should be a long delay between virus infection and the development of neoplasia.

Further uncertainties surround the molecular basis for transformation by EBV. Although recent studies suggest that specific regions of the viral DNA may be involved, there is no clear pattern and RNA transcripts derived from several different parts of the genome can be detected in transformed cells. Nevertheless, the interest aroused by EBV is such that these properties are now under intense investigation, greatly facilitated by the introduction of molecular cloning and DNA sequencing techniques. One consequence of these efforts is that the sequence of the entire 173 000 base-pair genome has now been derived so that the prospects for resolving the mysteries of EBV are bright.

4.3.5.2 Herpes Simplex Viruses. Two serotypes of herpes simplex virus, HSV-1 and HSV-2, are prevalent in the human population. The more common, HSV-1, is the causative agent of cold sores, the major route of transmission presumably being via oral secretions. HSV-2, on the other hand, is responsible for genital lesions and is a venereally transmitted agent whose incidence has increased remarkably in recent years, particularly in the USA. Both viruses are reportedly capable of transforming cells in culture, but these studies have been complicated by the fact that the various laboratories involved have used different methods for inactivating the virus, different recipient cells and different criteria for transformation. Nevertheless, it is clear that one or more specific subfragments of the viral DNA can induce some aspects of transformation, albeit with very low efficiency. Unfortunately, the transforming fragments differ in the two serotypes, the minimum piece of DNA required may be incapable of encoding a functional protein and the DNA is not always retained in the transformed cells. As a result, serious consideration is now being given to the idea that transformation by HSV-1 and HSV-2 may involve a type of 'hit and run' mechanism, where viral infection or the viral DNA itself induces some genetic alteration which persists whether or not the virus or the DNA remains in the cell.[19]

Most interest in herpes simplex viruses as tumorigenic agents stems from the association between HSV-2 infection and cervical cancer. Evidence indicating that the development of the disease might be influenced by the sexual promiscuity of both partners suggested that some venereally transmitted agent, such as HSV-2, might be involved. However, although there have been reports that cervical carcinoma patients show an above average titre and frequency of antibodies to HSV-2, other controlled studies imply that HSV is no more closely associated with neoplasia than other sexually transmitted infections, and indeed, that human papilloma virus may be a more likely candidate. Coupled with the conflicting and inconsistent reports regarding HSV macromolecules in tumour tissues, these findings make the status of HSV as a human cancer virus both doubtful and controversial.

4.3.5.3 Cytomegalovirus. Human cytomegalovirus (CMV), like EBV, usually establishes inapparent latent infections and has a world-wide distribution, with regional incidences varying between 20 and 98 per cent. However, congenital infection of the fetus or newborn can result in serious defects, and the virus is a major threat to immunosuppressed patients where it causes severe mononucleosis. Although its ubiquity again complicates the issue, CMV infections have been considered as contributory factors in a number of different malignancies, but particularly in Kaposi sarcoma. This is a variable tumour, probably of endothelial origin, and, although rare in the western world, it shows a clustered almost epidemic incidence in parts of equatorial Africa. By analogy to EBV and BL, the epidemiological association relates to the presence of CMV-encoded macromolecules and particles in the tumour tissues.[20] Outside Africa, Kaposi sarcoma is rare and mostly confined to elderly men of Jewish or Mediterranean origin. However, an aggressive form of the tumour has recently appeared among male homosexual patients suffering from acquired immune-deficiency syndrome (AIDS). Since CMV infection can both contribute to and be exacerbated by immune suppression, it is not clear what role the virus plays in either the cause or effect of this disease. Evidence for CMV infection is almost universal among homosexuals, but since AIDS patients are subject to opportunistic infections by a whole battery of agents, any one of these could also be implicated. Nevertheless, the detection of CMV DNA in tumour but not in normal tissue of the same patient supports the association established with the more classical cases of Kaposi sarcoma.

4.4 Summary, Speculations and Future Prospects

In situations where they can be shown to be oncogenic, viruses are the most potent carcinogens known. Both in whole animals and in tissue-culture systems, infection by a single virus particle is sufficient to induce a tumour or full morphological transformation, and in the latter situation only a small segment of the viral genome may be enough. Yet, with only a few exceptions, mostly in domestic animals, there is little evidence to suppose that the horizontal or vertical transmission of an infectious virus can be regarded as a direct cause of cancer in any naturally occurring animal populations, including humans.

Are there any prospects therefore that anti-viral prophylaxis would make any impact on cancer therapy? Clearly there are since, despite these negative generalisations, there are definite indications that in a number of specific human neoplasms prior infection by a virus, though not a direct cause, may be one of several factors involved in the aetiology. Thus virus infections may predispose cells to alterations inflicted by other environmental agents and may represent only one of several routes by which these effects may be achieved. In terms of prevention, measures might be directed against any one of these multiple factors, provided that they can be identified, but in many instances it may turn out that the virus is the most amenable to elimination. Where the virus also poses a major health problem, as with hepatitis B virus, there are obvious reasons to attempt anti-viral therapy, but with Burkitt's lymphoma, for example, the more logical course would be to eliminate the health risk of the suspected cofactor, malaria. With domestic animals, the measures taken to avoid viral carcinogenesis can include breeding of genetic resistance or the elimination of affected animals if their survival threatens the population as a whole, but with human subjects the choices are obviously more limited. Improved hygiene could potentially reduce or delay the spread of some viruses, and it seems feasible that early screening for viral antigen profiles or, for example, establishing the subtype of papillomavirus infections may have significant prognostic value in identifying individuals at risk of developing malignancy. However, the major option available remains immunisation. Unfortunately, a serious problem arises if attempts are to be made to develop effective vaccines against suspected human tumour viruses, in that many of them have been shown to have transforming potential only when they are rendered defective. Thus attenuated or inactivated virus preparations, as frequently used in immunisation programmes, may present a greater risk than the virus itself does in a natural context. Current efforts are therefore concentrating on the use of specific viral antigens or purified proteins as immunogens and the ability to express all or fragments of such proteins at high levels in genetically engineered prokaryotic systems offers tremendous advantages for this approach.

One of the obvious prerequisites for designing these genetically manipulated reagents is an intimate knowledge of the organisation and sequence of the viral genome. The large body of work devoted to the molecular characterisation of tumour viruses can therefore be justified along these lines, but even more so for the impact they have had on our understanding of eukaryotic gene expression in general. The tumour viruses have provided an enormous arsenal of molecular tools and pioneered many of the techniques now being applied to dissecting the genetic alterations associated with neoplasia, whether or not viruses are directly involved. Perhaps their most significant contribution has been to point to the existence of specific genes, or oncogenes, whose expression may be responsible for some of these changes.

In the DNA tumour viruses such oncogenes are strictly viral genes and all play significant and essential roles in the normal replicative cycle of the virus. Only under rare circumstances when their normal functions are subverted can these genes be seen to be oncogenic. In contrast, the 20 or so oncogenes contained within the genomes of transforming retroviruses are almost invariably tumorigenic when these viruses infect their natural hosts. Some researchers would claim that this is the only criterion by which an oncogene should be judged and that these *v-onc* genes are the only true examples. The reason for this is that the cellular, so-called *c-onc* equivalents from which they are derived are not normally oncogenic. While DNA transfection experiments can identify activated transforming genes within cellular DNA, no single cellular gene has been shown to have the capacity to convert a normal cell into a cancer cell. Such transfection assays have been traditionally performed on the immortalised NIH 3T3 cell line and the criticism has been that these cells cannot be regarded as normal and are already partially transformed. Moreover, few of the *c-onc* genes function in this assay. Considerable excitement has therefore been generated by the observation that combinations of oncogenes may act co-operatively in the induction of transformation.[21] Significantly, appropriate combinations transform primary cell cultures as opposed to cell lines and the resultant foci are tumorigenic when transplanted back in to animals. Although only a few pairs of genes have been tested so far, the implications are that genes such as adenovirus E1a, polyoma LT, and *myc* may encode functions akin to those required to establish an immortalised cell line (like NIH 3T3 cells), whereas others such as E1b, MT and *ras* are more directly responsible for other parameters of transformation. Since genes from heterologous sources (e.g. E1a + MT or *myc* + *ras*) can also co-operate in this assay, these data suggest (1) that the critical steps of transformation can be assigned to distinct separable genes and (2) that the products of these genes have intrinsic activities and function independently of one another.

How other oncogenes fit into this scheme remains to be seen, but there is now good evidence to implicate co-operative effects between oncogenes in naturally occurring tumours as well as in tissue-culture assays. An obvious example occurs in B-cell lymphomas induced by ALV where two distinct cellular genes appear to be implicated: the *c-myc* gene, presumably activated by provirus insertion, and the *B-lym* gene (Table 4.2C) detectable by DNA transfection assays. Similarly, in Burkitt's lymphoma, alteration of the *c-myc* locus as a result of chromosome translocation, can be accompanied by activation of either the *B-lym* or *N-ras* oncogenes. Add to this the potential role of EBV in immortalising the premalignant B-cell population and it is clear that this particular neoplasm may develop in multiple steps.

Such observations of co-operation between different oncogenes and between viruses and other environmental cofactors go a long way toward reconciling tumour virology with the multistage epidemiology of human cancers. Given the staggering rate of progress in our understanding of these genes and the genetic alterations required to make them oncogenic, the prospects of obviating these events no longer seem so daunting.

References

1. Temin, H.M. & Rubin, H. (1958) Characteristics of an assay for Rous sarcoma virus and Rous sarcoma cells in tissue culture. *Virology*, *6*, 669-88
2. Fried, M. (1965) Cell-transforming ability of a temperature-sensitive mutant of polyoma virus. *Proc. Natl. Acad. Sci. USA*, *53*, 486-91
3. Martin, G.S. (1970) Rous sarcoma virus: a function required for maintenance of the transformed state. *Nature (Lond.)*, *227*, 1021-3
4. Graham, F.L., van der Eb, A.J. & Heijneker, H.J. (1974) Size and location of the transforming region in human adenovirus type 5 DNA. *Nature (Lond.)*, *251*, 687-91
5. Varmus, H.E. (1982) Recent evidence for oncogenesis by insertion mutagenesis and gene activation. *Cancer Surveys*, *1*, 309-19
6. Stehelin, D., Varmus, H.E., Bishop, J.M. & Vogt, P.K. (1976) DNA related to the transforming gene(s) of avian sarcoma viruses is present in normal avian DNA. *Nature (Lond.)*, *260*, 170-3
7. Cooper, G.M. (1982) Cellular transforming genes. *Science*, *218*, 801-6
8. Leder, P., Battey, J., Lenoir, G., Moulding, C., Murphy, W., Potter, H., Stewart, T & Taub, R. (1983) Translocations among antibody genes in human cancer. *Science*, *222*, 765-71
9. Schwab, M., Alitalo, K., Varmus, H.E., Bishop, J.M. & George, D. (1983) A cellular oncogene (c-ki-*ras*) is amplified, overexpressed, and located within karyotypic abnormalities in mouse adrenocortical tumour cells. *Nature (Lond.)*, *303*, 497-501
10. Tabin, C.J., Bradley, S.M., Bargmann, C.I., Weinberg, R.A., Papageorge, A.G., Scolnick, E.M., Dhar, R., Lowy, D.R. & Chang, E.H. (1982) Mechanism of activation of a human oncogene. *Nature (Lond.)*, *300*, 143-9
11. Hunter, T. & Sefton, B.M. (1980) Transforming gene product of Rous sarcoma virus phosphorylates tyrosine. *Proc. Natl. Acad. Sci. USA*, *77*, 1311-15
12. Downward, J., Yarden, Y., Mayes, E., Scrace, G., Totty, N., Stockwell, P., Ullrich, A., Schlessinger, J. & Waterfield, M.D. (1984) Close similarity of epidermal growth factor receptor and v-*erb* B oncogene protein sequence. *Nature* (*Lond.*), *307*, 521-7
13. Popovic, M., Sarin, P.S., Robert-Guroff, M., Kalyanaraman, V.S., Mann, D.,

Minowada, J. & Gallo, R.C. (1983) Isolation and transmission of human retrovirus (human T-cell leukaemia virus). *Science*, *219*, 856-9
14. Botchan, M., Topp, W. & Sambrook, J. (1976) The arrangement of simian virus 40 sequences in the DNA of transformed cells. *Cell*, *9*, 269-87
15. Marion, P.L. & Robinson, W.C. (1983) Hepadna viruses: hepatitis B and related viruses. *Curr. Top. Microbiol. Immunol.*, *105*, 99-121
16. Beasley, R.B., Hwang, L.-Y, Lin, C.-C. & Chien, C.-S. (1981) Hepatocellular carcinoma and hepatitis B virus: a prospective study of 22,707 men in Taiwan. *Lancet*, *ii*, 1129-33
17. Ostrow, R.S., Bender, M., Niimura, M., Seki, T., Kawashima, M., Pass, F. & Faras, A.J. (1982) Human papillomavirus DNA in cutaneous primary and metastasized squamous carcinomas from patients with epidermodysplasia verruciformis. *Proc. Natl. Acad. Sci. USA*, *79*, 1634-8
18. Zur Hausen, H., Schulte-Holthausen, H., Klein, G., Henle, W., Henle, G., Clifford, P. & Santesson, L. (1970) EBV DNA in biopsies of Burkitt tumours and anaplastic carcinomas of the nasopharynx. *Nature (Lond.)*, *228*, 1056-8
19. Galloway, D.A. & McDougall, J.K. (1983) The oncogenic potential of herpes simplex viruses: evidence for a 'hit-and-run' mechanism. *Nature (Lond.)*, *302*, 21-4
20. Boldogh, I., Beth, E., Huang, E.-S., Kyalwazi, S.K. & Giraldo, G. (1981) Kaposi's sarcoma. IV. Detection of CMV DNA, CMV RNA and CMNA in tumour biopsies. *Int. J. Cancer*, *28*, 469-74
21. Land, H., Parada, L.F. & Weinberg, R.A. (1983) Cellular oncogenes and multistep carcinogenesis. *Science*, *222*, 771-8

Further Reading

Tooze, J. (ed.) (1981) *Molecular Biology of Tumor Viruses*, 2nd edn, part 1, *DNA Tumor Viruses*, Cold Spring Harbor Laboratory, New York

Varmus, H. & Levine, A.J. (eds) (1983) *Readings in Tumour Virology*, Cold Spring Harbor Laboratory, New York

Weiss, R., Teich, N., Varmus, H. & Coffin, J. (eds) (1982) *Molecular Biology of Tumor Viruses*, 2nd edn, part 2, *RNA Tumor Viruses*, Cold Spring Harbor Laboratory, New York

Wyke, J. & Weiss, R. (eds) (1984) Viruses in human and animal cancers: the current position. *Cancer Surveys*, *3*, No. 1, Oxford University Press, Oxford

5 CHEMICAL CARCINOGENESIS

D.H. Phillips

Contents

5.1 Introduction

It is apparent from studies of the incidence of cancer in different parts of the world that some striking regional differences exist. For most human cancers there is a lack of evidence to implicate known carcinogenic viruses as the critical factors responsible [the exceptions are the associations between hepatitis-B virus infection and liver cancer, herpesviruses and cervical cancer and Burkitt's lymphoma, and HTLV and T-cell lymphoma (see Chapter 4)]. Nor do genetic factors appear to play a major role in these geographical differences; studies on migrants reveal that, within two generations, i.e. once they have absorbed the habits and culture of their adopted nation, the migrants experience cancer incidences similar to those of the indigenous populations. Therefore the concept has emerged that the majority of human cancers is caused by environmental factors. Studies on chemical carcinogenesis originated from observations of unusually high incidences of cancers among workers in certain industrial occupations and from subsequent laboratory studies in which tumours were induced in animals by the same chemical mixtures to which the 'at-risk' groups of workers were exposed. To date more than 20 chemicals or chemical mixtures, and several industrial processes, are known to cause cancer in man as a result of occupational, medicinal or societal exposure (Table 5.1); approximately twice as many more are strongly suspected of doing so.[1] However, aside from small groups especially at risk and the strong association between tobacco smoking and lung cancer, the specific causes of the majority of human cancers are presently unknown. Nevertheless, it is possible that many cancers are caused by naturally occurring or synthetic chemicals present in the environment (e.g. in air, food, water, etc.) and are thus at least theoretically preventable.

The most direct way of controlling cancer is by the identification of the responsible carcinogens and their elimination from the environment or, where the latter is not feasible, a modification of their interaction with living cells such that they are rendered innocuous. The achievement of these goals requires not only rapid and economical methods of testing chemicals for carcinogenic activity (see Chapter 6) but also an

Table 5.1 Industrial Processes and Chemical Compounds Causally Associated with Cancer in Humans

A *Industrial processes*	Site of tumour
Auramine manufacture	Urinary bladder
Boot and shoe manufacture and repair (certain occupations)	Nasal cavity, urinary bladder
Furniture manufacture	Nasal cavity
Isopropyl alcohol manufacture (strong-acid process)	Paranasal sinuses
Nickel refining	Nasal cavity, lung
Rubber industry (certain occupations)	Urinary bladder
Underground haematite mining (with exposure to radon)	Lung

B *Chemicals and groups of chemicals*	*Exposure*	*Site of tumour, or type of malignancy*
4-Aminobiphenyl	Industrial	Urinary bladder
Analgesic mixtures containing phenacetin[a]	Medical	Renal pelvis
Arsenic and arsenic compounds[b]	Industrial	Lung, skin
Asbestos	Industrial	Pleura, peritoneum
Azathioprine	Medical	Skin, lymphoma
Benzene	Industrial	Leukaemia
Benzidine (4,4′-diaminobiphenyl)	Industrial	Urinary bladder
Betel quid	Societal	Buccal mucosa
N,N-Bis(2-chlorethyl)-2-naphthylamine (Chloronaphazine)	Medical	Urinary bladder
Bis(chloromethyl) ether	Industrial	Lung
Certain combined chemotherapies for lymphomas[a] (including MOPP[c])	Medical	Leukaemia, lymphoma
Chlorambucil	Medical	Leukaemia
Chromium and certain chromium compounds[a]	Industrial	Lung
Cigarette smoke	Societal	Lung, urinary bladder, pancreas
Conjugated oestrogens[a]	Medical	Endometrium
Cyclophosphamide	Medical	Urinary bladder, leukaemia
Diethylstilboestrol	Medical	Vagina
Melphalan (phenylalanine nitrogen mustard)	Medical	Leukaemia
Methoxsalen with ultraviolet A therapy (PUVA)	Medical	Skin
Mustard gas [bis(2-chloroethyl)sulphide]	Industrial	Respiratory tract
Myleran (1,4-butanediol dimethanesulphonate)	Medical	Leukaemia
2-Naphthylamine	Industrial	Urinary bladder
Soots, tars and oils[a]	Industrial	Skin, lung
Treosulphan (L-threitol 1,4-dimethanesulphonate)	Medical	Leukaemia
Vinyl chloride	Industrial	Liver mesenchyme

[a] Compound(s) responsible for the carcinogenic effect in humans cannot be specified
[b] Influence of other constituents in the environment cannot be excluded
[c] Procarbazine, nitrogen mustard, vincristine and prednisone
Adapted from International Agency for Research on Cancer (1982) *IARC Monographs on the Evaluation of the Carcinogenic Risk of Chemicals to Humans*, Suppl. 4, IARC, Lyon

understanding of their mechanism of action and the nature of the biochemical events critical to the development of tumours. This chapter describes the known types of chemical carcinogens, the pathways by which they are thought to exert their biological effects and some modulating influences (both enhancing and inhibitory) on the carcinogenic process.

5.2 Origins of Experimental Chemical Carcinogenesis

Observations concerning the association between cancer and exposure to specific materials were first made in the late eighteenth century. John Hill, a London physician, reported in 1761 six cases of nasal 'polypusses' among snuff-takers and cautioned against the excessive use of the substance.[2] In 1795 an association between clay-pipe smoking and carcinoma of the lips was reported in Germany by von Soemmering.[3] In 1775, in what is generally considered to be the first link between cancer and occupation, the London surgeon, Percival Pott, noted the high incidence of scrotal cancer among chimney sweeps and ascribed it to their exposure in pre-adolescence to soot.[4] As a direct result of this finding, preventative measures were introduced in the form of recommendations for daily baths and better protective clothing. In the last quarter of the nineteenth century the development of skin cancers was reported in workers in the tar[5] and shale oil[6] industries and among cotton-mill workers frequently exposed to crude lubricating oils used in spinning machines.[7] Also the occurrence of several cases of urinary bladder cancer among workers in a German aniline dye factory was tentatively attributed to their exposure to aromatic amines.[8]

These reports of tumours in humans exposed to particular chemical mixtures led to attempts to induce tumours in laboratory animals by treating them with the same agents. The first successful demonstration was by the Japanese workers, Yamagiwa and Ichikawa, who in 1915 reported the induction of tumours on the ears of rabbits by repeated applications of coal tar.[9] Shortly after, tumours were produced in mouse skin by the same method.[10] Intensive studies on the nature of the carcinogenic agent in coal tar followed, revealing that it was present in the high-boiling fraction, was free of nitrogen and sulphur and was probably a polycyclic aromatic hydrocarbon. Many compounds of this type were synthesised and their fluorescence spectra compared with those of the carcinogenic tar fraction; they were also tested for carcinogenic activity by repeated application to mouse skin. Dibenz[a,h]anthracene (Figure 5.1) was the first pure chemical found, in 1930, to produce tumours in these tests,[11] although its fluorescence spectrum was not identical with that of the tar fraction. Three years later the final purification of the carcinogenic material isolated from coal tar was achieved and the compound obtained identified as the pentacyclic aromatic hydrocarbon, benzo[a]pyrene (Figure 5.1).[12] (The letters 'a,h' in dibenz[a,h]anthracene, and 'a' in benzo[a]pyrene, indicate the positions at which the simple hydrocarbons anthracene and pyrene are substituted with additional benzene rings. The peripheral C–C bonds of the hydrocarbons are lettered a,b,c, etc., in a clockwise direction: thus dibenz[a,h]anthracene has the 'a' and 'h' positions of anthracene substituted with benzene rings.) Two legacies of these pioneering studies, conducted by Kennaway and his associates in London, are that polycyclic hydrocarbons, especially benzo[a]pyrene, are still the subject of extensive experimental studies, and mouse skin is still used as a convenient and sensitive tissue in which to test compounds for carcinogenic activity.

In the same year that the identity of the carcinogen in coal tar was announced, the induction of liver tumours was reported in rats fed 2′,3-dimethyl-4-aminoazobenzene, an aromatic amine that is a reduction product of the synthetic dye, scarlet red.[13] This was followed in 1938 by a report on the induction of urinary bladder cancer in dogs fed the industrial aromatic amine 2-naphthylamine.[14] Thus the induction of tumours in experimental animals by chemicals already suspected of being the cause of tumours among occupationally exposed humans provided an experimental basis for testing new compounds and for investigating the biochemical mechanisms involved in chemical carcinogenesis. One-hundred-and-sixty years after it was first implicated therefore the role of chemicals in the induction of tumours became amenable to laboratory investigation.

5.3 General Nature of Chemical Carcinogenesis

An early and important feature to emerge from studies of the experimental induction of tumours with chemicals was the long time-lag between treatment with the carcinogen and the appearance of tumours. In addition, repeated doses of active agents are often, although not always, required. The carcinogenic process appears to be multistep and is divisible into at least two qualitatively different steps; *initiation*, that appears to be a rapid and irreversible event, and *promotion*, a more protracted process that is at least partially reversible. In instances where repeated doses of a single chemical induce tumours, the substance is presumably acting as both an *initiator* and a *promoter* (thus as a *complete carcinogen*). Some agents act as promoters but appear to have little or no intrinsic initiating activity, whereas a few others have purely initiating activity. These concepts are discussed more fully in section 5.8. While a promoter is therefore a substance that increases the tumorigenic response to a carcinogen (or initiator) when applied after it, *cocarcinogens* are agents that increase the overall carcinogenic process when co-administered with a carcinogen but do not themselves have the ability to induce tumours.

It will be evident from the description in section 5.5 of the different classes of chemical carcinogens that no structure or partial structure is common to all these agents. Furthermore, they exhibit a wide range of chemical reactivity, from highly reactive (e.g. the alkylating agents) to somewhat inert (e.g. polycyclic aromatic hydrocarbons). However, the unifying concept that has emerged in the last 20 years is that the majority of chemical carcinogens, if not electrophiles themselves, are converted by metabolism into forms that are highly electrophilic.[15,16] It is these derivatives (*ultimate carcinogens*) that are responsible for the biological effects of the parent compounds (*precarcinogens*). The generation of an ultimate carcinogen may occur in several steps through intermediate formation of other non-electrophilic metabolites; these are termed *proximate carcinogens* or *procarcinogens*. Thus:

carcinogen (precarcinogen) → proximate carcinogen → ultimate carcinogen

A consequence of the electrophilicity of ultimate carcinogens is that they are capable of reacting with nucleophilic sites within the eukaryotic cell. It is thought that covalent modification of a cellular informational macromolecule, i.e. DNA, RNA or protein, by a chemical carcinogen is a necessary early event in the initiation phase of the carcinogenic process and such biomolecules contain a number of nucleophilic centres. Chemical mutagens, for which the critical cellular target is unequivocally DNA, are also, for the most part, strongly electrophilic in their ultimately reactive forms (exceptions are the intercalating non-binding frameshift mutagens and a number of base-analogue mutagens). The fact that the majority of chemicals that are carcinogenic also exhibit mutagenic activity toward bacteria when suitably activated by the inclusion of a metabolising system in the incubation mixture[17] (see section 5.4 and Chapter 6) suggests a strong formal relation between carcinogenesis and mutagenesis. This, and the fact that the extent of modification *in vivo* of DNA, but not RNA or protein, correlates well with carcinogenic potency for a series of closely related compounds,[18] lend strong support to the theory that DNA is the critical cellular target in chemical carcinogenesis. It should be emphasised that no direct unequivocal evidence for this has yet been obtained. Nevertheless, the interactions of chemical carcinogens with DNA have been intensively investigated and are described in section 5.6.

Thus present evidence suggests that most, and probably all, chemical carcinogens are, or are converted by metabolism into, electrophilic reactants that exert their biological effects by covalent interaction with

DNA. Chemical reactivity and mutagenic activity are properties characteristic of ultimate carcinogens; they may also be carcinogenic when administered to experimental animals, although in some cases the high reactivity of the compounds apparently results in their failure to reach the critical cellular target in sufficient concentrations to induce tumours. Proximate carcinogens, on the other hand, would be expected to show greater carcinogenicity than their parent compounds (i.e. precarcinogens) and also to exhibit enhanced mutagenic activity when tested in the presence of metabolising enzymes that would presumably convert them into their ultimate carcinogens. In fact these two properties of a metabolite of a chemical carcinogen are considered strong evidence that it is an intermediate in the pathway of metabolic activation of the precarcinogen. How and why metabolism of chemical carcinogens leads to their biological activation is considered in the next section.

5.4 Metabolism of Chemical Carcinogens: Some General Remarks

All living systems have to contend with unwanted environmental substances and food components that are of no nutritional value. Even if not acutely toxic, were such substances (especially lipid soluble compounds) to accumulate in the organisms they might well interfere with normal cellular functions. Successful living systems have developed a complex set of detoxication mechanisms that are versatile enough to detoxify a wide range of natural and synthetic chemicals.

In general, animals detoxify exogenous chemicals by metabolising them to more polar water-soluble derivatives that can be excreted readily.[19] The first stage (phase I metabolism) frequently involves the introduction of hydroxyl or epoxide groups by oxidative metabolism at nitrogen or carbon atoms; in some cases (e.g. with nitro compounds) this may be preceded by, or result solely from, reduction. These metabolites can then serve as substrates for a variety of conjugation mechanisms (phase II metabolism) resulting in the excretion of water-soluble conjugates in the faeces or urine. Thus most exogenous chemicals are metabolised and excreted in several steps, and some of these steps may involve the generation of reactive intermediates. In the vast majority of cases, such species are good substrates for the next stage of metabolism; occasionally, however, the reactive species are poor substrates for what would normally be the next enzyme in the detoxification process and, instead of being harmlessly and efficiently further metabolised, they may persist in the cell long enough to interact covalently with informational macromolecules and thus have the potential to initiate mutagenesis or carcinogenesis. Metabolic activation of chemical carcinogens can be viewed therefore as an unfortunate consequence of imperfections in a general-purpose system for the detoxication of foreign compounds.

The enzymes responsible for the metabolism of chemical carcinogens are located primarily bound to the endoplasmic reticulum in eukaryotic cells; in cell-fractionation studies, most activity is therefore found in the microsomal fraction. The initial introduction of oxygen into the compounds is usually catalysed by the family of NADPH-dependent cytochrome P-450 enzymes. The mechanism involves a two-electron transfer process and utilises molecular oxygen (O_2);[20] one oxygen atom is incorporated into the substrate and the other is reduced to water. Highest levels of these enzymes are found in the liver, although microsomal fractions prepared from many other tissues also contain enzyme activity; some activity is also located within the nuclear membrane of cells. The enzymes are inducible by a large number of compounds that serve as substrates for the system.[21] Other, non-cytochrome P-450, enzymes are involved in the initial oxygenation of some compounds, which will be discussed below in the context of the particular compounds involved. Microsomal reductions are mediated principally by flavoproteins.

Where the oxidation of compounds (aromatic or olefinic) initially produces epoxides, these can undergo conversion into dihydrodiols by epoxide hydrolases also located in the endoplasmic reticulum. In the case of the polycyclic hydrocarbons (section 5.5.1), some dihydrodiols can be substrates for a second round of oxidation by cytochrome P-450.

Many phase I metabolites are converted in a variety of conjugation reactions into phase II metabolites, including conjugates of glucuronic acid, sulphuric acid and glutathione. These reactions are catalysed by UDP-glucuronyltransferases, sulphotransferases and glutathione-*S*-transferases respectively. Normally, even with chemical carcinogens, these also will be detoxication processes, but there are instances, discussed below, where either the conjugate formed is an electrophile (e.g. some sulphuric acid esters) or a conjugate formed in one tissue is hydrolysed in another and a reactive intermediate regenerated.

For most chemical carcinogens, the overall metabolism is complex and involves both activation and detoxication pathways. Many factors can affect the balance between activation and detoxication and thus influence the carcinogenic process. In some cases these influences can account for the species and organ specificity of tumour induction observed with many chemical carcinogens. A number of modifiers of carcinogenesis are described later in this chapter.

Many systems are used experimentally to investigate the metabolism of chemical carcinogens. These range from whole animals to purified enzymes. In between these extremes, frequent use is made of tissue homogenates or microsomal fractions, as well as cell- and tissue-culture systems. Results obtained with these intermediate systems must be interpreted with caution as many of the vital components that comprise the complete system *in vivo* may be absent or the activity of other components may be elevated. This may be illustrated by the example of liver microsomal fractions prepared from pre-induced rodents. The use of such fractions *in vitro* frequently results in greater formation of biologically active metabolites of a carcinogen than occurs with microsomal fractions from uninduced animals, yet the effect *in vivo* of pre-induction on the production of tumours by a carcinogen is generally inhibitory, not enhancing.

5.5 Chemical Carcinogens

5.5.1 Polycyclic Aromatic Hydrocarbons

As has already been mentioned, polycyclic aromatic hydrocarbons were the first type of chemical carcinogen identified and as a group are still the subject of much current research.[22] They are formed by the incomplete combustion of organic material and are thus present in the atmosphere through the burning of fossil fuels and vegetation (it has been estimated that 900 tonnes of benzo[a]pyrene are released into the atmosphere in the USA annually), in cigarette smoke and in barbecued and smoked food, and they are strongly suspected of being a causative agent in the occurrence of cancer of the lung, skin and colon in man.

Some commonly studied polycyclic aromatic hydrocarbons are shown in Figure 5.1. In general, carcinogenic activity is limited to some

Figure 5.1 Some Polycyclic Aromatic Hydrocarbons

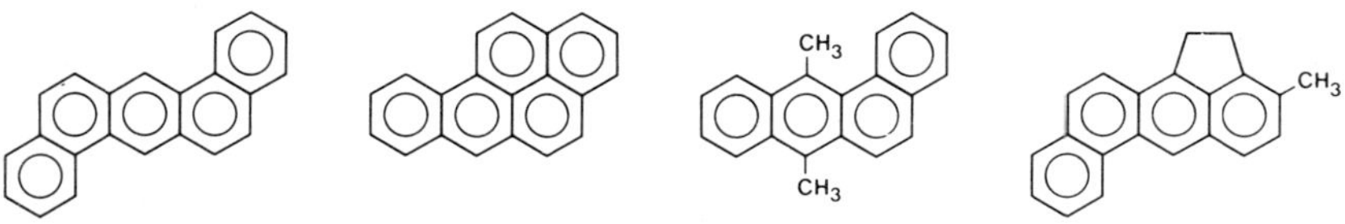

Dibenz[a,h]anthracene   Benzo[a]pyrene   7,12-dimethylbenz[a]anthracene   3-methylcholanthrene

Highly carcinogenic

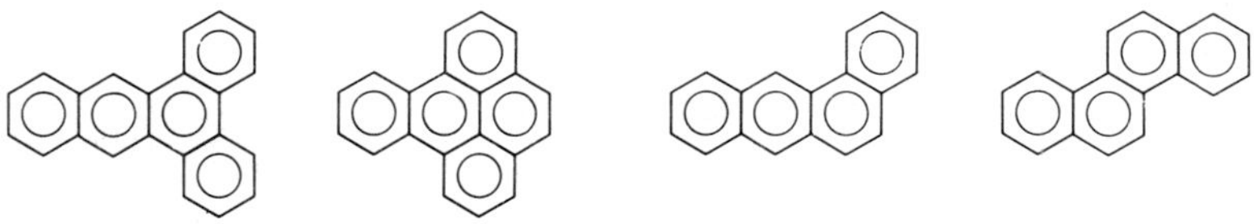

Dibenz[a,c]anthracene   Benzo[e]pyrene   Benz[a]anthracene   Chrysene

Weakly or non-carcinogenic

compounds with four, five or six condensed rings, and some heterocyclic compounds with similar ring structures are also active.[23] In the course of the early studies on polycyclic aromatic hydrocarbon carcinogenesis, several theories were proposed to account for the structure-activity relations observed. However, they were all based on the premise that the hydrocarbons themselves were responsible for the production of tumours, whereas it is now known that the compounds must first undergo metabolic activation to exert their biological effects. They are metabolised to a great many different derivatives, the major types being phenols, dihydrodiols and glutathione conjugates.[24] As all these metabolites can be postulated to arise from reactive epoxide intermediates, it was long suspected that an epoxide of some sort was the metabolite responsible for the biological activity of polycyclic aromatic hydrocarbons.

In the case of benzo[a]pyrene, the type of epoxide responsible became evident when it was found that the 7,8-dihydrodiol was metabolically activated to products that react extensively with DNA,[25] the mechanism being via epoxidation of the 9,10-double bond (Figure 5.2).[26] Biological evidence has shown that the 7,8-dihydrodiol (the proximate carcinogen) is highly carcinogenic and highly mutagenic toward bacteria in the presence of metabolising enzymes, whereas the 7,8-dihydrodiol-9,10-epoxide (the ultimate carcinogen) is highly electrophilic and mutagenic without further metabolism.[22] The reactivity of the diol-epoxide is due to the formation of a carbonium ion at the 10-position, adjacent to the so-called 'bay-region' of the molecule (Figure 5.2). Theoretical studies[27] have indicated

Figure 5.2 Major Pathway of Metabolic Activation of Benzo[a]pyrene (BP) (MO = mono-oxygenase; EH = epoxide hydrolase)

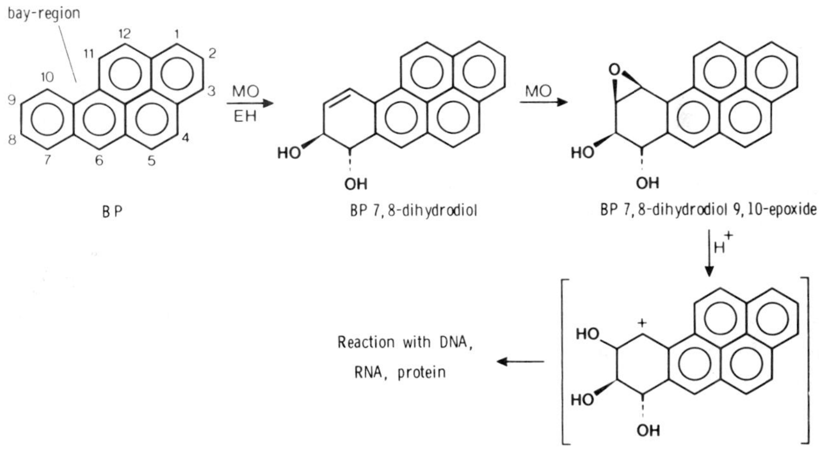

that for a large series of polycyclic aromatic hydrocarbons that possess a bay-region, carbonium ions formed at benzylic carbon atoms adjacent to the bay-region will be more reactive than those found at other parts of the molecules. The prediction that bay-region diol-epoxide formation is the major route of metabolic activation for carcinogenic polycyclic aromatic hydrocarbons has been confirmed for more than a dozen compounds.[21] In each case, the precursor dihydrodiol has high carcinogenic, mutagenic and cell-transforming activity when compared with the parent compounds and dihydrodiols formed at other parts of the molecules, and the bay-region diol-epoxides show higher mutagenic and electrophilic activity than isomeric non-bay-region diol-epoxides. Also in many cases the DNA adducts isolated from tissues or cells treated with the parent hydrocarbon are formed via bay-region diol-epoxide intermediates.[28]

The carcinogenic activity of a polycyclic hydrocarbon (or indeed any chemical carcinogen requiring metabolic activation) will depend not only on the reactivity of its putative ultimate carcinogen but also on the extent of formation of the latter and its ability to escape the normal detoxication mechanisms. Thus benzo[e]pyrene, whose bay-region diol-epoxide is highly reactive, is nevertheless a very weak carcinogen because very little of this derivative is formed by metabolism.

5.5.2 N-*Nitroso and Related Compounds*

5.5.2.1 Nitrosamines. Nitrosamines have been investigated for their carcinogenic and toxic properties since 1956 when the simplest alkylnitrosamine, *N*-nitrosodimethylamine (DMN, Figure 5.3), at the time an important industrial intermediate, was shown to produce liver tumours in rodents.[29] Since then more than 100 nitrosamines have been shown to be carcinogens in experimental animals, causing tumours mainly in the liver but also in a number of other tissues including lung, kidney, oesophagus and nasal cavities.[30] Examples of some nitrosamines to which humans have been exposed are shown in Figure 5.3. Exposure has occurred in a number of industrial settings, and nitrosamines have also been detected in beer, whisky, cheese, cooked bacon and tobacco smoke. However, possibly greater human exposure to nitrosamines occurs through their endogenous formation in the gastrointestinal tract where they arise from the nitrosation of secondary amines by inorganic nitrite.[31] While nitrite is widely used as a food preservative, particularly to cure meat, its use as a food additive probably only accounts for about 20 per cent of the daily intake of nitrite; a far greater source is the conversion in the saliva of nitrate (present in variable concentrations in vegetables and fruit) into nitrite. Although the ingestion of secondary amines and nitrate/nitrite in food would seem to be inavoidable, nitrosamine formation is inhibited by

Figure 5.3 Some Carcinogenic Nitrosamines to which Humans Have Been Exposed

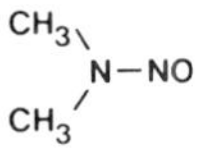	N-nitrosodimethylamine (dimethylnitrosamine) DMN	Leather tanning, rubber tyre manufacture, new car interiors, beer, herbicides, iron foundries
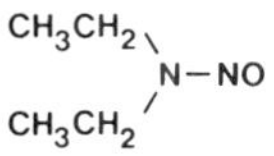	N-nitrosodiethylamine (diethylnitrosamine) DEN	New car interiors, whisky, iron foundries
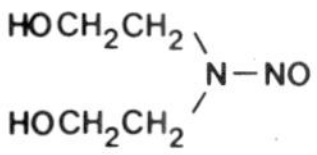	N-nitrosodiethanolamine NDELA	Cosmetics, lotions, shampoos, synthetic cutting fluids
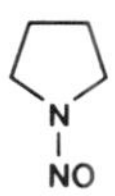	N-nitrosopyrrolidine NPYR	Cooked nitrite-cured bacon
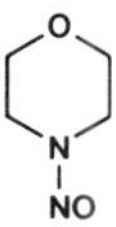	N-nitrosomorpholine NMOR	Leather tanning, rubber tyre manufacture, new car interiors

Occurrences (e.g. in cosmetics) are not universal and in many cases industrial processes have been modified to reduce or eliminate exposure.

vitamins E and C,[32] and the latter is added during the processing of pork for this purpose.

The first step in the metabolic activation of all aliphatic and cyclic nitrosamines is α-hydroxylation (Figure 5.4);[33] there is evidence for the involvement of several enzymes in this biotransformation, including cytochrome P-450s and flavin-dependent oxidases. The hydroxylated metabolite decomposes spontaneously to yield an aldehyde and an alkyl diazonium hydroxide intermediate, which breaks down further via an alkyl diazonium ion to a carbonium ion. There is some doubt as to which of these two electrophilic species is the ultimately reactive form, but in any case the alkylation of cellular macromolecules is likely to occur in a highly concerted manner because the reactive intermediates are extremely short-lived.

Pretreatment of mice with the enzyme inducer 3-methylcholanthrene decreases the carcinogenic activity of DMN[30] for liver but increases the incidence of lung and kidney tumours. Thus, this is an instance where enzyme induction results in enhanced metabolism via activation pathways relative to detoxication pathways, whereas the reverse is more usually the case.

Figure 5.4 Metabolic Activation of Dialkylnitrosamines

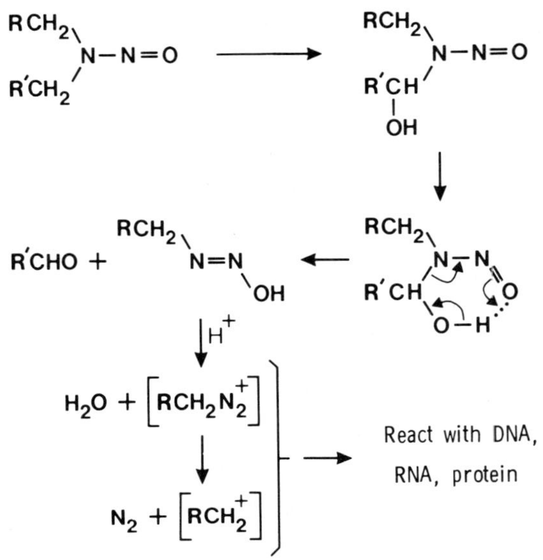

Figure 5.5 Examples of Carcinogenic Nitrosamides, Hydrazines and Triazenes

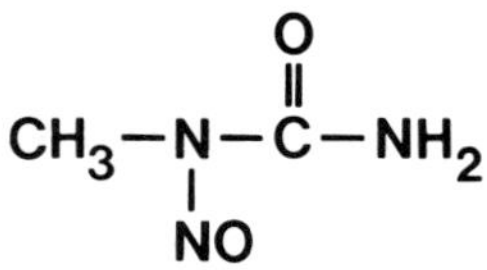

N-methylnitrosourea
MNU

$CH_3NH-NHCH_3$

1,2-dimethylhydrazine

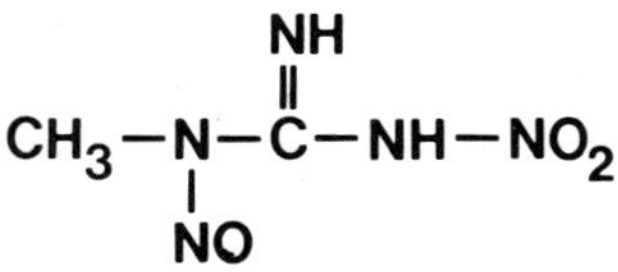

N-methyl-N'-nitro-
N-nitrosoguanidine
MNNG

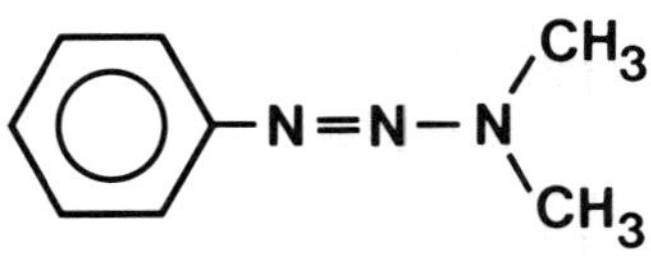

1-phenyl-3,3-dimethyl-
triazene

5.2.2.2 Dialkyltriazenes, Dialkylhydrazines and Nitrosamides. Several other classes of carcinogen are described here, their common feature being that they all appear to act through the generation of electrophilic species from an alkyl diazonium hydroxide.

Dialkyltriazenes, of which 1-phenyl-3,3-dimethyltriazene (Figure 5.5) is an example that causes tumours in rats in the central and peripheral nervous systems, kidney, uterus, ovary and heart,[30] are, like the nitrosamines, metabolically activated via hydroxylation of the alkyl groups. Reactive intermediates are then generated non-enzymically in the same manner as for the nitrosamines.

Dialkylhydrazines are a class of compounds that are also carcinogenic for many tissues in rodents including liver, lung, kidney and intestine.[30] 1,2-Dimethylhydrazine (Figure 5.5) is metabolised in several steps to methylazoxymethanol (see section 5.5.5.2), from which methyl diazonium hydroxide is again formed.

Nitrosamides are also active through similar reactive intermediates except that in these cases no metabolism is required. The compounds are unstable in aqueous environments and decompose to alkyl diazonium hydroxides from which a carbonium ion is generated. Although they are direct-acting carcinogens they frequently induce tumours in experimental animals in organs distant from the site of administration. Thus intravenous administration of *N*-methylnitrosourea (MNU) (Figure 5.5) to rats induces tumours in mammary gland, brain and nervous system.[34] But while intrarectal administration of *N*-methyl-*N*′-nitro-*N*-nitrosoguanidine (MNNG) (Figure 5.5) induces only colorectal tumours in rats, MNU similarly administered causes lung tumours and leukaemias as well.[35] Such organotropism (i.e. the selectivity whereby carcinogens induce tumours in some tissues, but not in others) may be explained partly by the efficiency of detoxication pathways of metabolism, but, as alkylated bases can be detected in the DNA of both susceptible and resistant tissues after administration of MNU (see section 5.6), it would appear that other factors, including rates of DNA repair (see sections 5.6 and 5.7), play an important role.

Although nitrosamides would generally not persist in the environment because of their chemical reactivity, they may be formed endogenously in the gastrointestinal tract through the reaction of amides with nitrite.

5.5.3 Aromatic Amines, Amides and Nitro Compounds

Evidence that exposure of industrial workers to several aromatic amines is associated with the development of cancer of the urinary bladder has stimulated interest in the mechanisms by which these and related chemicals induce tumours. They often cause tumours in experimental animals in organs distant from the site of administration and in most cases they require metabolic activation.[36] The initial step in the activation of aromatic amines and amides involves their oxidation to *N*-hydroxy derivatives.[37] Similarly activation of nitro-aromatic compounds is dependent on their reduction to hydroxylamines. An *N*-acetyltransferase-mediated conversion of arylamines into arylacetamides, and its reverse, a deacetylase-mediated conversion of arylacetamides into arylamines, have also been demonstrated.

Hydroxylation also occurs at aromatic ring positions but usually leads to phenolic products that are not carcinogenic. *N*-Oxidation may involve not only the cytochrome P-450 enzymes but also flavin-dependent oxidases.[38] *N*-Hydroxylation takes place readily only at certain amino-substituted ring positions. It is probably for this reason that 2-naphthylamine is a potent carcinogen and the 1-isomer is inactive, yet the synthetic *N*-hydroxy derivatives of both compounds are highly active.[35]

The *N*-hydroxy metabolites of aromatic amines and amides can undergo activation to electrophilic ultimate carcinogens in a variety of ways: these include protonation of the nitrogen atom of the hydroxylamine (occurring non-enzymically) with subsequent loss of water to give a nitrenium ion, oxidation of *N*-acyl-*N*-hydroxy derivatives to free radicals or esterification of the *N*-hydroxy group to give reactive esters.[35] These general mechanisms are outlined in Figure 5.6.

Figure 5.6 Possible Pathways of Metabolic Activation of Aromatic Amines, Amides and Nitro Compounds

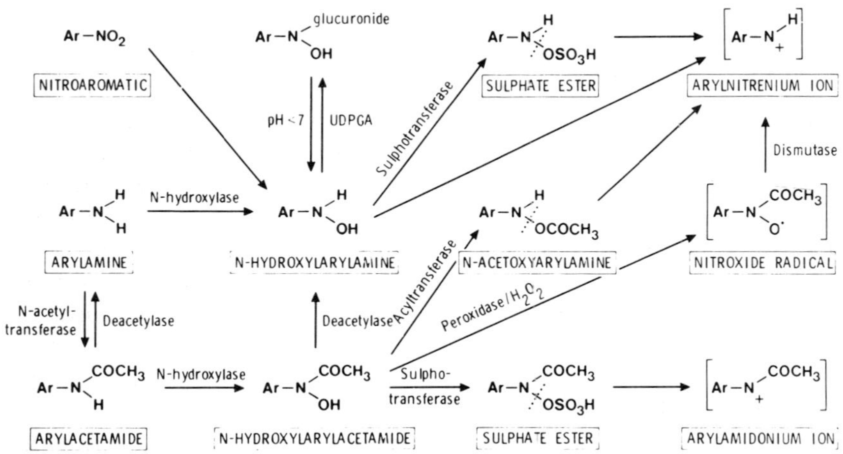

Examples of carcinogenic aromatic amines, amides and nitro compounds are shown in Figure 5.7; the pathways by which they are metabolically activated are described below.

Figure 5.7 Some Carcinogenic Aromatic Amines, Amides and Nitro Compounds

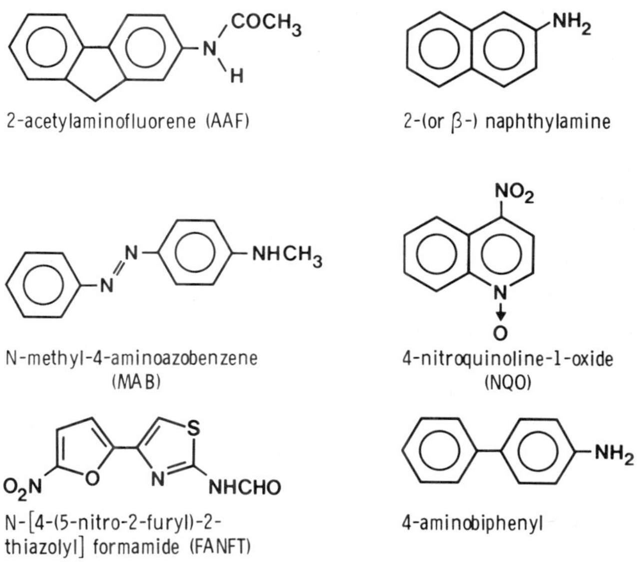

5.5.3.1 N-*Methyl-4-aminoazobenzene (MAB)*. This compound, which is a potent hepatic carcinogen in rodents, is also a proximate carcinogenic metabolite of the related *N,N*-dimethyl derivative.[36] MAB is *N*-hydroxylated by an NADPH-dependent flavoprotein and not by cytochrome P-450.[39] *N*-Hydroxy-MAB is converted into a sulphuric acid ester by a 3′-phosphoadenosine-5′-phosphosulphate (PAPS)-mediated sulphotransferase activity present in rat-liver cytosolic preparations and this ester is presumed to be the ultimate carcinogen.[40] *N*-Benzoyloxy-MAB has been used as a model electrophile and is a potent direct-acting mutagen, and *N*-hydroxy-MAB is mutagenic for *Salmonella typhimurium* strain TA98 (a frameshift mutagen detector) in the presence of metabolising liver preparations.[41] (A frameshift mutagen is one that, by intercalating between the bases in DNA, causes a shift in the reading frame when the DNA is transcribed.)

5.5.3.2 2-Acetylaminofluorene (AAF). This aromatic amide was originally developed as an insecticide, but never used because of the discovery of its powerful hepatocarcinogenicity in rats. It also induces tumours in rat bladder, kidney, intestine, ear duct, subcutaneous tissue and mammary gland, and in mouse bladder, liver and mammary gland.[36] Its

N-hydroxylation is catalysed by cytochrome P-450 in the liver to yield a metabolite that is more carcinogenic than the parent compound and is mutagenic for *Salmonella typhimurium* strain TA98 [the Ames test (see Chapter 6)], especially in the presence of metabolising enzymes from fortified liver fractions. There is strong evidence that the ultimate carcinogenic metabolite in the liver is a sulphuric acid ester. Thus, the hepatocarcinogenicity of *N*-hydroxy-AAF is enhanced by supplementing the diet of rats fed the carcinogen with inorganic sulphate, and reduced by administration of acetanilide, a compound that is excreted as a sulphuric acid conjugate and therefore depletes the sulphate pool.[42,43] Early data pointed to *N*-sulphoöxy-AAF as the major ultimate carcinogen, but an examination of the liver DNA adducts in rats and mice administered *N*-hydroxy-AAF has shown that only a minority of them contain the AAF moiety[44] (see section 5.6). The major adducts contain the deacetylated 2-aminofluorene (AF) moiety and recent data suggest that they arise via formation of another ultimate carcinogen, *N*-sulphoöxy-AF[45] (Figure 5.6). Some experiments suggest that other activated esters may also contribute to the carcinogenic process in liver.[45a]

In other susceptible tissues different pathways of activation of *N*-hydroxy-AAF are thought to occur, because sulphotransferase activity has not been detected in rat kidney, mammary gland, ear duct or subcutaneous tissue. An acyltransferase that catalyses the transfer of the *N*-acetyl group of *N*-hydroxy-AAF to the oxygen atom of the hydroxylamine to generate a strong electrophile (*N*-acetoxy-AF) is present in many tissues. Also peroxidases can generate a nitroxide free radical, a species that is electrophilic itself and that may generate *N*-acetoxy-AAF through dismutation.[35] This latter compound is widely used as a model electrophilic ester of AAF that reacts readily with nucleic acids and proteins at neutral pH, and is a potent mutagen.

A major metabolite of *N*-hydroxy-AAF in the liver is its *O*-glucuronide, which is excreted in the urine and bile and also transported in this form to other organs where it may be subjected to various enzymic reactions. The possible involvement of *O*-glucuronides of aromatic amines and amides in the induction of intestinal tumours in rats is indicated by the demonstration that bacterial flora in the gut can hydrolyse the glucuronide of *N*-hydroxy-AAF to generate the free hydroxylamine, which can undergo further activating metabolism in the colon.[35]

5.5.3.3 4-Nitroquinoline-1-oxide (*NQO*). Structure–activity studies on NQO (Figure 5.7) and its derivatives have indicated that the nitro group must be in the 4-position for carcinogenicity.[46] NQO is a potent carcinogen in the rat and mouse, with the tumours occurring usually at the site of administration by skin painting, subcutaneous injection, gastric intubation or oesophageal infusion. Although it possesses slight electrophilic reactivity, it is also metabolised to strongly electrophilic species. The formation of 4-hydroxyaminoquinoline-1-oxide at the site of subcutaneous injection of NQO in the rat has been demonstrated; this metabolite is converted into a reactive seryl ester by yeast and rat hepatoma cells containing seryl-tRNA transferase activity, and this ester has been proposed as the ultimate carcinogen *in vivo*.

5.5.3.4 2-Naphthylamine and 4-Aminobiphenyl. These compounds are both potent urinary bladder carcinogens for dogs and man.[1,36] They are readily *N*-hydroxylated in the liver and conjugates of their hydroxylamines are excreted in the urine. In addition to the formation of *O*-glucuronide conjugates already referred to, *N*-glucuronides, in which the glucuronic acid group is attached to the nitrogen atom of the hydroxylamine, are also formed and it has been proposed that these conjugates are intermediates in the induction of bladder tumours by the parent amines.[47] At pH less than 7, such as occurs in the bladder, *N*-glucuronides of the hydroxylamines are readily hydrolysed to generate reactive species, possibly nitrenium ions, that covalently bind to DNA. Thus the susceptibility of the urinary bladder to aromatic amine carcinogenesis would seem to result from the reactivation at acidic pH of a metabolite excreted from the liver as a detoxication product.

5.5.3.5 Nitrofurans. Many nitrofurans have antibacterial and antiprotozoan activity, and are widely used as human and veterinary medicines and as animal food preservatives;[48] a significant number of them have also been shown to possess carcinogenic activity in experimental animals. *N*-[4-(5-Nitro-2-furyl)-2-thiazolyl]formamide (FANFT, see Figure 5.7) is a potent bladder carcinogen in the dog, rat, mouse and hamster. Interestingly, replacement of the formylamino group with an acetylamino group, to give *N*-[4-(5-nitro-2-furyl)-2-thiazolyl]acetamide (NFTA), results in a compound that is carcinogenic for guinea-pig bladder but devoid of activity for this organ in the dog, rat and mouse. Instead, NFTA is a potent carcinogen for gall bladder and mammary gland in the dog, induces tumours primarily in mammary gland in the rat, and is potently leukaemogenic in the mouse.

Although the exact nature of the ultimately carcinogenic metabolites of nitrofurans is not presently known, a number of metabolic reactions have been characterised; among these is nitroreduction by a variety of enzyme systems including NADPH-cytochrome *c* reductase, xanthine oxidase and aldehyde oxidase. The formation of electrophilic intermediates capable of binding to cellular proteins and nucleic acids has been shown to be dependent on this nitroreduction. Most nitrofurans are mutagenic toward strains of *Salmonella typhimurium* only in the presence of metabolising enzymes. It would thus seem reasonable to suppose that one or more of the metabolic pathways implicated in the activation of aromatic *N*-hydroxy compounds, namely arylnitrenium ion formation, esterification or nitroxide ion formation, are involved in the metabolic activation of the carcinogenic nitrofurans.

5.5.4 *Direct-acting Carcinogens*

The carcinogenic alkylating agents comprise a diverse group of reactive chemicals that are electrophilic without metabolic activation and thus capable of direct reaction with nucleophilic sites in cellular macromolecules. In addition to the nitrosamides already mentioned, they include most of the alkylating agents used as anti-tumour drugs, strained lactones, some epoxides, imines and halogen derivatives[49] (Figure 5.8). These compounds usually induce tumours at or near the site of administration, i.e. in skin after topical application, forestomach after feeding and at the site of subcutaneous injections. It must be emphasised that most reactive chemicals are not carcinogenic for, while not

Figure 5.8 Some Carcinogenic Alkylating Agents and Sites of Tumour Induction

Compound		Species	Susceptible tissues	Route
β-propiolactone	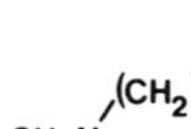	mouse rat	skin subcutaneous	topical s.c.
Propanesultone	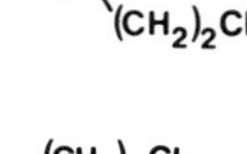	rat	subcutaneous	s.c.
Ethylene dibromide	$Br-CH_2-CH_2-Br$	mouse, rat	forestomach	oral
Methyl-bis(2-chloroethyl)-amine (Nitrogen mustard)	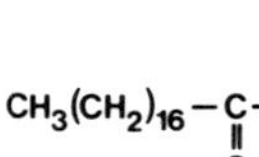	mouse rat	lung, subcutaneous, lymph gland subcutaneous	s.c. s.c.
Bis(2-chloroethyl) sulphide (Mustard gas)	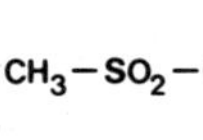	mouse	lung, lymph gland	s.c., i.v.
N-stearoylethyleneimine		mouse, rat	subcutaneous	s.c.
1,4-butanedioldimethane-sulphonate (Myleran)	$CH_3-SO_2-O-(CH_2)_4-O-SO_2-CH_3$	rat	subcutaneous	s.c.

necessarily requiring metabolic activation, they may nevertheless be subject to metabolic inactivation or may simply be too reactive to migrate in sufficient concentration within the mammalian cell to reach the critical cellular target.

Theoretically, the direct-acting carcinogens may alkylate DNA via the formation of a carbonium ion, that is by a monomolecular S_N1 reaction mechanism, or in a concerted bimolecular (S_N2) reaction via an intermediate complex involving both the substituted and displaced ligands (see Chapter 10 for a description of S_N1 and S_N2 mechanism). In practice the mechanism of most reactions is intermediate between S_N1 and S_N2, but in DNA alkylations the predominant mechanism of the reaction can be inferred to a certain extent from the sites of base substitution (see section 5.6).

5.5.5 *Naturally Occurring Carcinogens*

Naturally occurring carcinogens, in their broadest sense, can be taken to include viruses, ultraviolet and ionizing radiation, radioactive minerals and secondary metabolites of plants and micro-organisms. In this section, however, only plant and fungal products will be described.

The toxic properties of many plants have been known since antiquity. The evolution of the human diet, the use of plants for medicinal purposes and the avoidance of poisonous plants have been, until recent history, largely processes of trial and error. Toxic effects are usually immediate and therefore readily associated with the ingestion of the responsible material. However, the complexity of the human diet makes the detection by epidemiological studies of the carcinogenic activity (apparent only after a long time-lag) of any component of it extremely difficult. Only a relatively few natural products have been tested for carcinogenic activity in experimental animals,[50] and it is likely that many more, as yet untested, are carcinogenic. While the human diet may therefore contain small quantities of many carcinogens, proving an unequivocal link beween a known carcinogen in the diet and the incidence of a human cancer will still, in most cases, be a virtual impossibility.

5.5.5.1 Fungal Metabolites. Aflatoxin B_1 (Figure 5.9) is one of the most potent carcinogens known, although its carcinogenicity is strongly species-dependent; as little as 1 part in 10^9 in the diet of rats is sufficient to induce a significant incidence of hepatocellular carcinomas.[51] Closely related, although less carcinogenic, is aflatoxin G_1 (Figure 5.9). Both are metabolites of a cereal-contaminating fungus, *Aspergillus flavus*, whose toxic properties were originally discovered after the widespread poisoning of turkeys in England that had been fed mouldy peanut meal. The turkeys suffered fatal liver necrosis, and subsequent studies also revealed the carcinogenic activity of mould extracts. Members of the *Aspergillus*

Figure 5.9 Examples of Carcinogenic Fungal Metabolites

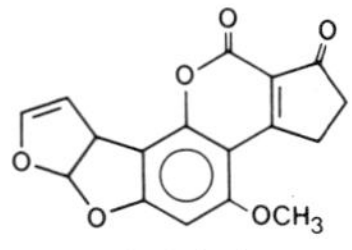

Aflatoxin B_1

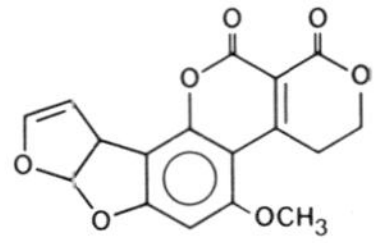

Aflatoxin G_1

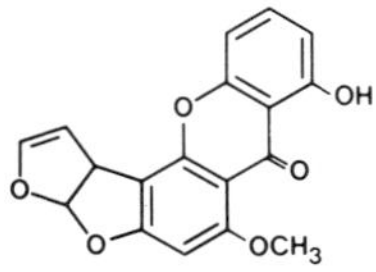

Sterigmatocystin

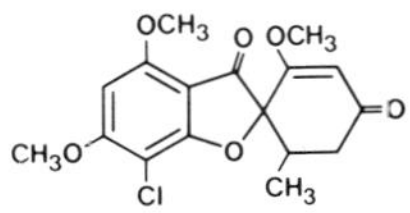

Griseofulvin

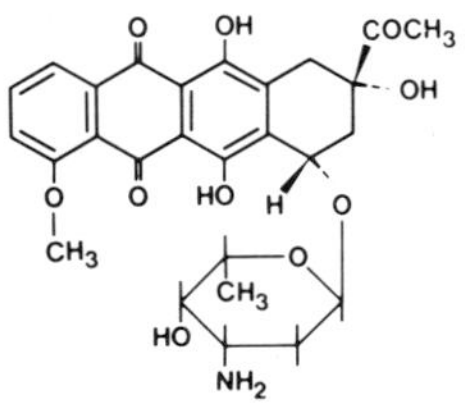

Daunomycin

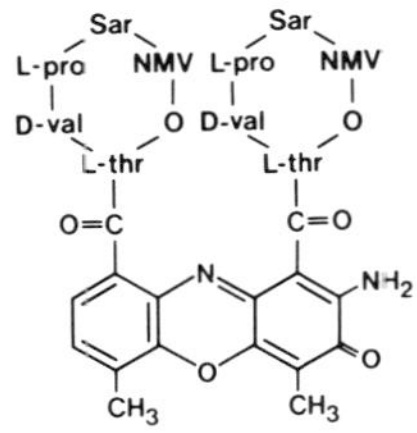

Actinomycin D

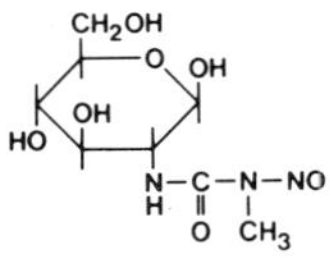

Streptozotocin

(L-Thr = L-threonine; D-Val = D-valine; L-Pro = L-proline; Sar = sarcosine; NMV = L-*N*-methylvaline)

family are known to occur widely in tropical regions and cereals stored in humid conditions are frequently contaminated. There is a strong association between human liver-cancer incidence in tropical regions and level of aflatoxin B_1 food contamination.[52,53] However, as hepatitis B virus infection is also endemic in areas with a high liver-cancer incidence, the question of whether chemical or viral agents, or both, are responsible remains a matter of controversy.

Aflatoxin B_1, in addition to being a potent hepatocarcinogen for the rat and many other species, also causes tumours at the site of application in mouse skin and rat subcutaneous tissue. Its ultimate carcinogen is the 2,3-epoxide[54-56] (see Figure 5.10). This dervative is too reactive to have been isolated as a metabolite or made synthetically; rather, its existence has been inferred from the nature of the covalent interaction of aflatoxin

B_1 with nucleic acids in animal tissues and from the recovery of the 2,3-dihydrodiol on hydrolysis of the nucleic acid adducts.

Another fungal furano metabolite, structurally similar to aflatoxin B_1, is sterigmatocystin (*Aspergillus versicolor*, *Penicillium luteum*)[50] (Figure 5.9). Its activity is about one-tenth that of aflatoxin B_1, and it induces hepatocellular carcinomas and stomach papillomas in the rat. Griseofulvin (Figure 5.9), a metabolite of *Penicillium griseofulvum*, is used as an orally administered drug to treat fungal skin infections in humans; when administered to mice as 0.5–1 per cent of their diet it induces liver tumours and also acts as a promoter for skin tumours initiated by 3-methylcholanthrene.

A number of other secondary metabolites of micro-organisms are also carcinogenic for rodents:[50] e.g. daunomycin (*Streptomyces peucetius*) (Figure 5.9), used as a human anti-tumour agent, causes kidney tumours when given to rats intravenously; actinomycin D (*Streptomyces parvalus*) (Figure 5.9), an anti-tumour agent, causes mesotheliomas in rats; streptozotocin (*Streptomyces achromogenes*) (Figure 5.9), an antibiotic and anti-tumour agent, is the 2-deoxy-D-glucose derivative of *N*-methyl-*N*-nitrosourea and induces tumours in the kidney, liver and pancreas of the rat (see Chapter 10 for further details of these compounds).

5.5.5.2 Plant Products. Alkenylbenzenes occur naturally in many plants and are constituents of many essential oils, herbs and spices widely used as food flavouring agents or in perfumery. Safrole (1-allyl-3,4-methylenedioxybenzene, Figure 5.10) is a major constituent of oil of sassafras and a minor component of several common herbs and spices. It is hepatocarcinogenic when fed to rats and mice as 0.5 per cent of their diet or when injected into newborn mice.[57] Thus, safrole is about six orders of magnitude less active than the most potent carcinogen, aflatoxin B_1. The related compounds estragole (1-allyl-4-methoxybenzene) and methyleugenol (1-allyl-3,4-dimethoxybenzene) are also carcinogenic for mice, but other naturally occurring alkenylbenzenes that are more highly methoxy- and/or methylenedioxy-substituted are mostly inactive.[57]

The major metabolites, and proximate carcinogens, of safrole and estragole in rodent liver are their 1′-hydroxy derivatives.[58,59] The major pathway by which these metabolites are converted into electrophilic species appears to be esterification at the 1′-position,[60,61] and there is strong evidence with 1′-hydroxysafrole for the involvement of the sulphuric acid ester. Thus the hepatocarcinogenicity of 1′-hydroxysafrole in the mouse is markedly reduced by the sulphotransferase inhibitor pentachlorophenol; also brachymorphic mice, congenitally deficient in the synthesis of PAPS, are resistant to tumour induction by 1′-hydroxysafrole compared with their phenotypically normal litter mates.[62]

Cycasin (Figure 5.10), a cycad plant product, is carcinogenic for the

Figure 5.10 Some Naturally Occurring Carcinogens and their Ultimate Carcinogens

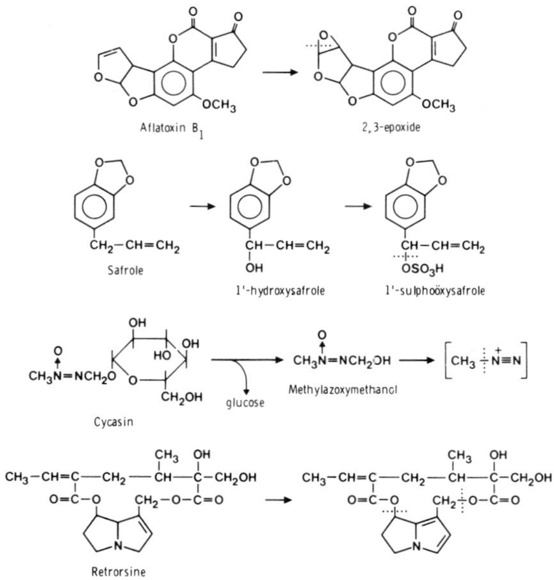

small and large bowel, liver and kidney of rats, but only when administered orally.[63] Although cycad plants are consumed by humans in some parts of the world, there is no direct evidence for their carcinogenic activity in man. Cycasin is the β-glucoside of methylazoxymethanol which, it will be recalled, is a proximate carcinogen of 1,2-dimethylhydrazine (see section 5.5.2.2). It has been demonstrated that intestinal bacteria are responsible for the β-glucosidase activity which hydrolyses cycasin to methylazoxymethanol,[64] from which is subsequently generated a reactive methylating species; hence the requirement for cycasin to be administered orally for it to induce tumours.

Another class of hepatocarcinogenic plant products are the pyrrolizidine alkaloids,[50] of which retrorsine (Figure 5.10) is an example. The presence of ester functions and a nuclear double bond are essential for carcinogenic activity in this class of compounds. Their enzymic dehydrogenation, by NADPH-dependent enzymes in liver microsomal fractions, converts the alkaloids to pyrrolic metabolites (dihydropyrrolizine esters) with electrophilic activity, and these are thought to be responsible for the carcinogenic activity of the parent compounds.[65,66]

5.5.6 Inorganic Carcinogens

A number of metal compounds have been shown to be carcinogenic;

these are listed in Table 5.2. For arsenic, chromium and nickel compounds, there is strong evidence for carcinogenic activity in man from studies of cancer incidence among workers in industries which use these materials (see Table 5.1), and laboratory studies with rodents have confirmed the carcinogenicity of several salts of chromium and nickel.[1] However, arsenic and its compounds are anomalous in this respect. They are known to be human skin and lung carcinogens after exposure through drugs, drinking water or pesticides, but in many studies designed to test the compounds in experimental animals, there is only limited inconclusive evidence for their carcinogenicity.[1] This is possibly the only case where laboratory studies on a compound, or group of compounds, have failed to support evidence for carcinogenic activity obtained from epidemiological studies on human populations.

Table 5.2 Carcinogenic Inorganic Chemicals

Arsenic salts
Beryllium metal and salts
Chromium (VI) salts and cobalt–chromium alloy
Lead salts
Cadmium salts
Nickel salts
Asbestos[a]
Cisplatin [*cis*-dichlorodiammine platinum (II)]

[a] Asbestos is a physical carcinogen (see section 5.6)

Short-term tests for the detection of carcinogens that have as their end-point mutation in bacteria (see Chapter 6) are largely unsuccessful in detecting activity in metals, with the exception of hexavalent chromium compounds. Whether this failure is due to technical deficiencies in the bacterial assays or fundamental differences in the mechanism of metal carcinogenesis is unclear. At very high doses, salts of arsenic, beryllium, lead and chromium (III) have been shown to cause chromosomal abnormalities and morphological transformation in cultured mammalian cells and, although metal cations are electrophilic, it is not clear whether or not direct interaction with nucleic acids occurs. On the other hand, some metal ions, including those of beryllium, cadmium, chromium, nickel and lead have been shown to decrease the fidelity of DNA polymerases *in vitro* and may thus cause genotypic changes in cells without directly interacting with DNA.[67]

Human occupational exposure to asbestos dust is associated with a high risk of lung carcinoma and mesothelioma; inhalation studies in rats, guinea pigs and monkeys can produce fibrotic lesions in the lung and pleura similar to those found in man. Also injection of any of the major forms of asbestos into the pleural cavity can cause mesotheliomas. There are several types of asbestos, of which chrysotile [white; $Mg_6(Si_4O_{10})(OH)_8$], amosite [brown; $(Fe_6Mg)Si_8O_{22}(OH)_2$] and crocidolite [blue; $Na(Fe_2^{3+}Fe_3^{2+})Si_8O_{22}(OH)_2$] are the most important commercially (all are fibrous silicates). However, carcinogenicity is dependent on particle size, not on chemical composition. The most active fibres are those greater than or equal to 10 μm in length and less than 0.5 μm in diameter; glass fibres of the same size are also carcinogenic. Such agents for which physical particle size is a more important factor in their tumorigenicity than chemical composition are termed *physical carcinogens*. Although fibre and foreign-body carcinogenesis has been extensively studied,[68] the molecular basis for the phenomenon remains unclear. One other point deserves mention here; cancer among asbestos workers who smoke cigarettes has a much higher frequency than can be accounted for by the additive effects of the two agents. This is an example of synergism between carcinogens.

Cisplatin [*cis*-dichlorodiammine platinum (II)] is a drug used as an anti-tumour agent (Chapter 10). Its cytotoxic properties are thought to be associated with its ability to form a complex with DNA. In common with many other DNA-damaging drugs it has been shown to be carcinogenic in rodents.[69]

5.6 Interactions of Carcinogens with DNA

When tumorigenic doses of chemical carcinogens are administered to experimental animals, modification of DNA typically occurs at frequencies of 1 in 10^5–10^7 nucleotides in the target tissue. Sensitive techniques are required to detect and analyse DNA modifications at these levels. Commonly, radiolabelled carcinogens, synthesised from radioactive precursors, are utilised to enable the binding to DNA to be detected by the decay of ^3H or ^{14}C radioactive isotopes.[70] More recently some other methods that do not require the synthesis and use of radioactive carcinogens have been adopted. For some compounds, e.g. polycyclic hydrocarbons, fluorescence spectroscopy is sufficiently sensitive to detect, and to provide limited structural information on, carcinogen–DNA adducts formed *in vivo*.[71] Antibodies to some adducts have been made[72-74] and they provide a sensitive means of detecting adducts where prior use of radiolabelled chemicals is not possible, e.g. in monitoring humans for industrial exposure to known carcinogens; a limitation of the method is that synthetically prepared pure carcinogen–DNA adducts are needed to make the antibodies and thus the technique is only applicable to known carcinogens whose interactions with DNA are already well characterised. Another recently developed technique uses enzymic ^{32}P-phosphorylation to 'post-label' non-radiolabelled adducts: DNA that contains bound carcinogen is hydrolysed to deoxyribonucleoside 3′-phosphates and then converted into 3′,5′-bisphosphates using [γ-^{32}P]ATP and T4 polynucleotide kinase, resolved chromatographically and the adducts detected by ^{32}P-decay.[75,76] The method has potential for examining large numbers of chemicals for DNA binding capability (and thus carcinogenic potential), without the need for synthesis of radiolabelled test compounds, and for monitoring human tissues for prior carcinogen exposure.

Other, less direct, methods are also used to detect DNA damage. A number of carcinogenic agents cause double- or single-strand breaks in DNA, which can be detected by monitoring changes in the dynamic properties of the DNA, e.g. by sedimentation velocity or electrophoretic mobility.[77] Adduct formation may lead to loss of the modified base from DNA, generating an apurinic/apyrimidinic site. Such lesions can be converted into strand breaks by alkali or nicking with an apurinic/apyrimidinic endonuclease and detected as above.[78] Also the induction of DNA repair processes (see section 5.7) has been used to infer the occurrence of DNA damage; thus, the occurrence of DNA synthesis under conditions in which normal proliferative replication is inhibited (unscheduled DNA synthesis, UDS), indicates the presence of reparable lesions in the DNA.

For the most part, covalent binding of chemical carcinogens to DNA involves modification of the purine and pyrimidine bases, although with some agents additional reaction with the phosphodiester linkages occurs. Interactions at N^2, N-3, O^6, N-7 and C-8 of guanine, N-1, N-3, N^6 and N-7 of adenine, O^2, N-3, N^4 and C-5 of cytosine and O^2 and O^4 of thymine (Figure 5.11) are known to occur, but the most reactive groups are the purine (i.e. guanine and adenine) nitrogens.

With the simple alkylating agents those that react by a predominantly S_N2 mechanism show a greater affinity for the ring positions on the bases than do those that react by a S_N1 mechanism. Thus the ratio of O^6/N-7 alkylation of guanine in DNA is 0.7 for *N*-ethylnitrosourea (ENU), 0.1 for MNU and 0.004 for methylmethanesulphonate (MMS). Similarly, with methyl-donating carcinogens that require metabolic activation, such as DMN and methylazoxymethanol, methylation occurs at N-7 and O^6 of guanine, but with all these agents additional minor adducts are evident[49,79] (Figure 5.11).

Polycyclic hydrocarbons react predominantly with the exocyclic 2-amino group of guanine residues in DNA *in vivo*.[22] Analysis of the reaction products of the 7,8-dihydrodiol-9,10-epoxide of benzo[a]pyrene with nucleic acids *in vitro* has demonstrated a number of minor adducts in addition to the major N^2 one, including those formed at O^6 and N-7 of

Figure 5.11 Sites of Interactions of Chemical Carcinogens with DNA Bases either after Administration of the Agents *in vivo* or after Reaction of their Ultimate Carcinogens with DNA *in vitro* (Underlining indicates the major adduct formed in most circumstances)

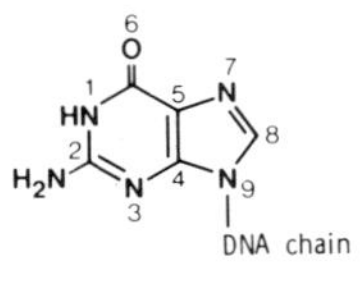

Guanine

N^2	Benzo[a]pyrene, Chrysene, 1'-hydroxysafrole, AAF, MAB, N-hydroxy-2-naphthylamine
N-3	ENU, MNU
O^6	N-hydroxy-1-naphthylamine, 1,2-dimethylhydrazine ENU, MNU, MNNG, MMS, DMN
N-7	ENU, MNU, MNNG, MMS, DMN, 1,2-dimethylhydrazine, Aflatoxin B_1, Bifunctional alkylating agents, Benzo[a]pyrene, MAB
C-8	AAF, MAB, Benzidine, N-hydroxy-2-naphthylamine

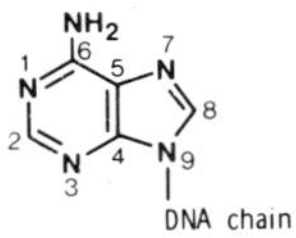

Adenine

N-1	ENU, MNU, MNNG, MMS, NQO, DMN
N-3	ENU, MNU, MNNG, MMS, 1,2-dimethylhydrazine
N^6	Benzo[a]pyrene, NQO, 1'-hydroxysafrole, N-hydroxy-2-naphthylamine

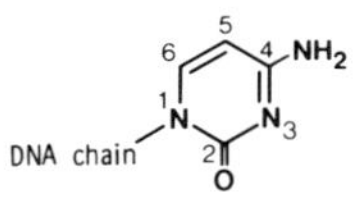

Cytosine

O^2	ENU
N-3	ENU, MNU, MNNG, MMS
N-4	Benzo[a]pyrene

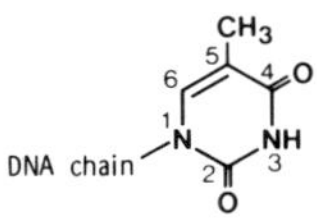

Thymine

O^2	ENU
O^4	ENU, MNU

guanine, N^6 of adenine and N^4 of cytosine.[80] It seems likely that some or all of these adducts are also formed *in vivo*.

The major site of interaction of the ultimate carcinogens of AAF is *C*-8 of guanine.[44,81] As has already been mentioned, the major adduct contains the deacetylated AF moiety (section 5.5.3.2) with attachment to guanine occurring through the nitrogen atom of the parent amide; a minor adduct is similarly formed that contains the intact AAF moiety. An adduct containing the latter species is also formed at N^2 of guanine, but in this case covalent binding occurs at the 3-position of the AAF aromatic nucleus.

Similarly the reactive species of the azo-dye MAB (section 5.5.3.1) interacts with DNA predominantly by covalent linkage of its amino group with *C*-8 of guanine but also through linkage of the aromatic ring to the 2-amino group of guanine.[82,83] Also, it has been shown that when *N*-benzoyloxy-MAB is reacted with DNA *in vitro*, adducts are formed that are lost from the DNA by depurination and that probably involve *N*-7 substitution of guanine.[84] If such adducts are formed *in vivo*, they will not be detected if they are lost from the DNA by depurination more rapidly than the DNA can be isolated from animal tissues. However, the prior formation of unstable adducts can often be inferred by the detection of apurinic/apyrimidinic sites in the isolated DNA.

Depurination after *N*-7 guanine substitution is a consequence of the destabilisation of the *N*-glycosidic bond by the positive charge introduced into the purine molecule by carbonium-ion addition; substitution at *N*-3 of guanine and adenine also has this effect.[85] With activated aflatoxin B_1, the main adduct formed *in vivo* is at *N*-7 of guanine. [56] As well as the adducts found in DNA isolated from treated tissues, the depurination species 2,3-dihydro-2-(guan-7-yl)-3-hydroxyaflatoxin B_1 could be detected in the urine of rats that had been fed the compound.[86]

The activated derivatives of 1′-hydroxysafrole and 1′-hydroxyestragole, presumed to be their sulphuric acid esters (see section 5.5.5.2) have been found in both cases to bind covalently to DNA in mouse liver mainly at N^2 of guanine, with minor involvement of N^6 of adenine.[60,61] In addition low levels of apurinic/apyrimidinic sites can be detected in DNA of cultured human cells after treatment with the model 1′-acetoxy derivatives, suggesting that *N*-7 guanine adducts may be formed;[87] reaction of the esters with DNA *in vitro* also generates apurinic/apyrimidinic sites.[88]

Surprisingly, the reactive metabolites of the *N*-hydroxy derivatives of 1- and 2-naphthylamine give entirely different spectra of adducts. With *N*-hydroxy-2-naphthylamine the adducts involve binding of the nitrogen atom to *C*-8 of guanine and of the 1-position of the aromatic ring to N^2 of guanine and N^6 of adenine.[89] On the other hand, *N*-hydroxy-1-naphthylamine modifies DNA at the O^6-position of guanine with binding occurring both to the nitrogen atom and to the 2-position of the aromatic ring.[90]

At the moment, our understanding of how damage to DNA by carcinogens results in tumour initiation is largely speculative. The structural consequences of DNA modification by a number of ultimate carcinogens have been studied. The N^2-guanine adduct of benzo[a]pyrene has the planar hydrocarbon molecule aligned in the minor groove of the DNA double helix;[91] computer analysis suggests that this conformation causes minimal distortion of the secondary structure of DNA.[92] Nucleic acids modified by *N*-acetoxy-AAF contain both *C*-8 and N^2 guanine adducts and it appears that the former causes much greater conformational changes than the latter, including changes from the right-handed B-form of poly[d(G-C)] to the left-handed Z-form.[93] It has been found that N^2 adducts are more persistent than *C*-8 adducts in liver DNA of rats administered AAF.[94] Tissue- and cell-culture studies have generally supported the concept that persistence of carcinogen–DNA adducts is a prerequisite for carcinogenesis. Thus adducts which alter the template function of DNA but do not produce a conformational distortion that is easily recognised and removed by DNA repair enzymes (see section 5.7) are most likely to be potentially carcinogenic lesions.[95]

In *Escherichia coli* it has been shown that DNA modification by benzo[a]pyrene diol-epoxide or aflatoxin B_1 epoxide can cause GC $\rightarrow$ TA transversions,[96,97] but the mechanism, indeed the particular adduct, through which they occur is not known. With the simple alkylating agents, however, the mechanism of point-mutation induction is better understood. One adduct-formation site of biological significance appears to be O^6 of guanine. It was first proposed, and later demonstrated in bacteria, that O^6-alkylated guanine can mispair during replication with thymine, resulting in a transition point mutation.[49,79] In relation to carcinogenesis by alkylating agents, O^6-ethylguanine is more persistent in the DNA of animals administered ENU than the initially predominant *N*-7-ethylguanine, and shows greater persistence in target tissues (e.g. rat brain) compared with non-susceptible tissues (e.g. rat liver).[98] However, one of the difficulties with attempting to correlate the occurrence of particular adducts with observed biological effects is that it is seldom possible to generate one type of adduct in the absence of other types, and the possibility that a minor adduct is more important cannot be excluded. Thus, when the persistence of O^6-ethylguanine and O^4-ethylthymine in hepatocyte DNA was investigated in rat liver after continuous administration of *N*-nitrosodiethylamine (DEN), the latter residues were found to be much more persistent than the former and therefore accumulated to a much greater extent.[99] This finding suggests that O^4-ethylthymine, not O^6-ethylguanine, could be the principal promutagenic lesion responsible for the induction of hepatocellular carcinomas by DEN.

To date, most studies have involved investigation of total cellular or nuclear DNA, but with the advent of techniques for analysing defined DNA sequences, it is likely that future studies will concentrate on carcinogen–DNA modification at particular sites in the genome. One other finding should be mentioned here: mitochondrial DNA, that accounts for less than 1 per cent of the DNA in mammalian cells, is modified by several carcinogens to a much greater extent than is nuclear DNA.[100,101] Whether this modification is relevant to carcinogenesis, or merely reflects the fact that mitochondria provide a favourable environment within the cytoplasm for lipophilic ultimate carcinogens, has yet to be established.

5.7 DNA Repair Mechanisms

Adducts deleterious to the DNA in living cells elicit processes by which the alteration is repaired and the genome restored to its original form.

Such processes serve to protect cells from the mutagenic, carcinogenic or lethal consequences of DNA damage and are essential to the maintenance of the cell and the stability of its genome.

The availability of defective mutant cells is often a useful means of delineating the components of a cellular process. For this reason most of our knowledge about DNA repair mechanisms is derived from studies on bacteria.[102] A great deal is known about the genetic control of DNA repair in *E. coli* from studies of many repair-deficient strains; much less is known about the process in eukaryotes because of the comparative paucity of mutant mammalian cells that are defective in DNA repair. However, one much-studied human cell-type is derived from patients with *Xeroderma pigmentosum* (XP). This autosomal recessive inherited syndrome is characterised by hypersensitivity to sunlight and ultraviolet, and a high incidence of skin carcinomas. Cultured fibroblasts from XP patients have been found to be defective in the removal of ultraviolet or carcinogen-induced lesions in their DNA.[103] Current evidence suggests that most XP cells are deficient in an early stage of excision repair (see below) but there are many complementation groups of XP cell lines[104] (hybrid cells of different complementation groups have normal repair capabilities) indicating that there are many components of the process that are deficient in XP cells. A number of other human hereditary disorders are thought to be related to DNA repair deficiency, and the responses to DNA-damaging agents of the cultured cells derived from patients with some of these have also been studied; these include *Ataxia telangiectasia*[105] (hypersensitivity to ionising radiation, high frequency of spontaneous chromosome aberrations), Bloom's syndrome[106] [hypersensitivity to ultraviolet, high frequency of sister chromatid exchange (SCE)] and Fanconi's anaemia[107] (hypersensitivity to DNA cross-linking agents, high frequency of spontaneous chromosome aberrations).

The most widely studied DNA lesion is the cyclobutane dimer, formed by the fusion of two adjacent pyrimidine bases (usually thymine) in a DNA strand after irradiation with ultraviolet. Although less well-studied, repair of DNA damage induced by other agents including the simple alkylating agents and more 'bulky' carcinogens such as AAF and benzo[a]pyrene has also been investigated.[102] In all these studies, however, one is usually monitoring repair in response to more than one type of DNA lesion, as it is seldom possible to generate a single type of DNA adduct or lesion with a particular agent.

The most general mechanism of repair is by the removal of the portion of the DNA containing the lesion and resynthesis of a 'repair-patch' by using the undamaged DNA strand as template[102] (see Figure 5.12). The first step involves recognition of the distortion to the DNA structure caused by the lesion and nicking of the DNA at or close to the lesion by an endonuclease; it is apparently this step that is deficient in most XP cells. Synthesis of a new stretch of DNA then occurs terminating with ligation of the new patch to the existing DNA. Repair-patch synthesis in response to damage by chemical carcinogens has been classified as '(ultraviolet) UV-like' ('long-patch'), where the patch size is approximately 30 nucleotides long (e.g. with *N*-acetoxy-AAF, aflatoxin B_1 epoxide, 1′-acetoxysafrole) or 'X-ray-like' ('short-patch') where the patch size is only three to four nucleotides long (e.g. with alkylating agents.)[102]

An apurinic/apyrimidinic site may be generated in DNA spontaneously or by the loss of an unstable carcinogen adduct (see section 5.6). Enzymic removal of a modified base by a *N*-glycosylase may also generate an apurinic/apyrimidinic site, and the enzymic removal of *N*-3-methyladenine, *N*-3-ethyladenine, O^6-ethylguanine and *N*-7-methylguanine from DNA has been demonstrated.[108] A base insertase may then simply replace the correct missing base, using the undamaged strand as template, without any strand-nicking (Figure 5.12). Alternatively, an apurinic/apyrimidinic endonuclease may recognise the lesion and nick the DNA so that repair can then proceed by the repair-patch synthesis pathway (Figure 5.12). It has been proposed, but not substantiated, that repair initiated by apurinic/apyrimidinic endonuclease proceeds by the short-patch pathway.[102]

Figure 5.12 Schematic Diagram Illustrating Mechanisms of Repair of Carcinogen–DNA Damage (■ = carcinogen-modified base; ▦ = newly incorporated bases or nucleotides; AP = apurinic/apyrimidinic)

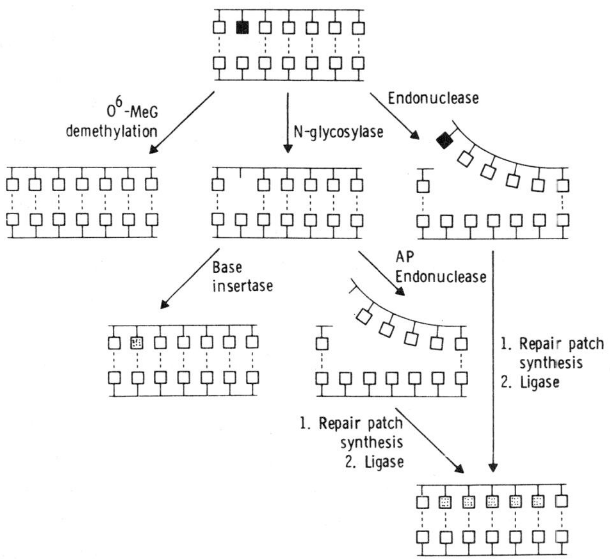

For some specific adducts, direct reversal of damage can occur without base or nucleotide loss. The classic example of this is the reversal of pyrimidine dimers by a photoreactivating enzyme.[109] More recently, an acceptor protein has been identified that removes the methyl group from O^6-methylguanine residues in DNA (Figure 5.12). This is not an enzyme, as each protein molecule acts only once. The activity, which is inducible, has been detected in rat and human liver[110,111] and may prove to be an important protective mechanism against potentially carcinogenic methylating agents, such as the nitrosamines.

When DNA replicating enzymes encounter a bulky adduct, they are apparently unable to copy DNA past the lesion and therefore leave a gap in the daughter strand. Thus, newly synthesised DNA in carcinogen-treated cells is initially of lower molecular weight than in untreated cells, but in time it increases to normal size. In bacterial and mammalian cells this is thought to occur by a pathway that involves strand displacement and branch migration, whereby a region of one newly synthesised daughter strand becomes the template for the other in place of the damaged region of parental DNA.[102] Recombination events may also occur in some circumstances, because adducts have been found in the daughter strand after DNA replication in cells that had previously been exposed to a carcinogen. There is a subgroup of XP patients (called 'XP variants') whose cultured cells are capable of normal levels of excision repair; however, they have been found to be deficient in daughter-strand gap closure.[112]

The postreplication events described are tolerance, rather than repair, processes, as the lesion is not removed from the DNA. However, once the DNA has been replicated and the daughter-strand gaps filled, excision repair can operate to remove the lesion. According to evidence so far obtained, the repair and tolerance mechanisms that have been identified in mammalian cells are error-free pathways, i.e. the damaged DNA is faithfully restored to its original state without the occurrence of mutations.[102] One mechanism of daughter-strand gap filling in bacteria, the inducible 'SOS' process,[113] is known to be error prone but there is, as yet, no evidence for the equivalent mechanism in mammalian cells.

5.8 Tumour Promotion and Multistage Nature of Carcinogenesis

Experimental evidence that non-carcinogenic agents could enhance tumour yield in response to a carcinogen was first obtained when tar painting on mouse skin was followed by mechanical wounding.[114] Subsequent studies demonstrated the same effect with a chemical agent, oil extracted from *Croton tiglium* L.[115] The active constituents of croton oil have been identified as phorbol esters, of which the most potent is

Figure 5.13 Structures of 12-*O*-Tetradecanoyl phorbol-13-acetate (TPA) and Teleocidin B

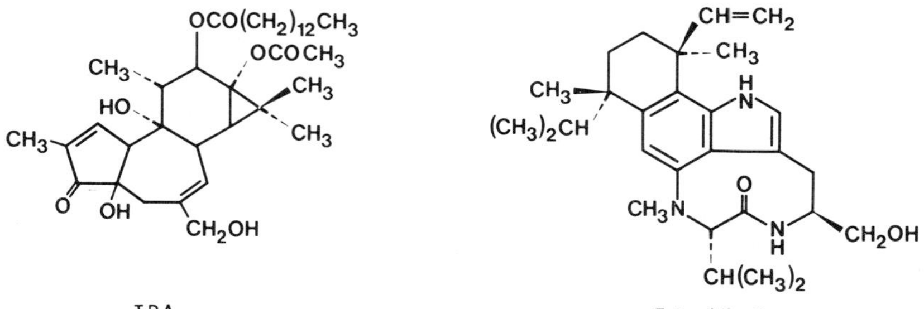

12-*O*-tetradecanoyl-phorbol-13-acetate (TPA) (Figure 5.13).[116,117] More recently, an equally potent agent, teleocidin B (Figure 5.13) has been isolated from mycelia of a strain of *Streptomyces*.[118] Agents that induce tumours only when administered after a carcinogen are called *promoters*.

Tumour promotion has been most extensively studied in mouse skin.[119] It has been established that treatment with a carcinogen (generally referred to in these circumstances as an *initiator*), at a dose not in itself sufficient to induce tumours, will nevertheless lead to tumours if followed by repeated doses of a promoting agent. Initiation appears to be an irreversible and rapid event; the start of promotion can be delayed for as long as a year without any diminution in the eventual tumour yield. In contrast, promotion is a slow protracted process that is to some extent reversible; multiple treatments with the promoting agent are required for many weeks and tumours will fail to develop if the time gaps between treatments are too long. The sequential nature of the process is indicated by the fact that if the promoter is administered before the initiator, no tumours develop (see Figure 5.14).[120]

When a chemical carcinogen is applied repeatedly to mouse skin, the usual type of tumours produced are carcinomas. According to the two-step initiation–promotion hypothesis, the chemical agent is acting as both an initiator and a promoter in these circumstances. When a single dose of initiator is followed by repeated doses of a promoter, the tumours produced are mainly benign papillomas, a few of which subsequently progress to malignant carcinomas. If the papillomas are treated with a second dose of initiator, however, a larger proportion of them becomes malignant.[121]

In certain instances, such as with urethane as initiator, enhanced production of tumours can occur when croton oil is applied before the

Figure 5.14 Protocol for Two-stage Carcinogenesis in Mouse Skin (I = single treatment with an initiating agent; P = multiple doses of a promoting agent)

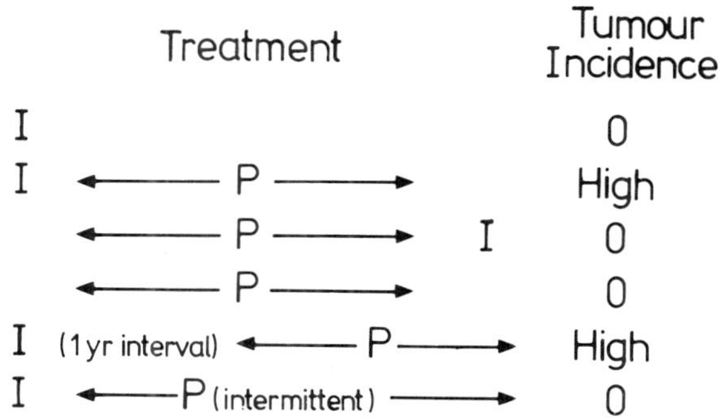

initiator.[122] Subsequent treatment with the promoting agent is still required, however, so the sequential requirement for initiation followed by promotion is not violated. The function of the croton oil pretreatment may be to stimulate more cells into mitosis and render them more susceptible to phenotypic alteration by the initiator.

The promotion stage of carcinogenesis can itself be subdivided experimentally into several stages. If croton oil is used in the early stages of promotion on mouse skin followed by turpentine for the later stages, the yield of papillomas is considerably higher than when the order of the agents is reversed, suggesting that turpentine influences later, but not early, events in promotion.[123] Also, there is the progression, already mentioned, of benign papillomas to malignant carcinomas.

In addition to mouse skin, protocols have been established that demonstrate the multistage nature of carcinogenesis in other tissues. Thus, a model system in liver has been used in which initiation is achieved by feeding AAF for limited periods, and promotion by subsequently feeding phenobarbitone for many months;[124] in the urinary bladder of the rat, sodium saccharin and sodium cyclamate act as promoters after initiating doses of MNU.[125]

Studies on the mechanism of tumour promotion have concentrated mainly on the effects of TPA. Extensive investigations by many workers have revealed that a somewhat bewildering number of cellular processes are affected by TPA.[126] Together with other promoters, it has been found to enhance cell proliferation, with associated stimulation of DNA, RNA, protein and phospholipid synthesis. The activity of ornithine decarboxylase in mouse skin and the levels of prostaglandins E and F are also increased. Such processes might promote tumours by increasing the proliferation of initiated cells, but it should be noted that not all mitogens are promoters. Proteinase activity is increased during promotion in mouse skin and histidase activity is decreased; these effects may also be relevant to the mechanism of promotion.

TPA also influences differentiation, although in some systems its effect is inhibitory, in others enhancing. The treatment of cells in culture with TPA has indicated a large number of effects on cell surfaces and membranes: these include increased phospholipid turnover, inhibition of the binding of epidermal growth factor to its receptors, decreased levels of LETS (*l*arge *e*xternal *t*ransformation *s*ensitive) protein and inhibition of metabolic co-operation between cells. Such cell-surface effects imply the interaction of TPA with some endogenous receptor and the protein which specifically binds to TPA has recently been found to have protein kinase activity.[127,128] Thus TPA may induce alterations in membrane protein and lipid phosphorylation by influencing the association of the TPA receptor with cellular membrane components.

It is generally believed that promoters do not interact with DNA, but it is now known that DNA damage in the form of strand breaks and chromosome aberrations is caused indirectly by TPA.[129] However, it still remains to be established which of the many observed effects of TPA and related agents are relevant to their promoting activity.

The fact that initiation is a rapid irreversible event is compatible with the concept that it occurs through the permanent alteration of the genetic material of the cell (see section 5.6). It has been argued that initiation of carcinogenesis does not proceed from simple point mutations because the frequency of transformation in cultured rodent cells is one to two orders of magnitude higher than the frequency of induction of mutations to specific markers, e.g. to drug resistance, in the same system.[130] Evidence from these and related studies favours a mechanism involving gene rearrangement. However, a point-mutation mechanism is implicated by recent findings from experiments with DNA transfection assays. The discovery that DNAs isolated from human and chemically induced animal tumours can transform NIH/3T3 mouse cells in culture has led to the identification of the activated genes responsible (oncogenes) and of their non-transforming alleles (proto-oncogenes) present in normal cells (see Chapter 4).[131] Many cases have been found in which the oncogenes differ from their respective proto-oncogenes by single base-pair substitutions.[132-134] Thus point mutations convert a non-transforming gene into a transforming one. Although this single-step mechanism might seem incompatible with the demonstrated multistage nature of carcinogenesis, it should be noted that the NIH/3T3 cell line is abnormal in that the cells have already acquired some of the characteristics of tumour cells (e.g. immortality). Therefore at present it seems reasonable to conclude that point mutations can account for one stage in the induction of tumours by carcinogens, while the reasons for the discrepancy between mutation and transformation frequencies remain a matter of controversy.

5.9 Hormonal Carcinogenesis

Hormones, particularly oestrogens (Figure 5.15), have been known to be carcinogenic for experimental animals for over 50 years.[135] More recently an association has been found between human cancer and both hormone replacement therapy and the use of oral contraceptives.[1,136,137] The general conclusions from animal studies are that: tumours of the pituitary, adrenal, cervix, endometrium, vagina, ovary, testis and breast can be induced by hormones; oestrogens and prolactin are important in the process, androgens and progestins less important; there are wide differences in species, strain and site susceptibility, with mice and rats being the most, monkeys the least, susceptible species.

Figure 5.15 Natural and Synthetic Carcinogenic Oestrogens

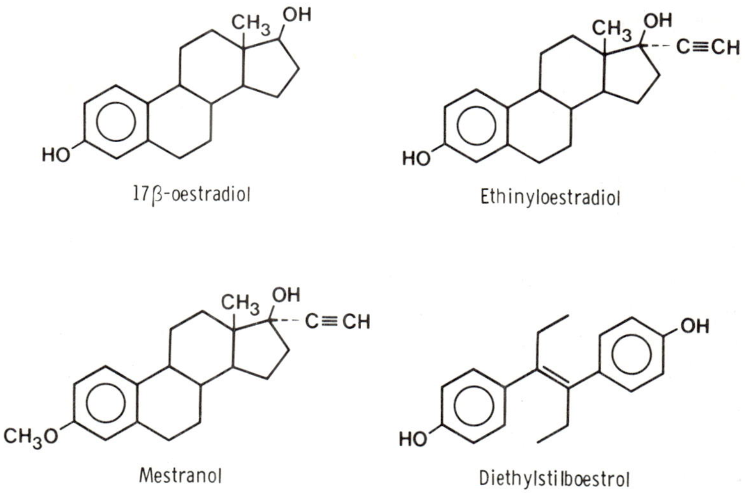

Usually, tumours are produced in animals only when the hormones are administered chronically for long periods, or when they are present in unphysiological amounts due to sustained changes in normal endocrine balances. Under such circumstances, the regulatory mechanism controlling cell proliferation and differentiation may be profoundly altered.

These characteristics, and their lack of activity in bacterial and mammalian cell-mutagenicity assays,[138] make it likely that hormones and hormone imbalances act primarily as promoters in the carcinogenic process. If this is indeed the case, then the initiating agent or agents in hormone-dependent cancer remains, for the most part, unknown (in certain animal models, e.g. mammary cancer in mice, an endogenous virus, the mouse mammary-tumour virus (MMTV), is thought to be the 'initiating' agent).[139] However, it remains unclear how carcinogenic hormones induce morphological and neoplastic transformation in cultured cells if they are purely tumour promoters. It is conceivable that carcinogenic hormones have weak initiating activity due to aberrant metabolism occurring only during hormonal imbalances; thus standard mutational assays would fail to detect such activity. There is some evidence for the microsome-mediated covalent binding of a number of steroidal hormones to proteins, but evidence of DNA binding is less clear.

A compound of particular interest is diethylstilboestrol (DES) (Figure 5.15), a synthetic non-steroidal oestrogen that has been widely studied since it was found to induce vaginal cancer in the adolescent daughters of women who had been given large doses of the compound during pregnancy to prevent miscarriage.[1] The drug has also been used as a growth promoter in cattle, although this practice is now discontinued. DES causes tumours in testis, ovary and cervix in the mouse, and promotes mammary tumours in mice carrying MMTV. It also causes kidney tumours in hamsters, and tumours in the bladder, pituitary gland and breast of rats. Like other hormonal carcinogens, DES is inactive in several bacterial and mammalian cell-mutation assays and induces morphological and neoplastic transformation of cells without causing any measurable gene mutation.[140] On the other hand, DES is mutagenic toward yeast,[141] and causes UDS in hamster embryo cells[142] and SCE in human fibroblasts,[143] activities suggestive of DNA damage. Indeed, low levels of DNA binding have been detected *in vivo*.[144] The induction of UDS by DES is dependent on the presence of metabolising enzymes,[142] and a number of metabolites of DES are electrophilic;[145] however, cellular transformation can occur in the absence of metabolising enzymes. Thus DES displays some properties characteristic of a chemical carcinogen and some of a hormonal carcinogen; which properties are the more relevant to the mechanism by which it produces tumours remains to be determined.

5.10 Modifiers of Carcinogenesis

The processes involved in carcinogenesis can be influenced by a number of host factors and several instances have already been mentioned in which different species of animals show different susceptibilities and organotropism to the same carcinogen. Furthermore, within a single species there can be marked strain differences in susceptibility, due in some cases to differences in carcinogen metabolism, but in other cases to other, as yet unknown, factors. Such genetic differences presumably exist within the human population. Identification of subpopulations particularly at risk would provide clinicians with a rationale for screening for early detection and for devising methods of cancer prevention.

In general, newborn animals are more susceptible than adults, due in part, perhaps, to normal growth providing a tumour-promoting stimulus in the affected organs and also to the slower metabolism of exogenous chemicals in newborns compared with adults. Age is also a factor in susceptibility among adult animals. Young adult female rats develop a high incidence of mammary tumours after a single oral dose of 7,12-dimethylbenz[a]anthracene (DMBA) (Figure 5.1), whereas older rats do not, yet, if DMBA is administered directly to the mammary gland, then both young and old rats develop high incidences of tumours.[146] In this case it seems that the more efficient detoxication of DMBA in the liver of mature rats provides the protecting influence.

With some carcinogens, the male of the species is more susceptible, with others, the female. In many cases these differences are due to the levels of metabolising enzymes; for example, sulphotransferase activity is greater in male rats than in females, and the males are more susceptible to AAF carcinogenesis.[147] Endocrine factors may also have an influence, although these are, for the most part, ill-defined.

It has been mentioned already (in section 5.4) that induction of metabolising enzymes *in vivo* usually results in increased detoxication vs activation and/or more rapid metabolism of the test compound. Also a number of specific enzyme inhibitors (e.g. pentachlorophenol, a sulphotransferase inhibitor) can act as anticarcinogens. A large number of less specific antioxidants have also been shown to inhibit carcinogenesis by a broad range of chemical carcinogens. These include vitamin C, the phenolic compounds butylated hydroxyanisole (BHA) and butylated hydroxytoluene (BHT) and non-phenolic compounds such as ethoxyquin, a widely used antioxidant commonly added to commercial animal feed. A number of components of the human diet (e.g. coumarins present in vegetables) have anticarcinogenic properties in experimental animals.[148]

Cocarcinogens, compounds that enhance the carcinogenic activity of other compounds but are inactive when administered alone, appear to act by increasing activation vs detoxication of the carcinogen. They are normally quite specific in their activity, e.g. non-carcinogenic polycyclic hydrocarbons can be cocarcinogens for carcinogenic ones, but not for other classes of carcinogens. An area that has received little attention is the possibility that some detoxication metabolites of a particular carcinogen may act as cocarcinogens for activated metabolites (i.e. proximate carcinogens).

As far as prevention of cancer in humans is concerned, diet may be the most important overall determining factor for the general population, excluding those individuals whose cancers are associated with tobacco smoking. Not only do humans ingest both carcinogens and anticarcinogens in their food, but other general factors can apparently increase (e.g. high-fat diet) or decrease (e.g. high-fibre diet) the risk of certain cancers. Diet is also important in determining the microbial population of the intestine, which can influence carcinogen metabolism.[149] But the biochemical basis of many dietary influences on tumour induction, and their complex interrelations have yet to be determined.

References

1. International Agency for Research on Cancer (1982) *IARC Monographs on the Evaluation of the Carcinogenic Risk of Chemicals to Humans*. Suppl. 4, IARC, Lyon
2. Redmond, D.E., Jr (1970) Tobacco and cancer: the first clinical report, 1761. *N. Eng. J. Med.*, *282*, 18-23
3. von Soemmering, S.T. (1795) De morbis vasorum abserventium corporis humani. Main, Frankfurt, p. 109
4. Pott, P. (1775) *Chirurgical Observations Relative to the Cancer of the Scrotum*, London, 1775. Reprinted in *Natl. Cancer Inst. Monogr.*, *10*, 7-13 (1963)
5. von Volkmann, R. (1875) *Beiträge zur Chirurgie*, Leipzig
6. Bell, J. (1876) *Edinb. Med. J.*, *22*, 135.
7. Henry, S.A. (1947) Occupational cutaneous cancer attributable to certain chemicals in industry. *Br. Med. Bull.*, *4*, 389-401
8. Rehn, L. (1895) Blasergeschwülste bei Fuchsin-Arbeitern. *Arch. Klin. Chir.*, *50*, 588-600
9. Yamagiwa, K. & Ichikawa, K. (1915) *Verh. Jpn. Path. Ges.*, 5, 142-48. See also Experimental study of the pathogenesis of carcinoma. *J. Cancer Res.*, *3*, 1-29 (1918)
10. Tsutsui. H. (1818) *Gann 12*, 17
11. Kennaway, E.L. & Hieger, I. (1930) Carcinogenic substances and their fluorescence spectra. *Br. Med. J.*, *ii*, 1044-6
12. Cook, J.W., Hewett, C.L. & Hieger, I. (1933) The isolation of a cancer-producing hydrocarbon from coal tar. *J. Chem. Soc.*, 395-405
13. Yoshida, T. (1933) Uber die serienweise Verfolgung der Veränderungen der leber bei der experimentellen Hepatomerzeugung durch o-Amidoazotoluol. *Trans. Jpn. Path. Soc.*, *23*, 636-8
14. Hueper, W.C., Wiley, F.H. & Wolfe, H.D. (1938) Experimental production of bladder tumors in dogs by administration of beta-naphthylamine. *J. Ind. Hyg. Toxicol.*, *20*, 46-84
15. Miller, J.A. & Miller, E.C. (1969) Metabolic activation of carcinogenic aromatic amines and amides via *N*-hydroxylation and *N*-hydroxy esterification and its relationship to ultimate carcinogens as electrophilic reactants. In E. Bergmann & E. Pullman (eds) *The Jerusalem Symposia on Quantum Chemistry and Biochemistry*, vol. 1, *Physiochemical Mechanisms of Carcinogenesis*, The Israel Academy of Sciences and Humanities, Jerusalem, pp. 237-61
16. Miller, J.A. (1970) Carcinogenesis by chemicals: an overview - G.H.A. Clowes memorial lecture. *Cancer Res.*, *30*, 559-76
17. Ames, B.N., Durston, W.E., Yamasaki, E. & Lee, F.D. (1973) Carcinogens are

mutagens: a simple test system combining liver homogenates for activation and bacteria for detection. *Proc. Natl. Acad. Sci. USA*, *70*, 2281-5

18. Brookes, P. & Lawley, P.D. (1964) Evidence for the binding of polynuclear aromatic hydrocarbons to the nucleic acids of mouse skin: relation between carcinogenic power of hydrocarbons and their binding to DNA. *Nature (Lond.)*, *202*, 781-4
19. Williams, R.T. (1959) *Detoxication Mechanisms*, 2nd edn, Chapman and Hall, London
20. Holtzman, J.L., Gillette, J.R. & Milne, G.W.A. (1967) The incorporation of ^{18}O into naphthalene in the enzymatic formation of 1,2-dihydronaphthalene-1,2-diol. *J. Biol. Chem.*, *242*, 4386-7
21. Conney, A.H. (1982) Induction of microsomal enzymes by foreign chemicals and carcinogenesis by polycyclic aromatic hydrocarbons: G.H.A. Clowes memorial lecture. *Cancer Res.*, 4875-917
22. Phillips, D.H. (1983) Fifty years of benzo[a]pyrene. *Nature (Lond.)*, *303*, 468-72
23. Dipple, A. (1976) Polynuclear aromatic carcinogens. In C.E. Searle (ed.) *Chemical Carcinogens*, ACS Monograph 173, American Chemical Society, Washington, DC, pp. 245-314
24. Sims, P. & Grover, P.L. (1974) Epoxides in polycyclic aromatic hydrocarbon metabolism and carcinogenesis. *Adv. Cancer Res.*, *20*, 165-274
25. Borgen, A., Darvey, H., Castagnoli, B., Crocker, T.T., Rasmussen, R.E. & Wang, I.Y. (1973) Metabolic conversion of benzo[a]pyrene by Syrian hamster liver microsomes and binding of metabolites to DNA. *J. Med. Chem.*, *16*, 502-6
26. Sims, P., Grover, P.L., Swaisland, A., Pal, K. & Hewer, A. (1974) Metabolic activation of benzo[a]pyrene proceeds by a diol-epoxide. *Nature (Lond.)*, *252*, 326-8
27. Jerina, D.M., Lehr, R.E., Yagi, H., Hernandez, O., Dansette, P.M., Wislocki, P.G., Wood, A.W., Chang, R.L., Levin, W. & Conney, A.H. (1976) Mutagenicity of benzo[a]pyrene derivatives and the description of a quantum mechanical model which predicts the ease of carbonium ion formation from diol epoxides. In F.J. de Serres, J.R. Fouts, J.R. Bend & R.M. Philpot (eds) In vitro *Metabolic Activation in Mutagenesis Testing*, Elsevier/North-Holland Biomedical Press, Amsterdam, pp. 159-78
28. Sims, P. & Grover, P.L. (1981) Involvement of dihydrodiols and diol epoxides in the metabolic activation of polycyclic hydrocarbons other than benzo[a]pyrene. In H.V. Gelboin & P.O.P. Ts'o (eds) *Polycyclic Hydrocarbons and Cancer*, vol. 3, Academic Press, New York, pp. 117-81
29. Magee, P.N. & Barnes, J.M. (1956) The production of malignant primary hepatic tumours in the rat by feeding dimethylnitrosamine. *Br. J. Cancer*, *10*, 113-22
30. Magee, P.N., Montesano, R. & Preussman, R. (1976) *N*-Nitroso compounds and related carcinogens. In C.E. Searle (ed.) *Chemical Carcinogens*, ACS Monograph 173, American Chemical Society, Washington, DC, pp. 491-625
31. Magee, P.N. (ed.) (1982) *Nitrosamines and Human Cancer*, Banbury Report No. 12, Cold Spring Harbor Laboratory, New York
32. Mirvish, S.S. (1977) *N*-Nitroso compounds: their chemical and *in vivo* formation and possible importance as environmental carcinogens. *J. Toxicol. Environ. Health*, *2*, 1267-77
33. Druckrey, H. (1973) Specific carcinogenic and teratogenic effects of 'indirect' alkylating methyl and ethyl compounds, and their dependency on stages of ontogenic developments. *Xenobiotica*, *3*, 271-303
34. Reddy, B.S., Cohen, L.A., McCoy, G.D., Hill, P., Weisburger, J.H. & Wynder, E.L. (1980) Nutrition and its relationship to cancer. *Adv. Cancer Res.*, *32*, 237-345
35. Weisburger, J.H. & Williams, G.M. (1982) Metabolism of chemical carcinogens. In F.F. Becker (ed.) *Cancer: a comprehensive Treatise*, vol. 1, *Etiology: Chemical and Physical Carcinogenesis*, 2nd edn, Plenum Press, New York, pp. 241-333
36. Clayson, D.B. & Garner, R.C. (1976) Carcinogenic aromatic amines and related compounds. In C.E. Searle (ed.) *Chemical Carcinogens*, ACS Monograph 173, American Chemical Society, Washington, DC, pp. 366-461
37. Miller, E.C., Miller, J.A. & Enomoto, E. (1964) The comparative carcinogenicities of 2-acetylaminofluorene and its *N*-hydroxy metabolite in mice, hamsters, and guinea pigs. *Cancer Res.*, *24*, 2018-26

38. Hlavica, P. & Hülsmann, S. (1979) Studies on the mechanism of hepatic microsomal *N*-oxide formation: *N*-oxidation of *N,N*-dimethylaniline by a reconstituted rabbit liver microsomal cytochrome P-448 enzyme system. *Biochem. J.*, *182*, 109-16
39. Kadlubar, F.F., Miller, J.A. & Miller, E.C. (1976) Microsomal *N*-oxidation of the hepatocarcinogen *N*-methyl-4-aminoazobenzene and the reactivity of *N*-hydroxy-*N*-methyl-4-aminoazobenzene. *Cancer Res.*, *36*, 1196-1206
40. Kadlubar, F.F., Miller, J.A. & Miller, E.C. (1976) Hepatic metabolism of *N*-hydroxy-*N*-methyl-4-aminoazobenzene and other *N*-hydroxy arylamines to reactive sulfuric acid esters. *Cancer Res.*, *36*, 2350-9
41. Yahagi, T., Degawa, M., Seino,Y., Matsushima, T., Nagao, M., Sugimura, T. & Hashimoto, Y. (1975) Mutagenicity of carcinogenic azo dyes and their derivatives. *Cancer Lett.*, *1*, 91-6
42. DeBaun, J.R., Smith, J.Y.R., Miller, E.C. & Miller, J.A. (1970) Reactivity *in vivo* of the carcinogen *N*-hydroxy-2-acetylaminofluorene: increase by sulfate ion. *Science*, *167*, 184-6
43. Weisburger, J.H., Yamamoto, R.S., Williams, G.M., Grantham, P.H., Matsushima, T. & Weisburger, E.K. (1972) On the sulfate ester of *N*-hydroxy-*N*-2-fluorenylacetamide as a key ultimate hepatocarcinogen in the rat. *Cancer Res.*, *32*, 491-500
44. Kriek, E. & Westra, J.G. (1979) Metabolic activation of aromatic amines and amides and interactions with nucleic acids. In P.L. Grover (ed.) *Chemical Carcinogens and DNA*, vol. 2, CRC Press, Boca Raton, FL, pp. 1-28
45. Lai, C.-C, Miller, E.C., Miller, J.A. & Liem, A. (1984) Strong evidence that *N*-sulfoöxy-2-aminofluorene is the major ultimate carcinogenic metabolite of *N*-hydroxy-2-acetylaminofluorene (*N*-HO-AAF) in infant male mice. *Proc. Am. Assoc. Cancer Res.*, *25*, 85
45[a]. Yamamoto, R.S., Williams, G.M., Richardson, H.L., Weisburger, E.K. and Weisburger, J.H. (1973). Effect of p-hydroxyacetanilide on liver cancer induction by *N*-hydroxy-*N*-2-fluorenylacetamide. *Cancer Res.*, *33*, 454-7.
46. Sugimura, T. (ed.) *The Nitroquinolines. Carcinogenesis: a Comprehensive Survey* vol. 6, Raven Press, New York
47. Kadlubar, F.F., Miller, J.A. & Miller, E.C. (1977) Hepatic microsomal *N*-glucuronidation and nucleic acid binding of *N*-hydroxy arylamines in relation to urinary bladder carcinogenesis. *Cancer Res.*, *37*, 805-14
48. Bryan, G.T. (ed.) (1978) *Nitrofurans*. (*Carcinogenesis: a Comprehensive Survey*) vol. 4, Raven Press, New York
49. Lawley, P.D. (1976) Carcinogenesis by alkylating agents. In C.E. Searle (ed.) *Chemical Carcinogens*, ACS Monograph 173, American Chemical Society, Washington, DC, pp. 83-244
50. Schoental, R. (1976) Carcinogens in plants and microorganisms. In C.E. Searle (ed.) *Chemical Carcinogens*, ACS Monograph 173, American Chemical Society, Washington, DC, pp. 626-89
51. Wogan, G.N. (1973) Aflatoxin carcinogenesis. *Methods Cancer Res.*, *7*, 309-44
52. Peers, F.G. & Linsell, C.A. (1973) Dietary aflatoxins and liver cancer - a population based study in Kenya. *Br. J. Cancer*, 27, 473-84
53. Peers, F.G., Gilman, G.A. & Linsell, C.A. (1976) Dietary aflatoxins and human liver cancer. A study in Swaiziland. *Int. J. Cancer*, *17*, 167-76
54. Swenson, D.H., Lin, J.-K., Miller, E.C. & Miller, J.A. (1977) Aflatoxin B_1-2,3-oxide as a probable intermediate in the covalent binding of aflatoxins B_1 and B_2 to rat liver DNA and ribosomal RNA *in vivo*. *Cancer Res.*, *37*, 172-81
55. Lin, J.-K., Miller, J.A. & Miller, E.C. (1977) 2,3-Dihydro-2-(guan-7-yl-3-hydroxy-aflatoxin B_1, a major acid hydrolysis product of aflatoxin B_1-DNA or -rRNA adducts formed in hepatic microsome-mediated reactions and in rat liver *in vivo*. *Cancer Res.*, *37*, 4430-8
56. Essigmann, J.M., Croy, R.G., Nadzan, A.M., Busby, W.F., Jr, Reinhold, V.N., Büchi, G. & Wogan, G.N. (1977) Structural identification of the major DNA adduct formed by aflatoxin B_1 *in vitro*. *Proc. Natl. Acad. Sci. USA*, *74*, 1870-4
57. Miller, E.C., Swanson, A.B., Phillips, D.H., Fletcher, T.L., Liem, A. & Miller, J.A. (1983) Structure-activity studies of the carcinogenicities in the mouse and rat of some

naturally occurring and synthetic alkenylbenzene derivatives related to safrole and estragole. *Cancer Res.*, *43*, 1124-34

58. Borchert, P., Miller, J.A., Miller, E.C. & Shires, T.K. (1973) 1′-Hydroxysafrole, a proximate carcinogenic metabolite of safrole in the rat and mouse. *Cancer Res.*, *33*, 590-600
59. Drinkwater, N.R., Miller, E.C., Miller, J.A. & Pitot, H.C. (1976) Hepatocarcinogenicity of estragole (1-allyl-4-methoxybenzene) and 1′-hydroxyestragole in the mouse and mutagenicity of 1′-acetoxyestragole in bacteria. *J. Natl. Cancer Inst.*, *57*, 1323-31
60. Phillips, D.H., Miller, J.A., Miller, E.C. & Adams, B. (1981) N^2 Atom of guanine and N^6 atom of adenine residues as sites for covalent binding of metabolically activated 1′-hydroxysafrole to mouse liver DNA *in vivo*. *Cancer Res.*, *41*, 2664-71
61. Phillips, D.H., Miller, J.A., Miller, E.C. & Adams, B. (1981) Structures of the DNA adducts formed in mouse liver after administration of the proximate hepatocarcinogen 1′-hydroxyestragole. *Cancer Res.*, *41*, 176-86
62. Boberg, E.W., Miller, E.C., Miller, J.A., Poland, A. & Liem, A. (1983) Strong evidence from studies with brachymorphic mice and pentachlorophenol that 1′-sulfoöxysafrole is the major ultimate electrophilic and carcinogenic metabolite of 1′-hydroxysafrole in mouse liver. *Cancer Res.*, *43*, 5163-73
63. Laqueur, G.L. (1970) Contribution of intestinal macroflora and microflora to carcinogenesis. In J.W. Burdette (ed.) *Carcinoma of the Colon and Antecedent Epithelium*, Thomas, Springfield, Il, pp. 305-13
64. Matsumoto, H., Nagata, Y., Nishimura, E.T., Bristol, R. & Haber, M. (1972) β-Glucosidase modulation in preweanling rats and its association with tumor induction by cycasin. *J. Natl. Cancer Inst.*, *49*, 423-34
65. Jago, M.V., Edgar, J.A., Smith, L.W. & Culvenor, C.C.J. (1970) Metabolic conversion of heliotridine-based pyrrolizidine alkaloids to dehydroheliotridine. *Mol. Pharmacol.*, *6*, 402-6
66. Mattocks, A.R. & White, I.N.H. (1973) Toxic effects and pyrrolic metabolites in the liver of young rats given the pyrrolizidine alkaloid retrorsine. *Chem.-Biol. Interactions*, *6*, 297-306
67. Sirover, M.A. & Loeb, L.A. (1976) Infidelity of DNA synthesis *in vitro*: screening for potential metal mutagens or carcinogens. *Science*, *194*, 1434-6
68. Brand, K.G. (1982) Cancer associated with asbestosis, schistosomiasis, foreign bodies, and scars. In F.F. Becker (ed.) *Cancer: a Comprehensive Treatise*, vol. 1, *Etiology: Chemical and Physical Carcinogenesis*, 2nd edn, Plenum Press, New York, pp. 661-92
69. Leopold, W.R., Miller, E.C. & Miller, J.A. (1979) Carcinogenicity of antitumor *cis*-platinum (II) coordination complexes in the mouse and rat. *Cancer Res.*, *39*, 913-18
70. Baird, W.M. (1979) The use of radioactive carcinogens to detect DNA modifications. In P.L. Grover (ed.) *Chemical Carcinogens and DNA*, vol. 1, CRC Press, Boca Raton, FL, pp. 59-83
71. Vigny, P. & Duquesne, M. (1979) Fluorimetric detection of DNA-carcinogen complexes. In P.L. Grover (ed.) *Chemical Carcinogens and DNA*, vol. 1, CRC Press, Boca Raton, FL, pp. 85-110
72. Müller, R. & Rajewsky, M.F. (1980) Immunological quantification by high-affinity antibodies of O^6-ethyldeoxyguanosine in DNA exposed to *N*-ethyl-*N*-nitrosourea. *Cancer Res.*, *40*, 887-96
73. Poirier, M.C., Santella, R., Weinstein, I.B., Grunberger, D. & Yuspa, S.H. (1980) Quantitation of benzo[a]pyrene-deoxyguanosine adducts by radioimmunoassay. *Cancer Res.*, *40*, 412-16
74. Haugen, A., Groopman, J.D., Hsu, I.C., Goodrich, G.R. Wogan, G.N. & Harris, C.C. (1981) Monoclonal antibody to aflatoxin B_1 modified DNA detected by enzyme immunoassay. *Proc. Natl. Acad. Sci. USA*, *78*, 4124-7
75. Gupta, R.C., Reddy, M.V. & Randerath, K. (1982) ^{32}P-Postlabeling analysis of non-radioactive aromatic carcinogen-DNA adducts. *Carcinogenesis*, *3*, 1081-92
76. Reddy, M.V., Gupta, R.C., Randerath, E. & Randerath, K. (1984) ^{32}P-Postlabeling test for covalent DNA binding of chemicals *in vivo*: application to a variety of aromatic carcinogens and methylating agents. *Carcinogenesis*, *5*, 231-43
77. Friedberg, E.C. & Hanawalt, P.C. (eds) (1981) *DNA Repair. A Laboratory Manual of*

Research Procedures, vol. 1, parts A and B, Marcel Dekker, New York
78. Kohn, K.W. (1979) DNA as a target in cancer chemotherapy: measurement of macromolecular DNA damage produced in mammalian cells by anti-cancer agents and carcinogens. *Methods Cancer Res.*, *16*, 291-345
79. Lawley, P.D. (1979) Approaches to chemical dosimetry in mutagenesis and carcinogenesis: the relevance of reactions of chemical mutagens and carcinogens with DNA. In P.L. Grover (ed.) *Chemical Carcinogens and DNA*, vol. 1, CRC Press, Boca Raton, FL, pp. 1-36
80. Cooper, C.S., Grover, P.L. & Sims, P. (1983) The metabolism and activation of benzo[a]pyrene. *Prog. Drug. Metab.*, *7*, 295-396
81. Visser, A. & Westra, J.G. (1981) Partial persistency of 2-aminofluorene and *N*-acetyl-2-aminofluorene in rat liver DNA. *Carcinogenesis*, *2*, 737-40
82. Lin, J.-K., Schmall, B., Sharpe, I.D., Miura, I., Miller, J.A. & Miller, E.C. (1975) *N*-Substitution of carbon 8 in guanosine and deoxyguanosine by the carcinogen *N*-benzoyloxy-*N*-methyl-4-aminoazobenzene *in vitro*. *Cancer Res.*, *35*, 832-43
83. Beland, F.A., Tullis, D.L., Kadlubar, F.F., Straub, K.M. & Evans, F.E. (1980) Characterisation of DNA adducts of the carcinogen *N*-methyl-4-aminoazobenzene *in vitro* and *in vivo*. *Chem.-Biol. Interactions*, *31*, 1-17
84. Tarpley, W.G., Miller, J.A. & Miller, E.C. (1982) Rapid release of carcinogen-guanine adducts from DNA after reaction with *N*-acetoxy-2-acetylaminofluorene or *N*-benzoyloxy-*N*-methyl-4-aminoazobenzene. *Carcinogenesis*, *3*, 81-8
85. Lawley, P.D. & Warren, W. (1976) Removal of minor methylation products 7-methyladenine and 3-methylguanine from DNA of *Escherichia coli* treated with dimethyl sulphate. *Chem.-Biol. Interactions*, *12*, 211-20
86. Bennett, R.A., Essigmann, J.M. & Wogan, G.N. (1981) Excretion of an aflatoxin-guanine adduct in the urine of aflatoxin B_1-treated rats. *Cancer Res.*, *41* 650-4
87. Phillips, D.H. & Hanawalt, P.C. (1982) Alkali-sensitive sites in DNA from human cells treated with ultraviolet light. 1′-acetoxysafrole or 1′-acetoxyestragole. *Carcinogenesis*, *3*, 935-40
88. Drinkwater, N.R., Miller, E.C. & Miller, J.A. (1980) Estimation of apurinic/apyrimidinic sites and phosphotriesters in deoxyribonucleic acid treated with electrophilic carcinogens and mutagens. *Biochemistry*, *19*, 5087-92
89. Kadlubar, F.F., Anson, J.F., Dooley, K.L. & Beland, F.F. (1981) Formation of urothelial and hepatic DNA adducts from the carcinogen 2-naphthylamine. *Carcinogenesis*, *2*, 467-70
90. Kadlubar, F.F., Miller, J.A. & Miller, E.C. (1978) Guanyl O^6-arylamination and O^6-arylation of DNA by the carcinogen *N*-hydroxy-1-naphthylamine. *Cancer Res.*, *38*, 3628-38
91. Geacintov, N.E., Gagliano, A., Ivanovic, V. & Weinstein, I.B. (1978) Electric linear dichroism study on the orientation of benzo[a]pyrene-7,8-dihydrodiol-9,10-oxide covalently bound to DNA. *Biochemistry*, *17*, 5256-62
92. Beland, F.A. (1978) Computer-generated graphic models of the N^2-substituted deoxyguanosine adducts of 2-acetylaminofluorene and benzo[a]pyrene and the O^6-substituted deoxyguanosine adduct of 1-naphthylamine in the DNA double helix. *Chem.-Biol. Interactions*, *22*, 329-39
93. Santella, R.M. & Grunberger, D. (1983) Induction of the base displacement or Z conformation in DNA by *N*-2-acetylaminofluorene modification. *Environ. Health Perspect.*, *49*, 107-15
94. Westra, J.G., Kriek, E. & Hittenhausen, H. (1976) Identification of the persistently bound form of the carcinogen *N*-acetyl-2-aminofluorene to rat liver DNA *in vivo*. *Chem.-Biol. Interactions*, *15*, 149-64
95. Grunberger,D. & Weinstein, I.B. (1979) Conformational changes in nucleic acids modified by chemical carcinogens. In P.L. Grover (ed.) *Chemical Carcinogens and DNA*, vol. 2, CRC Press, Boca Raton, FL, pp. 59-93
96. Eisenstadt, E., Warren, A.J., Porter, J., Atkins, D. & Miller, J.H. (1982) Carcinogenic epoxides of benzo[a]pyrene and cyclopenta[cd]pyrene induce base substitutions via specific transversions. *Proc. Natl. Acad. Sci. USA*, *79*, 1945-1949
97. Foster, P.L., Eisenstadt, E. & Miller, J.H. (1983) Base substitution mutations induced

by metabolically activated aflatoxin B_1. *Proc. Natl. Acad. Sci. USA*, *80*, 2695-8

98. Rajewsky, M.F., Augenlicht, L.H., Biessmann, H., Goth, R., Hülser, D.F., Laerum, O.D. & Lomakina, L. Ya. (1977) Nervous system-specific carcinogenesis by ethylnitrosourea in the rat: molecular and cellular aspects. In H.H. Hiatt, J.D. Watson & J.A. Winsten (eds) *Origins of Human Cancer*, Cold Spring Harbor Laboratory, New York, pp. 709-26
99. Swenberg, J.A., Dyroff, M.C., Bedell, M.A., Popp, J.A., Huh, N., Kirstein, U. & Rajewsky, M.F. (1984) *O*[4]-Ethyldeoxythymidine, but not *O*[6]-ethyldeoxyguanosine, accumulates in hepatocyte DNA of rats exposed continuously to diethylnitrosamine *Proc. Natl. Acad. Sci. USA*, *81*, 1692-5
100. Allen, J.A. & Coombs, M.M. (1980) Covalent binding of polycyclic aromatic compounds to mitochondrial and nuclear DNA. *Nature (Lond.)*, *287*, 244-5
101. Backer, J. & Weinstein, I.B. (1980) Mitochondrial DNA is a major cellular target for a dihydrodiol-epoxide derivative of benzo[a]pyrene. *Science*, *209*, 297-9
102. Hanawalt, P.C., Cooper, P.K., Ganesan, A.K. & Smith, C.A. (1979) DNA repair in bacteria and mammalian cells. *Ann. Rev. Biochem.*, *48*, 783-836
103. Cleaver, J.E. (1968) Defective repair replication of DNA in *Xeroderma pigmentosum*. *Nature (Lond.)*, *218*, 652-6
104. Bootsma, D. (1978) *Xeroderma pigmentosum*. In P.C. Hanawalt, E.C. Friedberg & C.F. Fox (eds) *DNA Repair Mechanisms*, Academic Press, New York, pp. 589-601
105. Paterson, M.C. (1978) *Ataxia telangiectasia*: a model inherited disease linking deficient DNA repair with radiosensitivity and cancer proneness. In P.C. Hanawalt, E.C. Friedberg, & C.F. Fox (eds) *DNA Repair Mechanisms*, Academic Press, New York, pp. 637-50
106. German, J. (1978) DNA repair defects and human disease. In P.C. Hanawalt, E.C. Friedberg & C.F. Fox (eds) *DNA Repair Mechanisms*, Academic Press, New York, pp. 625-31
107. Fujiwara, Y., Tatsumi, M. & Sasaki, M.S. (1977) Cross-link repair in human cells and its possible defect in Franconi's anemia cells. *J. Mol. Biol.*, *113*, 635-49
108. Rajalakshmi, S., Rao, P.M. & Sarma, D.S.R. (1982) Chemical carcinogenesis: interactions of carcinogens with nucleic acids. In F.F. Becker (ed.) *Cancer: a Comprehensive Treatise,* vol. 1, *Etiology: Chemical and Physical Carcinogenesis*, 2nd edn, Plenum Press, New York, pp. 335-409
109. Sutherland, B.M. (1978) Photoreactivation in mammalian cells. *Int. Rev. Cytol.*, *8*, (Suppl.) 301-34
110. Pegg, A.E. (1978) Enzymatic removal of *O*[6]-methylguanine from DNA by mammalian cell extracts. *Biochem. Biophys. Res. Commun.*, *84*, 166-73
111. Pegg, A.E., Roberfroid, M., von Bahr, C., Foote, R.S., Mitra, S., Bresil, H., Likhachev, A. & Montesano, R. (1982) Removal of *O*[6]-methylguanine from DNA by human liver fractions. *Proc. Natl. Acad. Sci. USA*, *79*, 5162-5
112. Lehmann, A.R., Kirk-Bell, S., Arlett, C.F., Paterson, M.C., Lohman, P.H.M., de Weerd-Kastelein, E.A. & Bootma, D. (1975) *Xeroderma pigmentosum* cells with normal levels of excision repair have a defect in DNA synthesis after UV-irradiation. *Proc. Natl. Acad. Sci. USA*, *72*, 219-23
113. Radman, M. (1975) SOS repair hypothesis: phenomenology of an inducible DNA repair which is accompanied by mutagenesis. *Basic Life Sci.*, *5A*, 355-67
114. Deelman, H.T. (1924) Die Entstehung des experimentellen, Teerkresbses und die Bedeutung der Zellenregeration. *Z. Krebsforsch.*, *21*, 220
115. Berenblum, I. (1941) The cocarcinogenic action of croton resin. *Cancer Res.*, *1*, 44-8
116. Hecker, E. (1968) Cocarcinogenic principles from the seed oil of *Croton tiglium* and from other Euphorbiaceae. *Cancer Res.*, *28*, 2338-49
117. Van Duuren, B.L. (1969) Tumour-promoting agents in two-stage carcinogenesis. *Prog. Exp. Tumor Res.*, *11*, 31-68
118. Fujiki, H., Mori, M., Nakayasu, M., Terada, M., Sugimura, T. & Moore, R. (1981) Indole alkaloids: dihydroteleocidin B, teleocidin, and lyngbyatoxin A as members of a new class of tumor promoters. *Proc. Natl. Acad. Sci. USA*, *78*, 3872-6
119. Berenblum, I. (1982) Sequential aspects of chemical carcinogenesis: skin. In F.F. Becker (ed.) *Cancer: a Comprehensive Treatise*, vol. 1, *Etiology: Chemical and*

Physical Carcinogenesis, 2nd edn, Plenum Press, New York, pp. 451-84
120. Boutwell, R.K. (1974) The function and mechanism of promoters of carcinogenesis. *CRC Crit. Rev. Toxicol.*, *2*, 419-43
121. Hennings, H., Shores, R., Wenk, M.L., Spangler, E.F., Tarone, R. & Yuspa, S.H. (1983) Malignant conversion of mouse skin tumours is increased by tumour initiators and unaffected by tumour promoters. *Nature (Lond.)*, *304*, 67-9
122. Pound, A.W. & Bell, J.R. (1962) The influence of croton oil stimulation on tumour initiation by urethane in mice. *Br. J. Cancer*, *16*, 690-5
123. Boutwell, R.K. (1964) Some biological aspects of skin carcinogenesis. *Prog. Exp. Tumor Res.*, *4*, 207-50
124. Peraino, C., Fry, R.J.M. & Staffeldt, E. (1971) Reduction and enhancement by phenobarbital of hepatocarcinogenesis induced in the rat by 2-acetylaminofluorene. *Cancer Res.*, *31*, 1506-12
125. Hicks, R.M., Wakefield, J. St J. & Chowaniec, J. (1975) Evaluation of a new model to detect bladder carcinogens or cocarcinogens. *Chem.-Biol. Interactions*, *11*, 225-33
126. Slaga, T.J., Sivak, A. & Boutwell, R.K. (eds) (1978) *Mechanisms of Tumor Promotion and Cocarcinogenesis*, vol. 1, Raven Press, New York
127. Castagna, M., Takai, Y., Kaibuchi, K., Sano, K., Kikkawa, U. & Nishizuka, Y. (1982) Direct activation of calcium-activated, phospholipid dependent protein kinase by tumor-promoting phorbol esters. *J. Biol. Chem.*, *257*, 7847-51
128. Ashendel, C.L., Staller, J.M. & Boutwell, R.K. (1983) Protein kinase activity associated with a phorbol ester receptor purified from mouse brain. *Cancer Res.*, *43*, 4333-7
129. Emerit, I. & Cerutti, P. (1982) The tumor promotor phorbol-12-myristate-13-acetate induces chromosomal aberrations in human lymphocytes via indirect action. In C.C. Harris & P.A. Cerutti (eds) *Mechanisms of Chemical Carcinogenesis*, Alan R. Liss, New York, pp. 495-7
130. Weinstein, I.B. (1981) Current concepts and controversies in chemical carcinogenesis. *J. Supramol. Struct. Cell. Biochem.*, *17*, 99-120
131. Weinberg, R.A. (1983) A molecular basis of cancer. *Sci. Am.*, *249* (5), 102-216
132. Tabin, C.J., Bradley, S.M., Bargmann, C.I., Weinberg, R.A., Papageorge, A.G., Scolnick, E.M., Dhar, R., Lowy, D.R. & Chang, E.H. (1982) Mechanism of activation of a human oncogene. *Nature (Lond.)*, *300*, 143-9
133. Reddy, E.P., Reynolds, R.K., Santos, E. & Barbacid, M. (1982) A point mutation is responsible for the acquisition of transforming properties by the T24 human bladder carcinoma oncogene. *Nature (Lond.)*, *300*, 149-52
134. Yuasa, Y., Srivastava, S.K., Dunn, C.U., Rhim, J.S., Reddy, E.P. & Aaronson, S.A. (1983) Acquisition of transforming properties by alternative point mutations within c-*bas/has* human proto-oncogene. *Nature (Lond.)*, *303*, 775-9
135. Lacassagne, A. (1932) Apparition de cancers de la mamelle chez la souris male, soumise a des injections de folliculine. *C.R. Acad. Sci. (Paris)*, *195*, 630-2
136. Pike, M.C., Henderson, B.E., Krailo, M.D., Duke, A. & Roy, S. (1983) Breast cancer in young women and use of oral contraceptives: possible modifying effect of formulation and age at use. *Lancet*, *ii*, 926-9
137. Vessey, M.P., Lawless, M., McPherson, K. & Yeates, D. (1983) Neoplasia of the cervix uteri and contraception - a possible adverse effect of the pill. *Lancet*, *ii*, 930-4
138. Drevon, C., Piccoli, C. & Montesano, R. (1981) Mutagenicity assays of estrogenic hormones in mammalian cells. *Mutat. Res.*, *89*, 83-90
139. Highman, B., Norvell, M.J. & Shellenberger, T.E. (1977) Pathological changes in female C3H mice continuously fed diets containing diethylstilbestrol or 17-beta-estradiol. *J. Environ. Pathol. Toxicol.*, *1*, 1-30
140. Barrett, J.C., Wong, A. & McLachlan, J.A. (1981) Diethylstilbestrol induces neoplastic transformation without measurable gene mutation at two loci. *Science*, *212*, 1402-4
141. Mehta, R.D. & Borstel, R.C. (1982) Genetic activity of diethylstilbestrol in *Saccharomyces cerevisiae*. Enhancement of mutagenicity by oxidizing agents. *Mutation Res.*, *92*, 49-61
142. Tsutsui, T., Degen, G.H., Schiffman, D., Wong, A., Maizumi, H., McLachlan, J.A.

& Barratt, J.C. (1984) Dependence on exogenous metabolic activation for induction of unscheduled DNA synthesis in Syrian hamster embryo cells by diethylstilbestrol and related compounds. *Cancer Res.*, *44*, 184-9

143. Rüdiger, H.W., Haenisch, F., Metzler, M., Oesch, F. & Glatt, H.R. (1979) Metabolites of diethylstilboestrol induce sister chromatid exchange in human cultured fibroblasts. *Nature (Lond.)*, *281*, 392-4
144. Lutz, W.K., Jaggi, W. & Schlatter, C.H. (1982) Covalent binding of diethylstilboestrol to DNA in rat and hamster liver and kidney. *Chem.-Biol. Interactions*, *42*, 251-7
145. Metzler, M. (1984) Diethylstilboestrol: reactive metabolites derived from a hormonally active compound. In H. Greim, R. Jung, M. Kramer, H. Marquardt & F. Oesch (eds) *Biochemical Basis of Chemical Carcinogenesis*, Raven Press, New York, pp. 69-75
146. Sinha, D.K. & Dao, T.L. (1980) Induction of mammary tumors in ageing rats by 7,12-dimethylbenz[a]anthracene: role of DNA synthesis during carcinogenesis. *J. Natl. Cancer Inst.*, *64*, 519-21
147. Miller, E.C. & Miller, J.A. (1981) Mechanisms of carcinogenesis. *Cancer*, *47*, 1055-64
148. Wattenberg, L.W. (1978) Inhibition of chemical carcinogenesis. *J. Natl. Cancer Inst.*, *60*, 11-18
149. Hill, M.J. (1980) Bacterial metabolism and human carcinogenesis. *Br. Med. Bull.*, *36*, 89-94

Further Reading

Becker, F.F. (ed.) (1982) *Cancer: a Comprehensive Treatise*, vol. 1, *Etiology: Chemical and Physical Carcinogenesis*, 2nd edn, Plenum Press, New York

Doll, R. & Peto, R. (1981) *The Causes of Cancer*, Oxford University Press, London

Grover, P.L. (ed.) (1979) *Chemical Carcinogens and DNA*, 2 vols, CRC Press, Boca Raton, FL

Hiatt, H.H., Watson, J.D. & Winsten, J.A. (eds) (1977) *Origins of Human Cancer*, 3 vols, Cold Spring Harbor, New York

Miller, E.C. (1978) Some current perspectives on chemical carcinogenesis in humans and experimental animals: presidential address. *Cancer Res.*, *38*, 1479-96

Miller, E.C. & Miller, J.A. (1979) Milestones in chemical carcinogenesis. *Sem. Oncol.*, *6*, 445-60

Searle, C.E. (ed.) (1976) *Chemical Carcinogens*, ACS Monograph No. 173, American Chemical Society, Washington, DC

Singer, B. & Grunberger, D. (1983) *Molecular Biology of Mutagens and Carcinogens*, Plenum Press, New York

6 TESTING FOR CARCINOGENS

J.M. Walker

Contents

6.1 Introduction

Epidemiological evidence strongly suggests that environmental factors are a major cause of cancer.[1,2] Although some of these factors are self-imposed (e.g. cigarette smoking) there are many other unintentional routes of carcinogen uptake, such as the diet, and by exposure to both natural and synthetic chemicals present in our environment (e.g. agricultural compounds, medicines, man-made pollutants, etc.). Since it is estimated that up to 80 per cent of all cancers are caused by environmental factors,[1] there is considerable justification for effort to be spent on the removal of these carcinogenic substances from our environment, as the identification and control of these substances should lead to a corresponding reduction in cancer incidence. Although epidemiology has proved, and is still proving, successful in identifying environmental factors that are carcinogenic in man, such identifications are necessarily retrospective, only being made after many people have been exposed, for many years, to the carcinogen. The need to test compounds for any carcinogenic effect, before their introduction into our environment, is still therefore a major and necessary requirement. Specific legislation requiring testing for the carcinogenic potential of new pharmaceutical products has existed for some time and such tests are an accepted part of the development of these compounds. However, until recently, similar legislation has not existed for many of the other synthetic compounds that are regularly introduced into our environment (e.g. fluorocarbons, polymers, etc.). Over the past 30 years the production of synthetic organic chemicals has increased tenfold in the USA[3] and it is not unrealistic to assume that increased production has resulted in a proportional increase in human exposure. Currently, over 50 000 synthetic compounds are estimated to be in more or less common use,[4] but unfortunately only a small fraction of these compounds were tested for carcinogenicity or mutagenicity before being introduced in to the environment, and it is not unreasonable to believe that at least some of these compounds are carcinogenic. Additionally, it is estimated that over a thousand new compounds are introduced into our environment each year.[4] Concern over this increasing exposure to man-made compounds has resulted in the introduction of guidelines in both the USA (Toxic Substances Control Act, 1976) and the UK (Health and Safety at Work Act, 1974) for monitoring industrial chemicals. These guidelines now require a more careful analysis of the carcinogenic potential of such compounds.

The traditional method for testing for carcinogenicity has involved animal studies. These tests involve treating animals with the compound (at different dose levels) for the life-span of each animal to determine if there are any differences in the time of appearance and in the number and types of tumour, between treated animals and controls. Such observations are made as a result of a detailed histopathological study of the animals, although the assessment and interpretation of results can sometimes be ambiguous, particularly when low-level effects are being assessed. A further complication arises from the difficulty in extrapolating data from animals to man. However, it is known that almost all the organic chemicals known to cause cancer in humans also cause cancer in experimental animals, which confirms the relevance of animal tests to the human situation.[5] The rat and mouse are usually used in carcinogenicity studies as they have a relatively short life-span (two years and 18 months respectively) and can be kept in the relatively large numbers (at least 500 per test) that are necessary to produce statistically significant results. These life-span studies obviously require a large expenditure of time, man-power and laboratory resources and consequently are very expensive, costing as much as £100 000–£500 000 over a three to four year period to test a single compound (completion of the pathological investigation can take one to two years). In recent years it has become increasingly apparent that this traditional method is not satisfying present-day demands for the rapid identification of carcinogens. For example, new drugs need to be assessed for possible carcinogenic effects as early and as quickly as possible during development of the drug to avoid further extensive and costly tests which would normally be carried out before marketing. Also, the testing of the many thousands of industrial and environmental chemicals to which humans are exposed, by animal studies, would be impossible both with respect to the cost and time involved. At present, detailed cancer tests are reported on only about 150 previously untested chemicals each year. It is therefore not surprising that considerable effort is being directed toward developing more rapid (short-term) and economical methods for screening potential carcinogens. At the time of writing, over 30 potential short-term tests have been described in the literature (see Refs 6–13 for reviews). Nearly all these tests are actually tests for mutagens, being based on the hypothesis that cancer results from a mutagenic event (somatic-mutation theory), an hypothesis that is supported by the observation that 90 per cent of the known animal carcinogens are mutagenic *in vitro*. Section 6.2 describes the basic principles behind some of the more successful short-term tests.

A number of studies have been carried out to compare the 'success-rates' of individual short-term tests in detecting known carcinogens and some of these observations, together with a discussion of the suitability and potential uses of short-term tests are described in section 6.3. Finally, a more direct approach to the problem of detecting carcinogens is described in section 6.4. This approach involves examining humans who are currently being exposed to compounds under suspicion (e.g. occupational exposure) and studying dose–response relations. Although these techniques are not as widely established as those used in animal studies or short-term tests, considerable progress is being made in the detection of human cellular damage after exposure to carcinogens.

6.2 Short-term Tests for Carcinogens

The tests described below are a cross-section of some of the more successful short-term tests that are currently being developed. The basic concept behind each test is given, together with a brief outline of the procedure for each test. However, it should be stressed that many modifications and adaptations of each approach have been described in the literature, and further variations can be expected. The relative merits of each test are not discussed as this is a complicated and involved area (but see section 6.2.3 and Table 6.1). In an attempt to mimic the situation *in vivo*, many tests *in vitro* include the use of an 'S9' mix. This is simply the supernatant from a rat-liver homogenate centrifuged at 9000 **g**, supplemented with glucose phosphate, NADPH and salts. This supernatant contains many of the metabolising enzymes that are thought to play a role in converting relatively harmless chemicals into potential carcinogens *in vivo* (see Chapter 5). The test compound is therefore often incubated in the presence of S9 mix before the test or, alternatively, the S9 mix is introduced with the test compound directly into the test system. To maximise the effect of the S9 mix, rats are often pretreated with an inducer such as arochlor, to stimulate the production of metabolising enzymes in the liver.

6.2.1 Ames Test

The Ames test[14,15] uses a strain of the bacterium *Salmonella typhimurium* which has a mutation (his^-) in a gene that codes for one of the enzymes involved in the synthesis of the amino acid histidine. As a result of this mutation, the bacterium is unable to synthesise its own histidine (which is required for protein synthesis) and will therefore only grow in a mineral nutrient medium if the medium is supplemented with histidine. If this strain is exposed to a mutagen, many different DNA sites, in different

Table 6.1 Performance of Some of the Short-term Carcinogenicity Tests Described in Section 6.2

Assay	No. of carcinogens tested	No. of non-carcinogens tested	Results: Sensitivity	Results: Specificity	Reference
Ames – 1	25	17	0.68	0.59	11
Ames – 2	58	62	0.91	0.93	6
Rec^- (*B.subtilis*)	25	17	0.84	0.47	11
Pol A	25	17	0.60	0.53	11
Inductest	24	16	0.50	0.75	11
Degranulation – 1	21	13	0.71	0.54	11
Degranulation – 2	58	62	0.71	0.71	6
Unscheduled DNA repair	15	10	0.60	0.50	11
Sister chromatid exchange	22	15	0.59	0.40	11
Cell transformation – 1	25	17	0.88	0.71	11
Cell transformation – 2	58	62	0.91	0.97	6
Micronucleus test	21	13	0.52	0.85	11
Sperm morphology	11	6	0.36	1.00	11
Sebaceous gland suppression	58	62	0.67	0.64	6

$$\text{Sensitivity} = \frac{\text{no. of carcinogens found positive}}{\text{no. of carcinogens tested}} \qquad \text{Specificity} = \frac{\text{no. of non-carcinogens found negative}}{\text{no. of non-carcinogens tested}}$$

bacteria, will be mutated. Ocassionally, a mutation will occur at exactly the same site as the his^- mutation, and this will restore the sequence of the DNA to its normal coding sequence. This mutation is called a reverse mutation (the mutation is his^- to his^+), and will restore the ability of the bacterium to synthesise its own histidine. Such a reversion can therefore be detected by the ability of the his^- bacterium to grow now on a medium lacking histidine. This reverse-mutation event will of course be a relatively rare one. However, as we are working with as many as 10^9 cells/ml of culture medium, such events will occur and they can be readily detected since each single his^+ revertant that is formed will grow to form a colony which is visible to the naked eye. (This ability to amplify a rare event represents one of the main advantages of working with bacterial systems.) To improve the sensitivity of this test system, other modifications have been made to this his^- strain. The strain contains an additional mutation, which causes partial loss of the outer lipopolysaccharide cell wall, thus making the cell more permeable to the test compound. A further mutation has been introduced that deletes the DNA excision-repair pathway which makes the strain more sensitive to DNA-damaging agents; a final refinement has been the introduction of an 'R-factor' plasmid which enhances the mutagenic response of the host bacterium while decreasing its sensitivity to any lethal effect of the mutagen.

For routine screening, the test strain (his^-, about 10^9 cells) is combined directly with both the compound being tested and some S9 mix, and the mixture plated on to a minimal agar medium. (The medium actually contains a trace of histidine which is sufficient to allow a small amount of growth of his^- strains leading to a cloudy appearance on the plate, but not growth to full colonies. This is done because some growth is needed for many mutagens to act.) Control plates are similarly prepared, but without the test compound. The plates are then incubated for two days at 37°C, after which time revertant colonies are counted. Control plates will always show some colonies due to spontaneous revertants. However, the number of mutagen-induced colonies are usually two or more orders of magnitude greater than the relatively low spontaneous incidence of revertants, and therefore easily detected.

5.2.2 *Inductest (*Prophage-induction Test*)*

A bacterium is said to be lysogenic when it carries in its own DNA, and in a dormant state, the DNA of a bacterial virus (a bacteriophage). In its dormant state, this DNA is known as a prophage. The most widely studied system is λ phage (a bacteriophage) in *Escherichia coli*. When lysogenic *E.coli* are subjected to any treatment which halts DNA replication, for example DNA damage, the prophage is induced, i.e. the phage DNA loops out of the bacterial DNA and directs the synthesis of proteins necessary to form the virus particle. This process is known as lysogenic induction. Within an hour the host cell lyses and releases phage particles. These phage particles will infect and destroy bacteria in their vicinity and are therefore easily detected as visible plaques produced on a lawn of indicator bacteria. This process has been adapted for use as a test for mutagens.[16,17]

The lysogenic tester bacteria, test compound, S9 mix and normal indicator bacteria, are mixed and spread on an agar plate. Control plates are similarly prepared but without the test compound. Both sets of plates are then incubated at 37°C for one day, after which time most of the plate is covered by a thick lawn of indicator bacteria. If the test compound is a mutagen, it will have caused lysogenic induction in tester bacteria, and these events will be detected as visible plaques on the indicator lawn, as already described. As with the Ames test, lysogenic tester bacteria have been constructed with additional mutations which enhance permeability to chemicals and sensitivity to DNA damage.

6.2.3 Tests for Detection of DNA Damage and Repair

Exposure of cells to mutagens will, by definition, produce chemical changes (damaged sites) on the DNA. Since cells contain enzyme systems which repair DNA damage, the mutation process results in the activation of these repair systems. Two of the more common repair systems, excision repair and postreplication repair, both involve, as part of their overall mechanism, DNA repair synthesis where a new piece of DNA is synthesised (unscheduled DNA synthesis) to fill a gap on one strand of the DNA. A number of tests have been developed that measure this unscheduled DNA synthesis. Other tests have been designed to detect either the physical damage to the DNA or the consequence of this damage in the absence of repair enzymes. Examples of each of these tests are given below. (See Refs 18 and 19 for general review articles on DNA damage/repair assays.)

6.2.3.1 Detection of Unscheduled DNA Synthesis in Mammalian Cells in vitro. This test measures unscheduled DNA synthesis in cultured human cells[20-22] (e.g. fibroblast cells are obtained from skin biopsies). This repair synthesis of DNA is measured by determining the amount of [^{3}H]thymidine incorporated in to cultured cells which have been treated with the test compound. An S9 mix is also included with the test compound. Since normal replicative DNA synthesis would interfere with this method ([^{3}H]thymidine would be incorporated into DNA during S-phase), the cells are grown either in the presence of hydroxyurea or in arginine-deficient medium, both of which inhibit replicative DNA synthesis. After exposure of the cells (2–5 h) to both [^{3}H]thymidine and the test compound, they are washed, trypsinised and portions taken for radioactivity counting. Alternatively, incorporation can be detected by radioautography of the cells. Increased incorporation of [^{3}H]thymidine by test cells over control cells indicates a positive result. One of the attractions of this method is the fact that one is working with human cells.

6.2.3.2 Bacterial rec$^-$ Assay and DNA Polymerase-deficient Assay. The bacterial rec$^-$ assay[23,24] uses mutants of *Bacillus subtilis* (rec$^-$) which are deficient in the so-called 'SOS' DNA repair system. Such bacteria are therefore particularly sensitive to damage by mutagens when compared with wild-type bacteria (rec$^+$) which do contain the repair system. Agar plates are prepared containing *B.subtilis* (rec$^-$) spores (germinating spores are more sensitive to DNA-damaging agents than are vegetative cells) and some S9 mix. Control plates are similarly prepared but using *B.subtilis* (rec$^+$). The test chemical is absorbed on to paper discs, a disc laid on both control (rec$^+$) and assay (rec$^-$) plates and the plates incubated at 37°C for 20–30 h. During this time the test compound diffuses into the gel. The concentration of the test compound used is the highest allowable concentration that produces no zone of growth inhibition around the filter disc on the control plate [i.e. there is (just) a non-toxic concentration around the disc]. This concentration is determined in a preliminary experiment. If the same concentration produces a zone of inhibition around the disc on the rec$^-$ plate, then this inhibition of growth is assumed to be due to cell death/inhibition of DNA replication caused by mutational effects of the compound.

The DNA polymerase-deficient assay[25,26] is based on a similar principle. The test uses two strains of *E.coli*, one deficient in DNA polymerase I (*E.coli* pol A$^-$) and a wild-type that contains the enzyme (*E.coli* pol A$^+$). Since DNA polymerase I plays an essential role in the DNA repair process exposure of these cells to agents which modify the DNA results in the preferential killing or inhibition of the pol A$^-$ strain due to its inability to repair the DNA damage. On the other hand, compounds that affect structures other than the DNA affect the two strains to the same extent.

6.2.3.3 Alkaline Elution Assay.[27,28] In this method cultured cells, previously grown in the presence of [^{14}C]thymidine (this is simply to aid monitoring of the DNA later), are grown in the presence of the test compound and S9 mix. The cells are then lysed, applied to a membrane filter and washed. A mat of DNA, free of most membranes, proteins and RNA, remains on the filter. An alkaline eluting solution (pH 12) is then slowly passed through the filter causing double-stranded DNA to unwind and elute from the filter at a rate proportional to the length of the single strands. The eluted DNA is monitored by its radioactivity and characteristic elution profiles can be obtained for various types of DNA damage. Depending on the elution profile obtained therefore various mutagenic effects can be detected. The DNA damage responsible for these various elution profiles includes single-strand breaks (large numbers of single-strand breaks cause rapid elution of DNA), alkylation of bases (chemical cleavage at 7-alkylguanine proceeds very quickly under the alkaline conditions used producing shorter than normal lengths of DNA) and interstrand cross-links (elution of cross-linked DNA is considerably retarded). A similar approach for determining DNA damage involves the separation of DNA fragments by centifugation on alkaline sucrose gradients.[29]

6.2.4 Degranulation of Endoplasmic Reticulum[6]

The membranes of the hepatic endoplasmic reticulum are usually considered to be of two forms, namely 'rough' or 'smooth' depending on whether or not they have bound ribosomes. This test is based on the observation that many carcinogens cause the degranulation (removal of ribosomes) from the rough endoplasmic reticulum (RER). Ribosome loss is monitored by determining RNA/protein ratios for membranes. This ratio is decreased in degranulated membranes due to the loss of RNA with the ribosomes. The method involves the use of membranes containing radiolabelled RNA. To achieve this, rats are starved to deplete pyrimidine precursors then injected with [^{14}C]orotic acid, and killed 17 h later. The livers are removed, homogenised and the radiolabelled RER isolated by sucrose-density-gradient centifugation. The test compound, with S9 mix, is then added to a portion of the RER preparation and incubated for about 2 h. Membranes are then separated from any free ribosomes by gradient centrifugation or gel filtration. The radioactivity in these membranes is then measured and the protein content determined by the Lowry method. Similar measurements are also made on control (sham-treated) samples of RER. A positive result is indicated by a decrease in the radioactivity (RNA)/protein ratio for the treated RER compared with the sham-treated RER.

6.2.5 Mammalian Cell Transformation[6,30]

This test investigates the ability of test compounds to transform cultured mammalian cells [e.g. baby-hamster kidney (BHK) cells]. Cells are grown in the presence of the test compound and S9 mix and then examined for malignant transformation. One very clear way of detecting transformation is to determine the ability of treated cells to induce malignancy in animals on injection (tumorigenicity). This, however, requires a considerable extension of the time-scale of the experiment. Fortunately, the ability of transformed cells to grow in semisolid agar correlates well with tumorigenicity, and it is this parameter that is now used to measure transformation. The number of colonies in soft agar (which is a measure of the number of transformed cells) produced by an aliquot of treated cells is counted and a measure of the transformation frequency determined by comparison with the number of surviving cells present in a same size aliquot of treated cells. An increase in the transformation frequency of treated cells over control cells is taken as a positive result.

6.2.6 *Sebaceous-gland-suppression Test*[6]

The sebaceous-gland test is based on a very early observation that, in experiments where repeated applications of carcinogens were made to the skin of animals to produce tumours, there was also a regression of the number of the recipients' sebaceous glands. This observation prompted an investigation into the usefulness of this phenomenon as a means of identifying potential carcinogens. Solutions of the test compound are applied twice a day for three days to the shaved dorsolumbar region of a group of mice, and then the animals killed four days after the cessation of treatment. Samples of test and control skin are removed, histological sections of the skin are prepared and then the percentage of sebaceous glands to hair follicles determined under the microscope (the hair follicles act as an 'internal' standard). A statistically significant suppression is taken as a positive result. Some compounds give as much as 90 per cent suppression. This test is particularly successful in detecting carcinogens from tobacco tars.[31]

6.2.7 *Sister Chromatid Exchange (SCE)*[32,33]

When DNA that contains mutation-induced lesions replicates during S-phase, exchange often occurs between the two daughter molecules. This can be observed at the cytological level as exchanges between the two sister chromatids of a metaphase chromosome and the observation can therefore act as an indicator of mutational events. However, to observe these exchanges, it is necessary to treat the cells so that the two sister chromatids can be differentiated. This is accomplished by exploiting the fact that DNA is a double molecule that replicates semiconservatively. Cells are allowed to undergo two replicative cycles in the presence of the thymidine analogue, 5-bromodeoxyuridine (BrdU), which will be incorporated into newly synthesised DNA strands in place of thymidine. After these two replicative cycles the resultant chromosomes will have, as a consequence of the semiconservative nature of DNA replication, one chromatid which has one chain substituted with BrdU, and one chromatid which has both chains substituted with BrdU (see Figure 6.1). These sister chromatids can then be made to stain differently from one another by using either Giemsa or a fluorescent dye. If any exchanges have occurred between sister chromatids, these exchanges will be identified as regions of differential staining within a chromatid (see Figure 6.2).

In practice, tissue-culture cells (e.g. Chinese hamster ovary, CHO) are grown through two replicative cycles (about 24 h) in the presence of both

Figure 6.1 Diagrammatic Representation Showing Two Cycles of DNA Replication in the Presence of BrdU

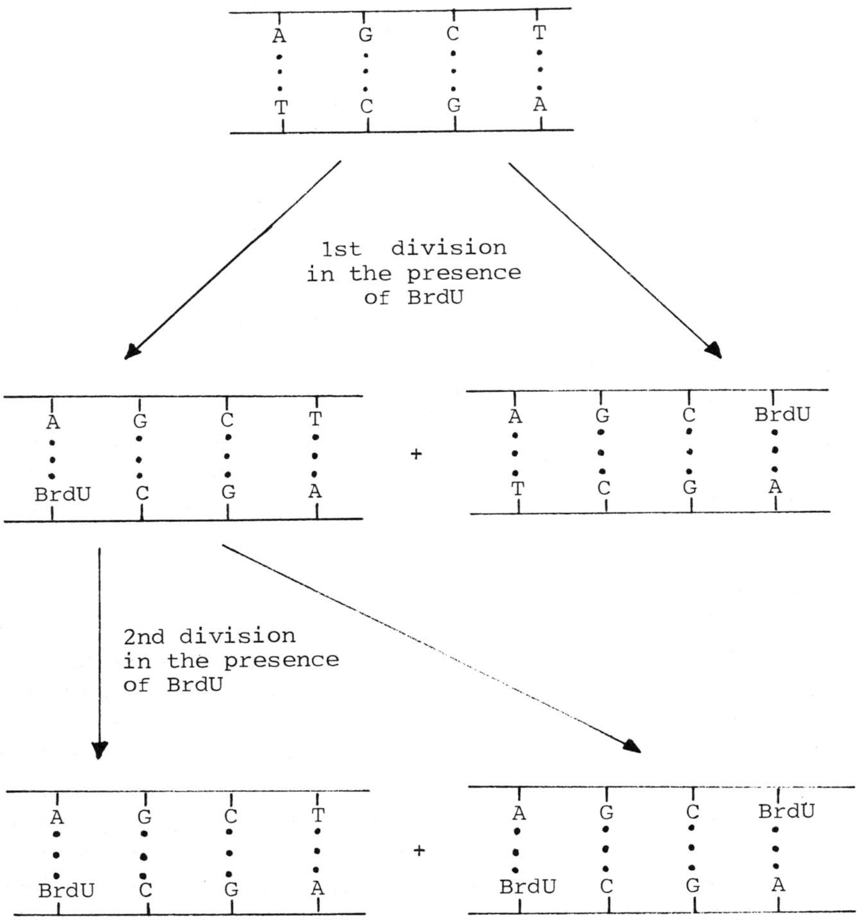

(The second replication is only shown for one chromosome.) The result is one chromatid which has one chain substituted with BrdU and the second with both chains substituted.

BrdU and the test compound. Colcemid is then added to the culture to cause cells to accumulate at mitosis (mitotic arrest) and these mitotic cells are then harvested. The cells are fixed, dried on a slide, stained as described already, and viewed. About 25 cells per sample are analysed. An increase in the mean SCE frequency per cell for treated cells, compared with that for control cells, is taken as a positive result.

Figure 6.2 Diagrammatic Representation of Sister Chromatid Exchange (SCE)

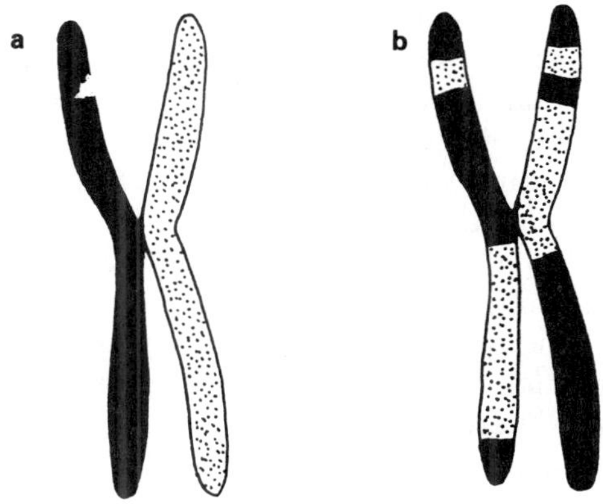

(**a**) Chromosome showing differential staining after two divisions in the presence of BrdU; (**b**) same chromosome showing SCE.

6.2.8 Micronucleus Assay[34,35]

The micronucleus assay is a cytogenetic screen *in vivo* that identifies agents which produce chromosome breaks. The test is based on the observation that cells with chromosomal breaks and/or exchanges often have aberrations in the distribution of chromatin during cell division. After division the daughter cells contain this displaced chromatin as distinct micronuclei in the cytoplasm, which can be observed under the microscope. The cell population tested consists of erythroblasts undergoing their final chromosome replication and mitosis before expulsion of the nucleus. Bone-marrow smears made 30–48 h after dosing yield a population of young enucleated erythrocytes that are known to have undergone mitosis during the time of treatment. These young erthrocytes are readily identifiable because they stain differently from mature erythrocytes and they are easily scored for the presence of micronuclei because they lack nuclei.

Mice are treated (peritoneal injection) with the test compound at 0 and 24 h, and then bone-marrow samples are taken (usually from the femur) at 48, 72 and 96 h, slides prepared and then scored for the presence of micronuclei. Increases in the frequency of micronuclei among treated animals over control-animal values are taken as an indication of treatment-induced genetic damage. The techniques of identifying and scoring micronuclei are relatively easy to master and the test itself is rapid and economical.

6.2.9 Sperm-morphology Assay[36]

The sperm-morphology assay (SMA) is a relatively simple test that examines the effect *in vivo* of compounds on the germ cells of male mice. The end-point in the SMA is the percentage increase of morphologically aberrant sperm in the treated series as compared with the control. Sperm head-shape and tail-attachment are the two parameters that are considered. Individual male mice are treated with the test compound. At various times after treatment (2, 4, 6 and 8 weeks), males are killed and sperm samples are collected from the vas deferens and/or the epididymis of each male. Samples are placed in sodium chloride solution, stained in eosin, slides prepared and counts of normal and abnormal sperm made at about ×800 magnification. The technique of distinguishing between normal and abnormal sperm can be easily mastered and results in consistent and reproducible data.

6.3 Evaluation of Short-term Tests

The Ames test has been the first short-term carcinogenicity test to gain widespread acceptance. The test has proved particularly successful in identifying carcinogens and has been well evaluated since its introduction by Bruce Ames in the mid-1970s.[14] One of the earlier studies showed the test to identify 90 per cent of known carcinogens and correctly to identify 87 per cent of known non-carcinogens.[37] Of all the short-term tests reported to date, the Ames test has proved to be by far the most successful. It has, for example, been instrumental in having the flame-retardant 'Tris BP' and the food additive AF_2 removed from the market. Both were positive in the Ames test and subsequently were demonstrated to cause cancer in experimental animals. AF_2 had in fact previously passed an animal test, but in the light of the Ames test result was more rigorously examined in animals and subsequently shown to be carcinogenic. Also, 16 out of 18 organic chemicals known or suspected as being human carcinogens are positive in the Ames test.[37] Since the Ames test always has some false positives and false negatives, some carcinogens will necessarily be passed by this test and some safe chemicals will be needlessly rejected, but the value of the system as a rapid first filter is undeniable. The general acceptance of the usefulness of the Ames test is shown by the fact that many drug companies now use the test as an initial screen of potential drugs. If the drug is positive in the Ames test it is discarded, if negative it is developed further and put through a full animal test before marketing. Additionally, the Ames test is now one of eight tests required under the Pesticide Act (USA) and is one of six tests required under the Toxic Substances Control Act (USA).

As well as the Ames test, a considerable number of other short-term tests have been described in the literature in recent years.[6-13] Like the Ames test, these tests all suffer from the disadvantage of giving some false-positive and some false-negative results (see Table 6.1), but some tests have proved sufficiently successful in identifying carcinogens to be considered worthy of further investigation. This proliferation of short-term tests has left both the scientist and the regulatory authorities with the problem of selecting the best and most useful tests for screening purposes. Consequently, a number of validation studies have been carried out (see, e.g. Refs 6 and 11), the basic purposes of which have been to assess the effectiveness with which short-term tests can detect carcinogens and non-carcinogens that have been previously classified on the basis of their activities in animal studies. An indication of the relative performances of some of these short-term tests is described in Table 6.1

A rigorous comparison of the performance of different systems is not easy to make at present since many of the available data describing the performance of individual tests have been developed in different ways in different laboratories. For example, the purity of the test chemical used, the protocols used and the criteria for judging a positive result can vary from laboratory to laboratory for a given test. Additionally, for many short-term assays experimental details have not been fully defined, and hence the test neither standardised nor investigated with a sufficient number of known mutagens/carcinogens. In the light of these observations larger comparative studies have been designed involving standardised strategies and protocol and these are currently in progress. However, it is still possible to draw some important general conclusions from the data presently available on short-term testing procedures.

(1) No single test is perfect, nor is one likely to be, considering the nature of the carcinogenic process. All miss at least 10 per cent of known carcinogens and they all give a significant percentage of positive results for known non-carcinogens. It is therefore unwise to base decisions on the results of any single test.

(2) The most effective tests are those that use bacteria, such as the Ames test. However, these bacterial tests fail to identify some carcinogens which other eukaryotic tests do identify, but these other tests in turn fail to identify chemicals which the bacterial tests detect. It is clear therefore that the use of more than one test can improve the ability to distinguish between carcinogen and non-carcinogen, and that the best approach for the future will therefore be to use a battery of complementary tests to investigate a given compound. The size and composition of the battery of assays that is best suited to screening carcinogens has yet to be decided. Nevertheless, it is generally agreed that in the near future results from a battery of validated tests will be of significant value in assessing the

carcinogenic potential of a chemical. Since some tests seem to give better results for one class of chemicals than for another, batteries of tests which have optimum performances for certain groups of chemical structures may well be developed in the future.
(3) At present, should a compound give positive results in more than one valid short-term test, operations involving the chemical should be immediately examined and human exposure to the compound reduced to as low a level as possible while long-term animal exposure studies are carried out.
(4) Future studies should pay particular attention to the metabolic activation aspect of the test. The introduction of a metabolic activation step has been an important refinement for many short-term tests. However, some chemicals appear to be less active in tests *in vitro* than they are *in vivo*. This suggests that the metabolic capabilities represented by the S9 mix are inadequate for some chemicals (e.g. some compounds are metabolized extrahepatically: 2,4-dinitrotoluene). Additionally, extrapolation from metabolism in a rat-liver homogenate to the complex metabolism in man is difficult and further research in this area is needed.
(5) The quantitative nature of short-term tests need to be investigated, i.e. is a chemical that comes up strongly in a test also a strong carcinogen? A knowledge of carcinogenic potential would be an important aid in human-risk assessment and would allow priorities to be established for dealing with these chemicals, since in practice one would treat a weak carcinogen differently from one which is highly carcinogenic. Although not perfect, there is a good relation between mutagenic potency in the Ames test and carcinogenic potency in animal tests.[38] However, such relations have yet to be established in other test systems and this is an area where further work is required.
(6) Encouraging though many of the short-term tests are, there is a need to move toward more mammalian-cell tests and tests *in vivo*. A tendency to rely totally on systems *in vitro* and non-mammalian systems in the evaluation of compounds should be avoided since systems *in vivo* and mammalian-cell systems are obviously more valid than non-mammalian ones in terms of extrapolation to man.

It is most likely that, when fully developed and validated, short-term tests will detect many environmental carcinogens and they therefore present a most promising way of ensuring that cancer-causing agents do not pass in to our environment. However, this does not mean that such compounds will necessarily have to be eliminated from our environment. In some cases, for example carcinogens identified in industrial processes that represent large commercial and financial investments, the appropriate protection of workers may represent a satisfactory solution to the problem. Indeed, such action has recently been taken for workers in the

PVC industry once the carcinogenic nature of vinyl chloride monomer was appreciated.

If other indispensable chemicals are identified as carcinogens in the future, it should again be possible to cope with these compounds by reducing human exposure to the safest levels possible rather than discontinuing the use of the compound. This will require the adoption of acceptable sets of safety standards implemented by stringent regulations.

6.4 Monitoring for Human Exposure to Carcinogens

Human exposure to carcinogens (e.g. alkylating agents, benzene, vinyl chloride, etc.) is commonly reflected by an increased number of chromosome aberrations, micronuclei formation and sister chromatid exchanges in the lymphocytes of those so exposed. Such parameters have already been used to monitor DNA damage *in vivo* in workers exposed to compounds under suspicion.[39] However, there is still a need for more sensitive and quantitative tests. It is now generally accepted that the interaction of a carcinogen with DNA is the major step in the initiation of the carcinogenic process (see Chapter 5). Many carcinogens are alkylating agents or are metabolised to alkylating species and, since many DNA alkylation products are well characterised, it is tempting to consider the possibility of monitoring the formation of alkylated bases as a measurement of carcinogen exposure. In practice, however, DNA from target cells is difficult to obtain in quantity and most of the reaction products are removed from the nucleic acid by normal repair processes. However, although the reaction of carcinogens with DNA is the critical reaction, reaction of alkylating agents with proteins also occurs. This led to the suggestion that the extent of reaction of carcinogens with proteins may reflect the extent of the crucial DNA reactions.[40] The measurement of reaction products with proteins would provide a simple means for monitoring exposure and determining the associated risk. Unlike DNA, alkylating damage in proteins is not repaired and therefore lesions are permanent throughout the lifetime of the protein. Haemoglobin is a particularly suitable protein to study since it is readily available from blood and has a long halflife which permits exposure to be detected many weeks after its occurrence. Since alkylating agents attack nucleophilic centres, the amino groups of the N-terminal amino acid, the -SH group of cysteine and the NH- group of the imidazole ring of histidine are among the groups that are attacked in proteins. Although the amounts of alkyl and acyl amino acids formed after exposure to a carcinogen are extremely small, analytical methods have been developed for measuring alkylation at each of these three sites. The most suitable procedure involves the use of combined gas chromatography/mass spectrometry.

The protein is first hydrolysed to its constituent amino acids which are then derivatised, separated by gas chromatography and detected by mass spectrometry. This method is capable of detecting as little as 1 nmol of alkylated amino acid/g of protein.[41] To date, most experiments of this nature have been carried out on animals, with encouraging results. For example, exposure to simple directly acting alkylating agents (e.g. ethylene oxide or propylene oxide) does indeed correlate with the amount of alkylated cysteine or histidine in haemoglobin.[41,42] Human studies are being carried out by various research groups on workers exposed to low levels of vinyl chloride (a potent carcinogen) and ethylene, ethylene oxide, propylene and propylene oxide (all compounds of undetermined human carcinogenicity). Although insufficient data have yet to be collected for an adequate assessment of this method, it would seem to have considerable potential for monitoring human exposure to carcinogens.

References

1. Cairns, J. (1975) The Cancer Problem. *Sci. Am.*, Nov., 64-78
2. Doll, R. (1980) The epidemiology of cancer. *Cancer*, *45*, 2475-85
3. Davies, D.L. & Magee, B.H. (1979) Cancer and industrial chemical production. *Science*, *206*, 1356-8
4. Fishbein, L. (1977) *Potential Industrial Carcinogens and Mutagens* (Publ. 560/5-77-005) US Environmental Protection Agency, Washington, DC
5. *Carcinogenicity Testing of Chemicals*, S. Goldberg (ed.) CRC Press, Boca Raton, FL (1980)
6. Purchase, I.F.H., Longstaff, E., Ashby, J., Styles, J.A., Anderson, D., Lefevere, P.A. & Westwood, F.R. (1978) An evaluation of six short-term tests for detecting organic chemical carcinogens. *Br. J. Cancer*, *37*, 873-903
7. Hollstein, M., McCann, T., Angelosanto, F.A. & Nichols, W.W. (1979) Short-term tests for carcinogens and mutagens. *Mut. Res.*, *65*, 133-226
8. Garner, C. (1980) Assessment of *in vitro* methods of detecting mutagens and carcinogens. *Chem. Ind.*, 1 Nov., 844-7
9. *Strategies for Short-Term Testing for Mutagens/Carcinogens*, B. Butterworth (ed.) CRC Press, Boca Raton, FL (1980)
10. Bridges, B.A., Butterworth, B.E. & Weinstein, I.B. (1983) *Banbury Report 13 – Indications of Genotoxic Exposure*, Cold Spring Harbor Laboratory, New York
11. *Evaluation of Short-term Tests for Carcinogens. Progress in Mutation Research*, vol. 1, F.J. de Serres & J. Ashby (eds) Elsevier/North-Holland, Amsterdam (1981)
12. *Short Term Tests for Chemical Carcinogens*, H.F. Stich & R.H.C. San (eds) Springer-Verlag, Heidelberg (1981)
13. Devoret, R. (1979) Bacterial tests for potential carcinogens. *Sci. Am.*, *241*, 40-9
14. Ames, B.N., McCann, J. & Yamasaki, E. (1975) Methods for detecting carcinogens and mutagens with the Salmonella/mammalian-microsome mutagenicity test. *Mut. Res.*, *31*, 347-64
15. Ames, B.N. (1979) Identifying environmental chemicals causing mutations and cancer. *Science*, *204*, 587-93
16. Moreau, P., Bailone, A. & Devoret, R. (1976) Prophage induction in *Escherichia coli* K12 ENVA UVR B. *Proc. Natl. Acad. Sci. USA.*, *73*, 3700-4
17. Speck, W.T., Santella, R.M. & Rosenkranz, H.S. (1978) An evaluation of the prophage induction (inductest) for the detection of potential carcinogens. *Mut. Res.*, *54*, 101-4

18. Williams, G.M. (1979) Review of *in vitro* test systems using DNA damage and repair for screening of chemical carcinogens. *J. Assoc. Off. Anal. Chem.*, *62*, 857-63
19. Swenberg, J.A. & Petzold, G.L. (1980) The usefulness of DNA damage and repair assays for predicting carcinogenic potential of chemicals. In B. Butterworth (ed.) *Strategies for Short-term Testing for Mutagens/Carcinogens*, CRC Press, Boca Raton, FL, pp. 77-86
20. San, R.H.C. & Stich, H.F. (1975) DNA repair synthesis of cultured human cells as a rapid bioassay for chemical carcinogens. *Int. J. Cancer*, *16*, 284-91
21. Stich, H.F., Kieser, D., Laishes, B.A., San, R.H.C. & Warren, P. (1975) DNA repair of human cells as a relevant, rapid and economical assay for environmental carcinogens. *Gann Monogr. Cancer Res.*, *17*, 3
22. Agrelo, C. & Amos, H. (1981) DNA repair in human fibroblasts. In F.J. de Serres & J. Ashby (eds) *Evaluation of Short-term Tests for Carcinogens. Progress in Mutation Research*, vol. 1, Elsevier/North-Holland, Amsterdam, pp. 528-32
23. Kada, T. (1981) The DNA damaging activity of 42 coded compounds in the Rec-Assay. In F.J. de Serres & J. Ashby (eds) *Evaluation of Short-term Tests for Carcinogens. Progress in Mutation Research*, vol. 1, Elsevier/North-Holland, Amsterdam, pp. 175-82
24. Kada, T., Tulikawa, K. & Sadaie, Y. (1972) *In vitro* and host mediated 'rec-assay' procedures for screening chemical mutagens. *Mut. Res.*, *16*, 165-74
25. Leifer, Z., Hyman, J. & Rosenkranz, H. (1981) Determination of genotoxic activity using DNA polymerase-deficient and -proficient *E.coli*. In H.F. Stich & R.H.C. San (eds) *Short Term Tests for Chemical Carcinogens*, Springer-Verlag, Heidelberg, pp. 127-39
26. Rosenkranz, H. & Leifer, Z. (1980) Determining the DNA-modifying activity of chemicals using DNA-polymerase-deficient *E.coli*. In F.J.de Serres (ed.) *Chemical Mutagens: Principles and Methods for Their Detection*, vol. 6, Plenum Press, New York, pp. 109-47
27. Swenberg, J.A., Pelzold, G.L. & Harbach (1976) *In vitro* DNA damage/alkaline elution assay for predicting carcinogenic potential. *Biochem. Biophys. Res. Commun.*, *72*, 732-8
28. Kohn, K. (1978) DNA as a target in cancer chemotherapy. Measurement of macromolecular DNA damage produced in mammalian cells by anti-cancer agents and carcinogens. *Methods Cancer Res.*, *16*, 291-345
29. Stich, H.F. & Koropatruck, D. (1977) The adaption of short term assays for carcinogens to the gastrointestinal system. In E. Farber (ed.) *Pathophysiology of Carcinogenesis in Digestive Organs*, University of Tokyo Press, Tokyo/Park Press, Baltimore, pp. 121-34
30. Styles, J. (1977) A method for detecting carcinogenic organic chemicals using mammalian cells in culture. *Br. J. Cancer*, *36*, 558
31. Healey, P., Mawdsley-Thoms, L. & Barry, D. (1970) Short-term tests for evaluating potential carcinogenic activity of tobacco condensates. *Nature* (*Lond.*), *228*, 1006
32. Perry, P. & Evans, H. (1975) Cytological detection of mutagen-carcinogen exposure by sister chromatid exchange. *Nature (Lond.)*, *258*, 121-5
33. Evans, E. & Mitchell, A. (1981) Effects of 20 coded chemicals on sister chromatid exchange frequencies in cultured Chinese hamster cells. In F.J. de Serres & J. Ashby (eds) *Evaluation of Short-term Tests for Carcinogens. Progress in Mutation Research*, vol. 1, Elsevier–North-Holland, Amsterdam, pp. 538-50
34. Schmid, W. (1976) The micronucleus test for cytogenetic analysis. In A. Hollander (ed.) *Chemical Mutagens: Principles and Methods for the Detection*, vol. 4, Plenum Press, New York, pp. 31-53
35. Salamone, M., Heddle, J. & Katz, M. (1981) Mutagenic activity of 41 compounds in the *in vivo* micronucleus test. In F.J. de Serres & J. Ashby (eds) *Evaluation of Short-term Tests for Carcinogens. Progress in Mutation Research*, vol. 1, Elsevier/North-Holland, Amsterdam, pp. 686-97
36. Wyrobek, A. & Bruce, W. (1975) Chemical induction of sperm abnormalities in mice. *Proc. Natl. Acad. Sci. USA*, *72*, 4425-9
37. McCann, J., Choi, E., Yamasaki, E. & Ames, B. (1975) Detection of carcinogens as mutagens in the Salmonella/microsome test: assay of 300 chemicals. *Proc. Natl. Acad. Sci. USA*, *72*, 5135-9

38. Meselson, M. & Russell, K. (1977) Comparisons of carcinogenic and mutagenic potency. In H. Hiatt, J. Watson & J. Winsten (eds) *Origins of Human Cancer*, Cold Spring Harbor Laboratory, New York, pp. 1473-81
39. Hollstein, M., McCann, J., Angelosanto, F. & Nichols, W. (1979) Short-term tests for carcinogens and mutagens. *Mut. Res.*, *65*, 133-226
40. Ehrenberg, L. & Osterman-Golkar, S. (1980) Alkylation of macromolecules for detecting mutagenic agents. *Teratog. Carcinog. Mutagen.*, *1*, 105-27
41. Osterman-Golkar, S., Farmer, P.B., Segerback, D., Bailey, E., Calleman, C.J., Svensson, K. & Ehrenberg, L. (1983) Dosimetry of ethylene oxide in the rat by quantitation of alkylated histidine in hemoglobin. *Teratog. Carcinog. Mutagen.*, *3*, 395-405
42. Bailey, E., Connors, T.A., Farmer, P.B., Gorf, S.M. & Rickard, J. (1981) Methylation of cysteine in haemoglobin following exposure to methylating agents. *Cancer Res.*, *41*, 2514-17

Further Reading

Bridges, B.A., Butterworth, B.E. & Weinstein, I.B. (1983) *Banbury Report 13 – Indications of Genotoxic Exposure*, Cold Spring Harbor Laboratory, New York

Butterworth, B. (ed.) (1980) *Strategies for Short-term Testing for Mutagens/Carcinogens*, CRC Press, Boca Raton, FL

de Serres, F.J. & Ashby, J. (eds) (1981) *Evaluation of Short-term Tests for Carcinogens. Progress in Mutation Research*, vol. 1, Elsevier/North-Holland, Amsterdam

Goldberg, S. (ed.) (1980) *Carcinogenicity Testing of Chemicals*, CRC Press, Boca Raton, FL

Hollstein, M., McCann, T., Angelosanto, F.A. & Nichols, W.W. (1979) Short-term tests for carcinogens and mutagens. *Mut. Res.*, *65*, 133-226

Stich, H.F., & San, R.H.C. (eds) (1981) *Short Term Tests for Chemical Carcinogens*, Springer-Verlag, Heidelberg

7 MOLECULAR APPROACHES TO DIAGNOSIS OF CANCER

P. Thomas

Contents

7.1 Nature of the Problem

Many cancers can be controlled with existing methods of therapy provided that they are treated early enough. These cancers include the major killers, cancers of the lung, breast and gastrointestinal tract. For example, while no significant change has occurred in death rates from colorectal cancer over the last 50 years, the American Cancer Society has estimated that early diagnosis and prompt treatment could save a majority of the 50 000 Americans who die annually of the disease. Methods for early detection of cancers are therefore of great benefit to the patient and, while not providing a cure in themselves, make the patient more curable.

The two main areas of early detection are a general population screening for the presence of tumour and screening of patients at high risk. This includes the monitoring of patients who have been apparently successfully treated for cancer to detect recurrence and thereby to initiate further therapy.

7.1.1 Population Screening

The ideal tumour marker could be used as a screen for the general population and would detect only those who had cancer even in its earliest stages. These patients would be further investigated, treated and hopefully cured. Unfortunately, no such test exists, although many claims to early cancer tests have been made. Such screening is fraught with difficulty and some of the problems have been outlined by Bagshaw.[1] The main problems arise with false-positive tests. For example, in a test that is 90 per cent accurate, the false-positive rate (the percentage of people who do not have the disease but have a positive test) and the false-negative rate (the percentage of people who have the disease but have a negative test) are both 10 per cent. The false-negative results are acceptable as 90 per cent of people with cancer will be identified. The false-positive rate, however, creates far greater problems. If the prevalence of cancer is 2.5 per 1000 patients screened then a 10 per cent false-positive rate would give 39 false positives for every true positive. Thorough investigation of these patients would be time consuming, costly and may carry a risk to the healthy subjects. For a general screen this false-positive rate is clearly not acceptable. Nevertheless, some population screens for patients at risk are felt to be useful. The Pap test for cervical cancer in women, the stool Guaiac test for the detection of occult blood from colorectal cancer and the education of the public to the signs and symptoms of cancer, should all show their effects in decreased death rates or prolonged survival times. The yearly biochemical blood or urine test for cancer, however, still eludes us.

7.1.2 Detection of Tumour Recurrence

It is in the area of detecting residual tumour after surgery or therapy and in the earlier detection of tumour recurrence that biochemical markers make their greatest impact. These markers can serve to give the oncologist information about prognosis and the effectiveness of therapy. Probably the nearest to the ideal tumour marker available is human chorionic gonadotropin (HCG) used to monitor gestational cancers. α-Fetoprotein is another marker routinely used to monitor hepatoma and certain gestational cancers and carcinoembryonic antigen (CEA) is now the most widely used of the tumour markers and, while not always behaving ideally, can be used for monitoring a range of the most common cancers. Another example, acid phosphatase, is used for the investigation of cancer of the prostate. These four markers will be discussed later in greater detail to show the effectiveness of biochemical monitoring of cancer even in the presence of relatively high false-positive and false-negative rates. Many other examples of tumour markers appear in the cancer literature but most of these have yet to gain general acceptance as effective monitors of tumour progression or remission.

7.1.3 Limits of Tumour Detection

7.1.3.1 Physical Methods. These are usually classified as radiology and include X-ray, computerised tomography (CT scan), nuclear magnetic resonance (NMR scans) and various isotopic methods that include liver-spleen scan, bone scans, etc. In general the best resolution obtainable under ideal conditions is the detection of a tumour between 0.5 and 1 cm in diameter. This represents about 1 g of tissue or 10^{12} tumour cells. Advances in physical methods of detection occur all the time but it is unlikely that these limits of detection will be significantly improved on in the foreseeable future.

7.1.3.2 Biochemical Methods. A biochemical test, and this includes immunochemical assay procedures, should be the most sensitive way to detect the presence of a tumour. If a cancer produces a unique substance and this substance finds its way to urine or blood, and a test sensitive enough to detect nanogram quantities or less is available, then it would be theoretically possible to detect the prescence of a single tumour cell. However, to date no unique or tumour-specific substance has been found that allows this type of sensitivity or specificity. Tests such as those for HCG or CEA in plasma are capable of detecting small amounts of tumour, but many factors influence the levels of tumour markers (see section 7.2).

7.1.4 Ideal Tumour Marker

This would be a molecular substance produced by all tumour cells that distinguishes them from normal cells. Its production must be directly related to tumour mass and it must be found in sera or urine allowing a test suitable for automation to be produced. The test would detect cancer reliably and early enough for curative therapy. False positives or negatives would not occur. Even if these criteria are met there are still problems associated with such a general cancer test as the site of the tumour would still be unknown. More useful perhaps would be a battery of tests each obeying the above criteria but specific for a particular organ site such as lung, colon or breast.

7.2 Biochemical Methods for Cancer Detection

The biochemical monitoring of cancer has become a practical proposition since the development of highly specific methods for measuring substances in biological fluids. The major advance was the development of the radioimmunoassay (RIA) and later the enzyme-linked immunoassay (EIA) methods. These procedures are capable of quantitatively detecting down to picogram (10^{-9}g) per millilitre amounts of substances in biological fluids provided that specific antibodies are available. The RIA relies on the competition for binding to the antibody between the purified substance that has been radiolabelled (in the case of proteins such as HCG or CEA with ^{125}I) and the substance in the biological fluid. Unbound radiolabel is separated from bound radiolabel and the amount of bound radioactivity is inversely proportional to the amount of material in the sample. The EIA works slightly differently. The antibody is bound to a solid support (e.g. a nylon bead) and is incubated with the sample. Any antigen present binds to the antibody. A second antibody conjugated to an enzyme (often peroxidase) is incubated with the bead and reacts with the bound antigen. Incubation of the bead with a chromogenic substrate for the enzyme results in colour development which is proportional to the amount of antigen in the sample. A related procedure used in tumour diagnosis is immunohistochemical staining of tissue sections to detect specific antigens. This involves incubating a paraffin-embedded tissue section with a specific antibody followed by incubation with an anti-antibody conjugated to an enzyme (again often peroxidase) and then the section is incubated with a substrate that gives an insoluble coloured product. The deposition of the product on the section indicates the presence of the antigen. Diagrammatic representations of the RIA, EIA and immunoperoxidase procedures are shown in Figure 7.1. Many variations of these procedures exist and are used to increase sensitivity, decrease background, etc. (More detailed accounts of these procedures are given; see Refs 2–4.)

How well do existing methods for detection of cancer stand up to the criteria of the ideal marker? A number of examples of useful tumour monitors will be discussed in the next section. A list of these markers and some of their properties are shown in Table 7.1.

Table 7.1 Some Biochemical Tests in Clinical Use for Detection and Monitoring of Cancer

Substance (in serum)	Structure	Mol. wt.	Method of assay	Use
Human chorionic gonadotropin (HCG)	Glycoprotein	46 000 subunit 16 000 subunit 30 000	RIA, EIA	Gestational cancers
Carcinoembryonic antigen (CEA)	Glycoprotein	180 000	RIA, EIA	Wide range, including cancer of colon, breast, lung, pancreas and ovary
α-Fetoprotein (AFP)	Glycoprotein	70 000	RIA, EIA	Hepatoma, gestational cancers
Acid phosphatase	Glycoprotein	102 000	Spectrophotometric RIA, EIA	Prostatic cancer
Calcitonin	Peptide	3500	RIA	Medullary cancer of the thyroid, breast cancer?
β_2-Microglobulin	Protein	11 800	RIA, EIA	Lymphoma, multiple myeloma

RIA = radioimmunoassay; EIA = enzyme-linked immunoassay

7.2.1 Hormones

7.2.1.1 Human Chorionic Gonadotropin (HCG). This glycoprotein hormone was the first useful tumour marker to be described and is probably the most specific. It is usually detected in serum by RIA and is used in the management of patients with trophoblastic tumours. However, its specificity still falls far short of our ideal tumour marker as it is found in large quantities throughout the three trimesters of pregnancy and its measurement forms the basis of early pregnancy tests. It is also produced by some non-trophoblastic tumours as an ectopic product.

Figure 7.1 Diagrammatic Representations of Assay Methods used in Tumour Diagnosis

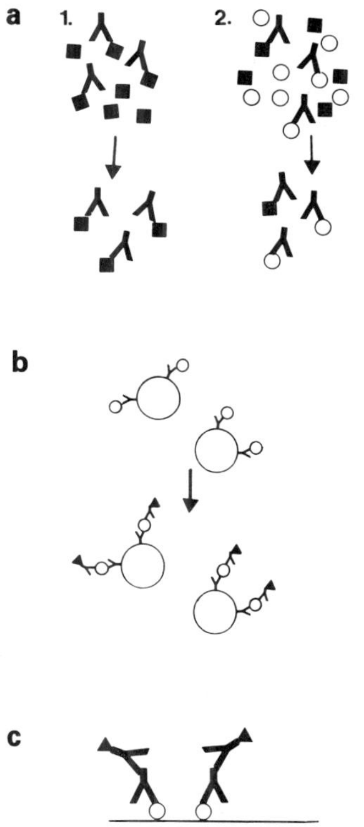

(**a**) Radioimmunoassay. (1) Antibody (⅄) is incubated with radiolabelled antigen (■) and bound antigen is separated from free antigen usually by precipitation with a second antibody, 50 per cent saturated ammonium sulphate or polyethylene glycol. (2) The procedure is repeated in the presence of the solution to be tested containing unlabelled antigen (○). Free antigen is separated from bound as before. The reduction in the number of counts bound can be compared with a standard curve constructed from known concentrations of unlabelled antigen to determine the antigen concentration in the unknown sample. (**b**) Enzyme immunoassay. Antibody (⅄) bound to a solid support, e.g. a nylon bead, binds antigen (○). The bead is incubated with a second antibody (conjugated to an enzyme, e.g. peroxidase or alkaline phosphatase) which also binds to the antigen (X). The bead is incubated with a substrate for the enzyme to develop a colour which is read spectrophotometrically. The amount of antigen present in a sample is proportional to the colour developed. By substituting a radiolabelled second antibody for the antibody enzyme conjugate this system can also be used as a radioimmunoassay. (**b**) Immunohistochemical staining. Antigen (○) present on a tissue section on a slide is incubated with antibody (⅄) and then with an anti-antibody enzyme conjugate (peroxidase, alkaline phosphatase or glucose oxidase) (X). The section is then incubated with a substrate for the appropriate enzyme that produces an insoluble coloured product. Microscopic examination of the section will then reveal the presence and location of the antigen in the tissue. There are a number of variations of this procedure, including incubating the section directly with an antibody conjugate against the required antigen, and a number of procedures for amplifying the reaction (see Ref. 3).

Nevertheless it has proved to be extremely useful in screening patients at risk for choriocarcinoma and those requiring treatment can be easily identified. For example, a persistently elevated HCG level in a young woman without the other signs of pregnancy would be highly suspicious for malignancy. The concentration of HCG in serum also reflects the tumour burden. This has been demonstrated both in human patients with cancer and in nude mice xenografted with human choriocarcinoma. Serum HCG is thus a good predictor of prognosis in patients with trophoblastic tumours and mortality rates tend to increase with elevations in HCG. Serum HCG may also be used for the primary diagnosis of testicular cancer and choriocarcinoma. The other major use of serum HCG measurements is in monitoring these cancers after therapy. Serum HCG measurements are able to detect residual tumour well below the volume detectable by radiological methods. Immunoperoxidase staining of, for example, testis for HCG can reflect biochemical changes in the cells preceding morphological changes and it is therefore useful in the identification of testicular tumours.

7.2.1.2 Calcitonin and Other Hormones. Calcitonin is a peptide hormone produced by the thyroid gland and is detected in serum by RIA. This hormone is used to monitor a specific but rare tumour, medullary carcinoma of the thyroid. It was from such a tumour that the hormone was first isolated. However, its specificity is not limited to thyroid cancer; other tumours, particularly carcinoma of the breast, have been reported to produce the hormone ectopically.

Other hormones have also been put forward as possible monitors of tumour progression. Peptide hormones such as adrenocorticotropic hormone (ACTH) and growth hormone can be produced by tumours of the adrenal and pituitary respectively. Again these tumours are rare and the hormones may also be produced ectopically by other cancers. Except in specific cases therefore these hormones are not useful as general monitors of cancer. When they are produced ACTH and growth hormone may manifest themselves clinically by producing the syndromes associated with their excess production.

7.2.2 Fetal Antigens

7.2.2.1 Carcinoembryonic Antigen (CEA). Carcinoembryonic antigen (CEA) is the most widely used test for cancer. Both RIA and EIA procedures are available commercially for its measurement. It is also arguably the most useful of the tumour markers in that it is produced by the most common cancers, those of the colon, breast, lung and pancreas. CEA is a glycoprotein with a mol. wt. of approx. 200 000; 50–60 per cent of the molecule by weight is carbohydrate. Its clinical usefulness and the factors affecting its concentration in plasma have been reviewed recently by Zamcheck.[5] He refers to CEA as the prototype cancer marker. A discussion of the factors affecting its clinical use will be applicable to other glycoprotein markers both present and future. Since its discovery by Gold and Freedman in 1965 CEA has been the most studied of the tumour markers. It is found in all cancers of epithelial cell origin, in fetal gut and in the normal colon. Initially CEA was thought to be tumour-specific and a potential screen for colon cancer. This was found not to be the case as patients with inflammatory conditions of the bowel, liver disease and smokers were found to have elevated plasma levels. Also only 50–60 per cent of patients with cancer of the colon have elevated levels and the percentage positive with breast and lung cancer is even lower. CEA is now used largely as an aid to diagnosis but more importantly to follow patients' responses to therapy. After surgical removal of a colon cancer, for example, an initially elevated CEA can fall to normal. Following such a patient on the basis of CEA can detect recurrences of the tumour sometimes as much as 18 months before other clinical symptoms appear. Some surgeons have used increasing CEA levels as the basis for a second operation (second look). Tumour is usually found in such cases but is often inoperable. CEA can also be used to monitor patients receiving chemo- or radio-therapy and in general a decrease in plasma CEA indicates a response.

The reasons for the 'false positives' and 'false negatives' have been a subject of much study. It is only recently that the factors affecting the circulating levels of CEA have been defined.[5,6] The concentration of any substance in the plasma depends not only on the rate at which it enters but also on the rate at which it leaves. When the rate of entry equals the rate of exit then a steady-state concentration is achieved. Many factors affect CEA entry into the circulation, including the following. (1) Differentiation of the tumour; well-differentiated tumours produce the largest amounts of CEA and undifferentiated ones may produce no CEA at all. (2) Amount of tumour. (3) Amount of invasion of normal tissue. This allows access to the blood supply. Under normal circumstances CEA produced by the colon is secreted into the lumen and very little enters the circulation. Only when CEA-producing cells invade normal tissue can the CEA be secreted in any quantity into the blood. (4) Site of tumour. If a metastasis is present at a highly vascular site, e.g. the liver, then not only does the tumour have a better supply of nutrients but access of secreted products to the blood supply is increased. Once CEA enters the blood stream, however, it is rapidly removed by the action of the liver. Rapid uptake by the liver macrophages (Kupffer cells) occurs followed by transfer to the parenchymal cells where it is degraded; both these processes are receptor-mediated. A small amount of CEA also passes directly into the bile by a paracellular mechanism. Recently, it has been shown that variations in the carbohydrate structure of CEA can affect hepatic clearance.[7] CEA containing large amounts of sialic acid is not removed from the circulation as efficiently as those with lesser amounts. These many factors can be used to explain some of the extraordinary variations in CEA levels that can occur in patients with similar tumour loads.

The detailed investigations carried out on CEA should be followed for all new tumour markers if their use is to be optimised. Many false claims have been made regarding the usefulness of tests for cancer. In the colorectal area a new test should be superior to or directly complement the CEA assay before large sums of money are spent on its development.

Over the almost 20 years since the first report on CEA, many new developments in tumour diagnosis have occurred including the use of antibodies to tumour markers for radioimmunodetection and therapy. The initial studies in these areas used CEA as their model or prototype marker and these studies will be discussed in more detail later in the chapter. Many reports of potentially useful cancer diagnostic tests now use CEA for comparison and will probably continue to do so. CEA while being far from our ideal tumour marker is, nevertheless, the most useful that is currently available for the detection and monitoring of the common cancers.

7.2.2.2 α-Fetoprotein (AFP). α-Fetoprotein is a sialoglycoprotein of mol. wt. approx. 70 000. It is present in fetal and neonatal sera and in amniotic fluid of many species. It is elevated in sera of pregnant women and when grossly elevated can be a sign of fetal abnormalities. RIA is available for its detection. It is present only at low levels in adult sera, although it is the major serum protein in the fetus. AFP is found elevated in the sera of a majority of patients with hepatocellular carcinoma. Like CEA in colonic cancer serum levels of AFP increase with the progress of liver cancer and fall on surgical removal of the tumour. AFP is also elevated in patients with testicular teratocarcinoma but not seminoma and thus compliments the HCG test. AFP is synthesised by the fetal liver and is also synthesised by the adult regenerating liver. As with CEA, elevations of AFP have been reported in liver disease including cirrhosis and hepatitis. Many of the cautions over the use of CEA will also apply to AFP. The detailed mechanisms by which AFP enters and is removed from the circulation have not been elucidated. Nevertheless, the measurement of this oncofetal antigen is very useful in monitoring both liver and testicular cancers.

7.2.2.3 Other Antigens. Over the past 20 years many other substances, largely glycoproteins or glycolipids, have been advocated as serum tests for cancer. Few of them have had any proven clinical utility. Some are still being actively studied, for example the pancreatic oncofetal antigen

(POFA), which shows some specificity for pancreatic cancer,[8] and the gross cystic disease fluid protein (GCDFP-15), a glycoprotein with a subunit of 15 000 daltons which can be used for monitoring patients with metastatic breast cancer.[9] Recently two commercial tests utilising monoclonal antibodies have been marketed. While not approved for clinical use they are being sold for research purposes. One of these, CA 19-9, is a test for gastrointestinal cancer that is designed to complement the CEA assay and detects certain carbohydrate structures, specifically sialyated lacto-*N*-fucopentaose II found on certain glycoproteins and glycolipids. Paired with CEA the CA 19-9 RIA has shown an increased capability for prediction of recurrence and can permit improved prognostic monitoring. The second test CA 125 can be used to monitor the course of epithelial ovarian cancer and detects a high-molecular-weight glycoprotein that is expressed in coelomic epithelium during embryonic development. Measurement of this glycoprotein in sera has been shown to aid the monitoring of response to therapy of patients with epithelial ovarian cancer.[10]

With the interest of industry in diagnostic tests and in monoclonal antibody technology we can expect more tests to be available in the future. They will all need to be examined very carefully against properly staged patients and compared with known diagnostic aids such as CEA before their worth can be fully assessed.

7.2.3 Enzymes

7.2.3.1 Prostatic Acid Phosphatase. The relation between increased serum acid phosphatase and prostatic adenocarcinoma has been known for almost 50 years and the prostate produces an acid phosphatase distinct from that found in other cells. A number of different assay procedures have been used for its measurement. The colorimetric assay for enzyme activity with substrates such as *p*-nitrophenol phosphate has some problems in that no substrate is specific for the prostatic isoenzyme. However, the prostatic enzyme is sensitive to tartrate, whereas the other acid phosphatases are not. Nevertheless, the use of tartrate to inhibit prostatic acid phosphatase is only semispecific. More recently immunoassay procedures have become available for the measurement of the prostate-specific enzyme. There is, however, a false-negative rate of 20 per cent and thus acid phosphatase cannot be used independently to diagnose cancer of the prostate. A variety of other cancers can also release an acid phosphatase that is similar in antigenic properties and enzyme activity to the prostatic acid phosphatase; however, the serum levels of this enzyme are seldom very high. A further source of false-positive results are patients with benign conditions of the prostate. Even with these limitations for screening or diagnosis, serial acid phosphatase measurements in patients with cancer of the prostate are useful in determining response to therapy and it is now well established that the measurement of acid phosphatase is essential to the work-up of patients with suspected prostatic carcinoma. A recent review[11] described in greater detail the chemistry and the application of acid phosphatase to prostatic cancer.

7.2.3.2 Glycosyltransferases and Other Enzymes. Because of the changes that are known to occur on the surfaces of cells during malignant transformation (see Chapter 2) and the possible role of sialic acid in the production of metastases, a number of sugar-transferring enzymes have been studied as possible tumour markers. These include serum galactosyltransferase isoenzyme II for pancreatic cancer and serum sialytransferase and fucosyltransferase for breast cancer. None of these enzymes is useful for early detection though they may have some clinical utility in patient monitoring. Other enzymes are used for specific purposes; alkaline phosphatase and glutamyltranspeptidase will give some indication of the presence of bone or liver metastases. More recent work has shown that measurement of the lysosomal enzyme β-*N*-acetylhexosaminidase may be useful in distinguishing benign from malignant extrahepatic biliary obstruction. These enzymes are not cancer-specific and can only be used in conjunction with other clinical evidence. (For a review dealing more fully with enzymes in cancer see Ref. 12.)

7.2.4 Miscellaneous Proteins

Many other plasma and urinary proteins have been used to attempt to diagnose or monitor cancer. Urinary Bence-Jones proteins are used for diagnosis of multiple myeloma.[13] Serum ferritin has been used as a marker for breast and other cancers.[14] Acute-phase glycoproteins, α_1-acid glycoprotein, C-reactive protein and haptoglobin have also been advocated for monitoring of cancer.[15] Tissue polypeptide antigen (TPA) has been used to monitor patients with a variety of cancers and Holyoke and Chu[16] concluded that TPA is as sensitive as CEA for most tumours except colon cancer. β_2-Microglobulin is increased in patients with lymphoma and multiple myeloma and its serum levels can be related to tumour load, prognosis and disease activity. None of these examples are specific for cancer and a diagnosis cannot be based on their serum or urine levels alone. As long as the limitations of these markers are appreciated they can all be useful in certain situations and some have found wide enough clinical application (not exclusively in the cancer area) that commercial kits for their measurement are available utilising both EIA and RIA methodologies.

Monoclonal antibodies produced to surface antigens of leukaemia cells have proved useful in grouping leukaemias into various subsets. This information can be used in deciding the most appropriate therapy and can also have prognostic significance.[17]

7.2.5 Other Tests

A number of other tests that do not involve measuring a protein in body fluids are used clinically in the detection and management of cancer. The measurement of polyamines such as spermidine and putrescine have some diagnostic value. They are not a general test for cancer even though they may be an index of cell proliferation. They may be measured by using an amino acid analyser and their measurement in cerebrospinal fluid is used in the diagnosis of medulloblastoma.

A very useful and simple test for colon cancer is the detection of occult or hidden blood in the stool. This is a general screening test for populations at risk, for example patients over 50 years old, patients with a history of polyps and women with a history of genital or breast cancer. Colon cancer is usually asymptomatic in its early stages and one of the earliest indications is small amounts of blood in the stool. The Guaiac paper test is an inexpensive way of detecting occult blood. The basis of the test is the oxidation of phenolic compounds present in 'Guaiac' (guaiaconic acids) to quinones resulting in a blue colour according to the following scheme:

1. Haemoglobin + $H_2O_2 \rightarrow 2H_2O + O_2$
2. O_2 + Guaiac (colourless) $\rightarrow$ oxidised Guaiac (blue)

This reaction is possible because the haematin portion of haemoglobin can function as a pseudoenzyme (similar to peroxidase). This test is not specific for cancer and false positives may be obtained for many reasons including red meat in the diet, bleeding haemorrhoids, other diseases of the gastrointestinal tract such as gastritis, diverticulitis and bleeding ulcers. Certain medications and improper specimen collection can also cause false positives. Occasional false-negative tests may also be obtained. This test is only used therefore for populations at risk. Further diagnostic procedures are performed on every patient with a positive test. In spite of these limitations the stool Guaiac test has become generally accepted and its use has probably saved or extended many thousands of lives.

7.3 Radioimmunolocalisation of Cancer

A more recent development in the detection of cancer has been to use radiolabelled antibodies against tumour-associated antigens to localise tumours within the patient. The method involves injection of radiolabelled (^{131}I) antibody and a search for the tumour by using external scintiscanning. The first successful studies were carried out in humans using antibodies to CEA by Goldenberg and his colleagues.[18] Initial experiments in animals had shown that human colonic cancers established as heterografts in hamsters could be visualised by this procedure. The localisation of tumours in humans is not as clear cut as in experimental animals. Tumour deposits can only be detected if allowance is made for non-tumour radioactivity by injecting a second non-localising marker (^{125}I-labelled immunoglobulin) and the two images obtained subtracted by computer. Varying success with these methods has been reported but generally both primary and secondary tumours can be visualised provided that they are greater than 1–2 cm in diameter. The presence of even large amounts of antigen in the circulation seems to make little difference to the success of the method. The main emphasis with radioimmunolocalisation has been to use CEA, AFP or HCG as the target antigens. Other targets are now being studied and the use of monoclonal antibodies should result in an expansion of these studies. The few investigations with monoclonal antibodies to CEA in patients have given similar results to targeting with polyclonal antisera. However, studies with monoclonals in experimental animal systems have shown some advantages over the polyclonal antisera, including better tumour to normal tissue ratios of radioactivity. A great deal of effort is also being expended to improve the resolution of the procedure and a number of approaches are being used. Changing the radioactive isotope from ^{131}I to ^{123}I because of its better dosimetry, or ^{111}In because of its suitability for detection by conventional gamma-cameras and intracellular accumulation, may give better resolution. Improvements are being made in the scanning technique and in methods for background radioactivity subtraction. Attempts have been made to more rapidly clear unbound antibody from the circulation by using free or liposome-entrapped second antibody. Antibody fragments (Fab) have been used and can localise tumour as effectively as whole antibody. New antigen targets are being sought and one productive area of study is the determination of the type of antigen target most suited for localisation. Would better success be obtained by using a membrane-bound antigen rather than secreted substances such as CEA, AFP or HCG? As well as being used for tumour localisation, antibodies against tumour-associated antigens are being used to target therapeutic agents to the cancer. This will be discussed in detail in Chapter 10. The field of immunolocalisation and that of immunotargeting of therapy is in its infancy and major advances in both areas are likely to occur over the next few years.

7.4 The Future

The examples of clinically useful tests for cancer described above demonstrate the lack of specificity in cancer detection. However, these tests used properly are all useful and if specificity were the sole criterion there would be no tests for cancer. No doubt in the future the search for specific tests will continue and with the expanding use of monoclonal antibodies and the use of recombinant DNA technology the chances of finding tumour-specific molecules (assuming such exist) are better than ever before.

The recent discovery of oncogenes and their products (see Chapter 4) should cause a great expansion in the effort to determine if these proteins can be used for early detection of cancer or for diagnosis of premalignant states. More effective antibodies to new markers, possibly membrane-bound, should improve radioimmunolocalisation. Similarly the use of human monoclonal antibodies in place of the mouse monoclonals now in general use should reduce the problems of immune responses to the injected antibodies. Research on the presently available markers such as CEA will also continue with the aim of improving their use. Studies of the factors affecting their plasma concentrations could lead to ways of increasing their levels in blood perhaps by blocking their metabolism. These studies could lead to earlier detection of recurrence. Research in cancer detection has expanded greatly over the past ten years and should continue to expand. Further advances in this area should have a substantial effect on survival rates for many of the common cancers.

References

1. Bagshawe, K.D. (1983) Tumour markers – Where do we go from here? *Br. J. Cancer*, *48*, 167-75
2. Harris, C.C., Yolken, R.A. & Hsu, I.-H. (1982) Enzyme immunoassays: applications in cancer research. In H. Busch & L.C. Yeoman (eds) *Methods in Cancer Research*, vol. 20, Academic Press, New York, pp. 213-43
3. Primus, F.J. & Goldenberg, D.M. (1982) Functional histopathology of cancer: a review of immunoenzyme histochemistry. In H. Bush & L.C. Yeoman (eds) *Methods in Cancer Research*, vol. 20, Academic Press, New York, pp. 139-82
4. Hunter, W.M. (1973) Radioimmunoassay. In D.M. Weir (ed.) *Handbook of Experimental Immunology*, Blackwell Scientific Publications, London, pp. 2-36
5. Zamcheck, N. (1981) The expanding field of colorectal cancer markers: CEA the prototype. *Cancer Bull.*, *33*, 141-51
6. Thomas, P. & Zamcheck, N. (1983) Role of the liver in clearance of excretion of circulating carcinoembryonic antigen (CEA). *Dig. Dis. Sci.*, *28*, 216-24
7. Thomas, P., Toth, C.A., Byrn, R.A., Saravis, C.A., Loewenstein, M.S. (1983) Five patients with markedly elevated CEA blood levels and impaired clearance of circulating CEA. *Clin. Res.*, *31*, 696A
8. Gelder, F.B., Reese, C.J., Moossa, A.R., Hall, T. & Hunter, R. (1978) Purification, partial characterization and clinical evaluation of a pancreatic oncofetal antigen. *Cancer Res.*, *38*, 313-24
9. Haagensen, D.E. (1981) Biochemical relationship between gross cystic disease and breast carcinoma. In C.D. Haagensen, C. Bodian & D.E. Haagensen (eds) *Breast Carcinoma Risk and Detection*, W.B. Saunders, Philadelphia, pp. 300-38
10. Bast, R.C., Klug, T.L., St John, E., Jenison, E., Niloff, J.M., Lazarus, H., Berkowitz, R.S., Leavitt, T., Griffiths, C.T., Parker, L., Zurawski, V.R. & Knapp, R.C. (1983) A radioimmunoassay using monoclonal antibody to monitor the course of epithelial ovarian cancer. *N. Engl. J. Med.*, *309*, 883-7

11. Choe, B.-K. & Rose, N.R. (1982) Prostatic acid phosphatase: a marker for human prostatic adenocarcinoma. In H. Busch & L.D. Yeoman (eds) *Methods in Cancer Research*, vol. 19, Academic Press, New York, pp. 199-232
12. Balinksy, D. (1980) Enzymes and isoenzymes in cancer. In S. Sell (ed.) *Cancer Markers*, Humana Press, Clifton, NJ, pp. 191-224
13. Soloman, A. (1980) Monoclonal immunoglobulins as biomarkers of cancer. In S. Sell (ed.) *Cancer Markers*, Humana Press, Clifton, NJ, pp. 57-87
14. Marcus, D.M. & Zinberg, N. (1975) Measurement of serum ferritin by radioimmunoassay, results in normal individuals and patients with breast cancer. *J. Natl. Cancer Inst.*, *55*, 791-5
15. Cooper, E.H. & Stone, J. (1970) Acute phase reactant proteins in cancer. *Adv. Cancer Res.*, *30*, 1-44
16. Holyoke, D. & Chu, T.M. (1979) Tissue polypeptide antigen. In R.B. Herberman & K.R. McIntire (eds) *Immunodiagnosis of Cancer*, Marcel Dekker, New York, pp. 513-21
17. Blackstock, R. & Humphrey, G.B. (1982) Surface marker in the characterization of leukemias. In H. Busch & L.D. Yeoman (eds) *Methods in Cancer Research*, vol. 19, Academic Press, New York, pp. 3-51
18. Goldenberg, D.M, Deland, F. & Kim, E. (1978) Use of radiolabeled antibodies to carcinoembryonic antigen for the detection and localization of diverse cancers by external photoscanning. *N. Engl. J. Med.*, *298*, 1384-8

8 HORMONES AND CANCER

M.G. Rowlands

Contents

8.1 Introduction

Hormones are chemical 'messengers' that are synthesised in specialised body tissues called endocrine glands. In response to certain stimuli, hormones enter the circulation and are carried to their target tissues. In the target tissues, specific metabolic activities are stimulated or inhibited by the hormones, producing the required response to the original stimuli. On a structural basis, hormones fall into three categories, namely amino acid derivatives (thyroid hormones and catecholamines), peptide hormones [luteinising hormone (LH) and follicle-stimulating hormone (FSH)] and the steroid hormones (sex hormones and glucocorticoids). In addition, hormones differ in their mechanism of action. The steroids and thyroid hormones interact with their target tissues by entering the cells, whereas peptides and catecholamines act at the cell surface.

At present, there is no evidence to suggest that hormones can initiate the transformation of a normal human cell into a tumour cell. However, it has long been known that certain cancers are hormone-dependent. The association between endocrine organs and abnormal or neoplastic growth in humans was recognised before hormones were discovered. During the late nineteenth century, castration was reported as a means of reducing the size of enlarged prostate glands in men[1] and removal of the ovaries (oophorectomy) caused regression of breast tumours.[2]

The involvement of endocrine function, in particular, the action of the steroid hormones, in malignant disease arises because the tumours developing from hormone-responsive tissue retain the ability of the parent cell to respond to hormones. The most common tumours to exhibit this hormone dependency are those of the reproductive system, such as the breast and prostate.

The traditional methods of treating such endocrine-related cancers has been either the surgical ablation of the endocrine glands or the addition of steroid hormones, such as oestrogens, to antagonise the action of the endogenous hormones. Ablative techniques are associated with the usual complications of major surgery and significant mortality and morbidity, whereas administration of oestrogens leads to fluid retention and cardiovascular problems.

During the last two decades, renewed interest in endocrine therapy has arisen due to two developments. First, the elucidation of the biosynthetic routes and mechanism of action of steroid hormones has helped to define the role of these hormones in tumour growth. For example, the discovery of oestrogen receptors in breast cancer, has led to a greater understanding of the disease and conferred selectivity on the administration of endocrine therapy. Secondly, new agents have been discovered that offer effective and safer alternatives to the conventional methods of treatment. An example is the anti-oestrogen, tamoxifen (see later) that elicits similar clinical responses in patients with advanced breast cancer as the traditional therapy of oophorectomy, but without the morbidity of surgery and with minimal toxicity.

This chapter will describe the classical hormone-dependent cancers and the methods of treatment both past and present. Finally, some of the directions that endocrine therapy may progress in the future will be discussed.

8.2 Biosynthesis and Mechanism of Action of Steroid Hormones

The five classes of steroid hormones are glucocorticoids, mineralocorticoids, androgens, oestrogens and progestins. The structure of the chief naturally occurring member of each class is shown in Figure 8.1. All steroid hormones are derived from cholesterol by a series of enzymic reactions, which either cleave fragments of the molecule at certain positions or introduce hydroxyl groups at several sites on the steroids. The major steps in steroidogenesis are shown in Figure 8.2.

The glucocorticoids, synthesised by the adrenal cortex, are concerned with the metabolism of lipids, proteins and carbohydrates and also have an anti-inflammatory function. Because glucocorticoids cause involution of lymphoid tissue, they are employed in the treatment of leukaemias and lymphomas. Mineralocorticoids are also produced in the adrenal cortex and they function in the regulation of transepithelial sodium transport. The androgen, testosterone, is predominantly secreted by the testis, although a small proportion of circulating androgens derive from the adrenals. The hormone is responsible for the maintenance of the male sex characteristics. Its target tissues are the prostate and seminal vesicles. Within prostatic tissue testosterone is converted by the 5α-reductase enzyme into 5α-dihydrotestosterone, which is the active androgen within the cell.

The female sex hormones, oestrogens and progestins, are secreted by the ovaries in premenopausal women. They are required for the growth and maintenance of the breast, vagina and uterus and the development of the secondary sexual characteristics. Progestins prepare the body for

Figure 8.1 Structural Formula of Major Classes of Steroid Hormones [A = cortisol (glucocorticoids); B = aldosterone (mineralocorticoids); C = progesterone (progestins); D = oestradiol (oestrogens); E = testosterone (androgens)]

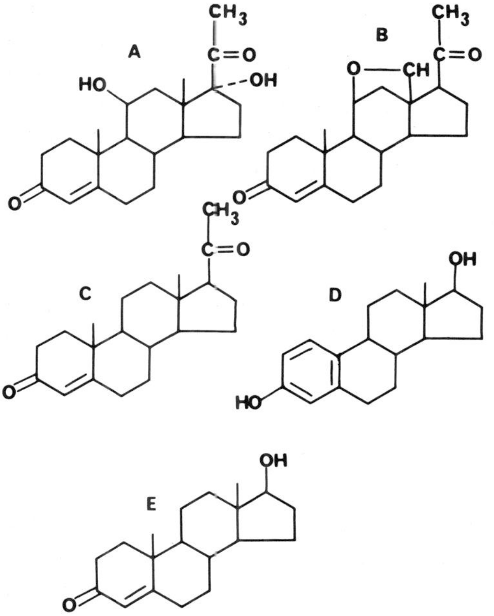

Figure 8.2 Pathways of Steroidogenesis

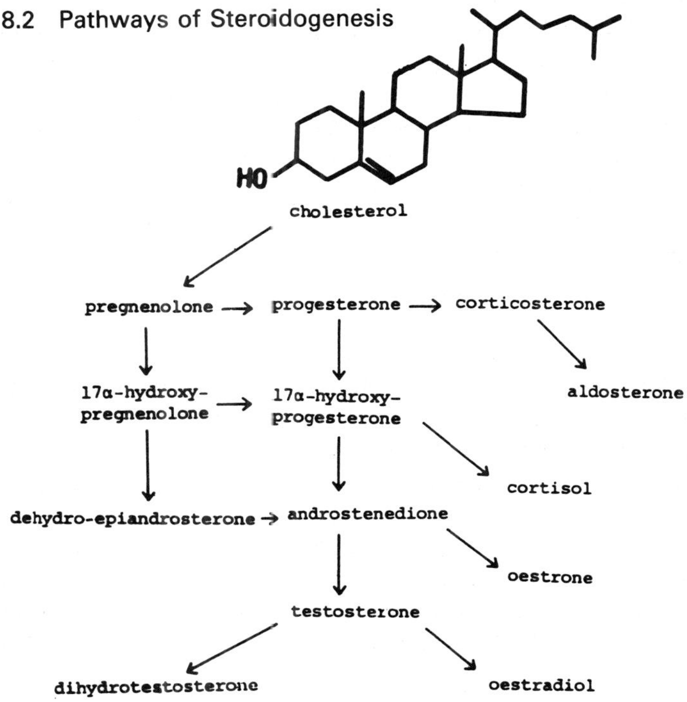

pregnancy. In postmenopausal women, ovarian sex-hormone production is insignificant and androgens produced by the adrenals are converted into oestrogens in peripheral tissues, such as muscle and adipose tissue. Furthermore, the breast-tumour cell possesses the ability to synthesise oestrogens.

Gonadal synthesis of steroid hormones is regulated by the peptide hormones, luteinising hormone (LH) and follicle-stimulating hormone (FSH), secreted by the anterior region of the pituitary. These in turn are controlled by the luteinising-hormone releasing hormone (LHRH) peptide produced by the hypothalamus. Secretion of LHRH, LH and FSH is inhibited by increased blood levels of steroid hormones so that a negative-feedback system operates to control testicular and ovarian steroidogenesis.

There appears to be a similarity in the basic mechanism of action whereby the different classes of steroid hormone induce their specific actions on target tissues. The steroids are released from the relevant endocrine gland into the circulation and bind to specific plasma transport proteins. These carrier proteins or steroid-hormone binding globulins (SHBG) appear not to be involved in the expression of hormone action, because the biological potency of the gonadal steroids is inversely related to the fraction present in serum bound to SHBG. They may have some role in protection against fluctuations in hormone levels and they may contribute to the metabolism of the hormones. Therefore, the uptake of steroid into the cell depends on the concentration of free steroid in the plasma as opposed to the concentration of steroid bound to protein.

In the early 1960s evidence accumulated to support the concept of the presence of steroid-hormone receptors within the target tissues. Experiments with radioactive-labelled steroids demonstrated that these lipophilic molecules penetrated the cells of all tissues, but were retained only in the target tissues. This selective retention of hormones was demonstrated by using tritiated oestrogens in rat uterus and vagina,[3] human mammary tumours[4] and experimental mammary tumours induced in rats by the administration of the carcinogen, dimethylbenzanthracene.[5] The receptors are protein molecules found free in the supernatant fraction of the cell. They are localised in the hormone's target tissues and possess high binding affinity for the respective steroid. On binding, the hormone causes significant changes to the protein receptor. This temperature-dependent activation process is not clearly understood, but may involve a conformational change of the receptor or the loss or gain of a protein subunit. The result of this activation is the translocation of the hormone-receptor complex to the nucleus and its interaction with the DNA-chromosomal protein complex (chromatin). This leads to a stimulation of DNA transcription and subsequently nuclear RNA synthesis. The nuclear RNA is processed, in a mechanism which is unclear, to produce messenger RNA, that is transported to the cytoplasm for translation on ribosomes to new protein products. These protein products elicit the response induced by the steroid hormone.[6] A view of the mechanism of steroid-hormone action is shown in Figure 8.3. This model for the action of steroid hormones was based on work with oestrogens, but the evidence supports the view that this basic mechanism is applicable to most steroids. The above model is very simplistic and many questions remain unanswered. For example, multiple binding sites with differing affinities have been discovered for 17β-oestradiol in target cells, but the function of each site is unknown. It is unresolved whether the activation of the steroid-receptor complex occurs in the cytoplasm or in the nucleus. Unoccupied nuclear receptors have been detected. Of importance to cancer therapy is the fact that some cells containing receptors are unresponsive to the appropriate hormone. This unresponsiveness may be due to the structurally defective receptor or a lack of the translocation step into the nucleus. Much effort is being applied to answering these questions and obtaining a better understanding of steroid-hormone action.[7,8]

Figure 8.3 Mechanism of Steroid-hormone Action

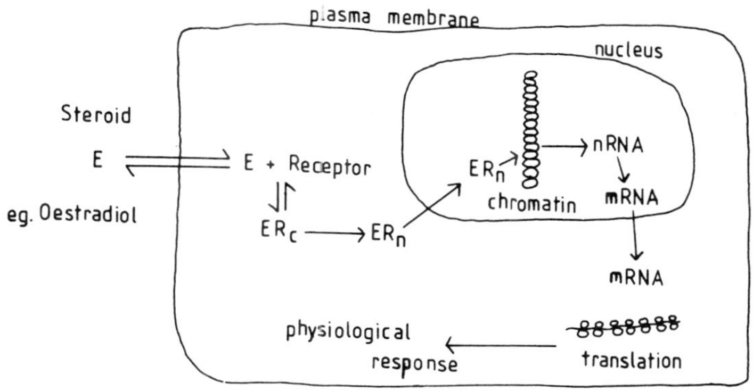

ER_c = cytoplasmic oestrogen-receptor complex; ER_n = nuclear oestrogen-receptor complex; nRNA = nuclear RNA; mRNA = messenger RNA.
Details are described in the text.

8.3 Breast Cancer

The demonstration by Beatson in 1896 that oophorectomy in premenopausal patients with breast cancer with advanced disease caused significant regression of the tumour was promptly confirmed.[2] However, the response was of a short duration, up to a year, and only about one-third of patients responded. At that time, surgeons were searching for a curative treatment rather than palliation, so oophorectomy was not widely adopted. In the early 1940s renewed interest developed in endocrine therapy when Huggins showed the benefit of orchidectomy for men with metastatic prostate carcinoma and oophorectomy, either by surgery or radiation, became standard treatment for breast cancer in premenopausal women.[9] Recurrence was common and the failure to cure the disease was thought to be due to another organ taking over the role of the ovary in secreting the hormones. This was suggested to be the adrenals. Recent work has shown that the adrenals synthesise androgens that are converted into oestrogens in other tissues of the body, including the tumour itself. At the time, synthetic glucocorticoids were unavailable, so the removal of the adrenals would have led to an Addisonian crisis due to insufficiency of the adrenal hormones, such as glucocorticoids. By 1952 replacement glucocorticoids were available and adrenalectomy was shown to produce good remissions in postmenopausal women. Hypophysectomy was undertaken to suppress oestrogen and adrenal-hormone synthesis in oophorectomised women. In addition, animal experiments had shown that the growth of some mammary tumours may be dependent on the pituitary hormones, growth hormone and/or prolactin. However, the clinical results achieved with hypophysectomy were similar to those with adrenalectomy.

As an alternative to these ablative procedures, additive hormone therapy was employed. In the 1940s tumour regression in patients was observed after the administration of pharmacological doses of androgens, such as testosterone, to antagonise the action or synthesis of oestrogens. Long-term use of the active male hormone leads to virilising side-effects. Haddow and co-workers in 1944 showed that large doses of oestrogens could cause tumour regression.[10] The most commonly used synthetic oestrogen is diethylstilboestrol, but the mechanism of action is unknown. Remission rates were around 35 per cent, but prolonged use led to uterine bleeding, cardiovascular toxicity, nausea and vomiting. With all these procedures, either additive or ablative, the response rate remained about one in three patients and the response was of a temporary nature.

Therefore, in the past, the conventional sequence of endocrine therapy for breast cancer was oophorectomy for premenopausal patients and additive therapy for postmenopausal patients. Patients responding to these first-line treatments were considered for bilateral adrenalectomy or hypophysectomy as second-line treatment on relapse. Such therapeutic options have lengthened the survival of patients that respond but, as described, they are all associated with considerable morbidity.

Within the last decade, the discovery of the oestrogen receptor within hormone-dependent primary breast-tumour cells and the development of effective non-toxic hormonal drugs has revolutionised the treatment and understanding of breast cancer. The inability in the past to identify potential responders to endocrine therapy has meant exposing a large proportion of patients to the risks of additive or ablative procedures without any therapeutic benefit. In seeking to avoid these side-effects and problems, increasing attention is being paid to the receptor status of breast tumours as a basis for patient selection. Steroid-receptor assays are based on the binding of a radioactive-labelled steroid to tissue slices or cytosol proteins. The oestrogen receptor is measured either by sucrose-density-gradient centrifugation or by a dextran-coated-charcoal technique. Both provide reliable and sensitive assay systems.

Analysis of the primary tumours has shown that 50–60 per cent of all tumours contain measurable quantities of the oestrogen receptor.[11] Patients whose tumours contained little or no oestrogen receptor rarely responded to endocrine therapy, whereas 60 per cent of patients with receptor-positive tumours responded. The remaining third of receptor-positive tumours that failed to respond to hormone manipulation may not contain a fully functional oestrogen-receptor system. Although the oestrogen receptor is the most widely used assay system, at the time of writing, attention has also focused on the progesterone receptor as a means of selecting tumours that will respond to endocrine therapy. Measurement of the progesterone receptor may provide a more accurate marker of hormonal responsiveness, because synthesis of this receptor within reproductive tissues depends on oestrogen stimulation, therefore it is a product of an intact oestrogen-receptor system. Patients whose tumours contain both oestrogen and progesterone receptors have an 80 per cent response rate to endocrine therapy. At present, the assay for progesterone receptors is not as sensitive or reliable as the technique for the oestrogen-receptor determination; however, much effort is being applied to rectify this situation.

Nowadays, the most commonly used hormonal agents for advanced breast cancer are tamoxifen (Nolvadex) and aminoglutethimide (Orimeten) (see Figure 8.4). Aminoglutethimide was originally designed as an anticonvulsant in the late 1950s, but was withdrawn in 1966 because adrenal insufficiency occurred on prolonged administration. This toxic action proved to be a blockage of adrenal steroidogenesis mainly by inhibition of the enzymic conversion of cholesterol into pregnenolone. This reaction is catalysed by the cytochrome P- 450-dependent desmolase (cholesterol side-chain cleavage) enzyme complex. Consequently, aminoglutethimide was suggested as an alternative to adrenalectomy for treating metastatic breast disease in postmenopausal patients. Because the inhibition is at an early stage in the steroid biosynthetic pathway (see Figure 8.2), the production of glucocorticoids is suppressed, and therefore the reflex rise in ACTH caused by blockage of cortisol synthesis overcomes the drug-induced adrenal suppression. Prevention of this adrenal escape phenomenon is achieved by simultaneous administration of replacement glucocorticoids. Clinical trials, with a regimen of aminoglutethimide and hydrocortisone, showed an objective response rate of 32 per cent in unselected patients and 52 per cent in patients with oestrogen-receptor-positive tumours. The mean duration of response was about 18 months. The most frequent responses were in patients with bone and soft-tissue metastases, and, in addition, many patients obtain pain relief. The side-effects of the drug are dose-dependent and include lethargy, dizziness and skin rash. Clinical results are similar to those achieved with surgical adrenalectomy or hypophysectomy, but without the morbidity and significant mortality (4–15 per cent) associated with the ablative techniques. Further biochemical studies demonstrated that aminoglutethimide inhibits other cytochrome P-450-dependent steroid-hydroxylation enzymes. The most potent action of the drug was shown by Thompson and Siiteri to be on the aromatase enzyme complex which converts androstenedione and testosterone into oestrone and oestradiol respectively (see Figure 8.2).[12] This is now assumed to be the clinically relevant site of action of aminoglutethimide. In postmenopausal women the main source of oestrogens is provided by the aromatase enzyme in the peripheral tissues. On the other hand, aminoglutethimide is ineffective for premenopausal women due to its inability to suppress ovarian aromatisation.[13,14]

Figure 8.4 Structural Formula of Drugs used for Endocrine Therapy of Breast Cancer [A = aminoglutethimide (Orimeten); B = tamoxifen (Nolvadex); C = danazol (Danol)]

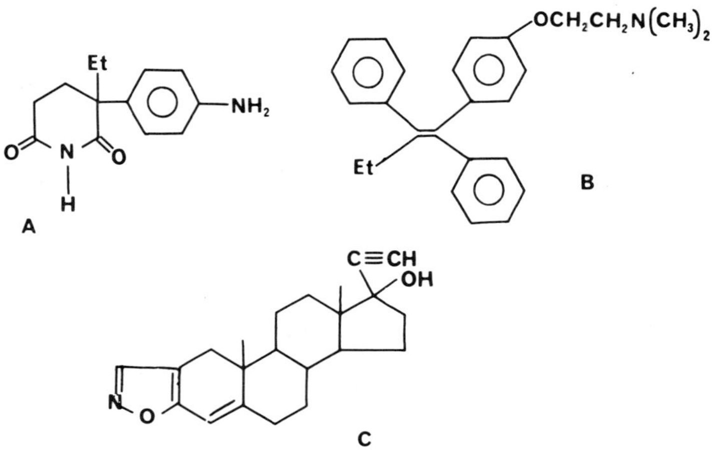

Tamoxifen was not specifically designed for cancer therapy, but was one of a group of non-steroidal anti-oestrogens developed as potential antifertility drugs. The discovery of the oestrogen receptor in breast tumours focused attention on the importance of oestradiol for tumour viability and the known capability of anti-oestrogens to block oestradiol binding stimulated their evaluation for advanced breast cancer. The initial clinical trials were in 1971 and the drug is now in widespread use. An objective response rate of 35 per cent was found in unselected postmenopausal patients and 60 per cent for oestrogen-receptor-positive tumours. Responses were most frequent in soft-tissue metastases. The mean duration of response was around 20 months. In addition, tamoxifen gives response rates in premenopausal women with breast cancer that are similar to oophorectomy. Side-effects of the drug are mild and uncommon. Occasional nausea and vomiting, oedema, weight gain, vaginal bleeding and hot flushes are encountered. Tamoxifen competes with oestradiol for the high-affinity cytoplasmic oestrogen receptor and binds to this receptor. The resulting receptor-anti-oestrogen complexes translocate into the nucleus and tend to remain there much longer than the oestradiol-receptor complexes, thereby suppressing the replenishment of cytosol receptors and rendering the cells less responsive to oestradiol. Synthesis of DNA and mitosis are suppressed and cell death occurs. Tamoxifen elicits responses in some oestrogen-receptor-negative tumours and the intracellular accumulation of tamoxifen is in excess over the receptor concentration. Tamoxifen may therefore influence receptor-independent functions. Furthermore, an oestrogen-independent anti-oestrogen binding site, distinct from the oestrogen receptor, has been isolated, but its function remains uncertain.[7,15,16]

Because of its low toxicity, tamoxifen is widely employed as the initial endocrine therapy for postmenopausal women with advanced breast cancer. Aminoglutethimide is regarded as second-line therapy because a significant proportion of postmenopausal patients who relapse on tamoxifen then go into remission with aminoglutethimide. This does not apply when aminoglutethimide is used as primary treatment. Both agents are being evaluated in adjuvant situations with surgery for treatment of primary breast cancer and the early results appear encouraging.

8.4 Prostate Cancer

In the 1940s Huggins and his associates demonstrated that deprivation of the androgenic hormones by orchidectomy had therapeutic benefit for patients with metastatic prostate cancer.[9] Furthermore, he showed that the adrenals were a secondary source of androgens and adrenalectomy, as is the case with advanced breast cancer, was useful in treating castrated men. In parallel with this work, the administration of large doses of oestrogens, in particular the synthetic oestrogen, diethylstilboestrol, was shown to counteract the endogenous androgens and offer an alternative to the ablative techniques. After 40 years, orchidectomy and oestrogen therapy remain the standard regimens for treating prostate cancer. With either procedure, the rate of response is around 50 per cent with a duration of about 18 months. However, as with all endocrine therapy, recurrence of the tumour often happens. Both normal and cancerous prostate glands synthesise large quantities of the enzyme, acid phosphatase. Metastases of the tumour lead to elevated blood levels of the enzyme and a favourable response to therapy is indicated by decreased levels of the enzyme. Both therapies are associated with considerable side-effects. Orchidectomy has the morbidity of surgery and psychological trauma, and oestrogen therapy is associated with problems of cardiovascular disorders, gynaecomastia, nausea and vomiting. Some toxicity can be alleviated by lower dosages, but gynaecomastia remains a considerable problem. The demonstration that adrenalectomy was beneficial in castrated men with advanced prostatic cancer prompted clinical trials with aminoglutethimide as a medical adrenalectomy. The most recent trial reported a 38 per cent objective response with a mean duration of eight months.[17] Progestational agents, such as medroxyprogesterone acetate and megestrol acetate (see Figure 8.5), have been employed in a small number of trials. Both agents lowered plasma testosterone levels and were well tolerated. However, despite some remissions, further information is required before these alternative treatments can be assessed.

Figure 8.5 Structural Formula of Progestins [A = megestrol acetate (Megace); B = medroxyprogesterone acetate (Provera)]

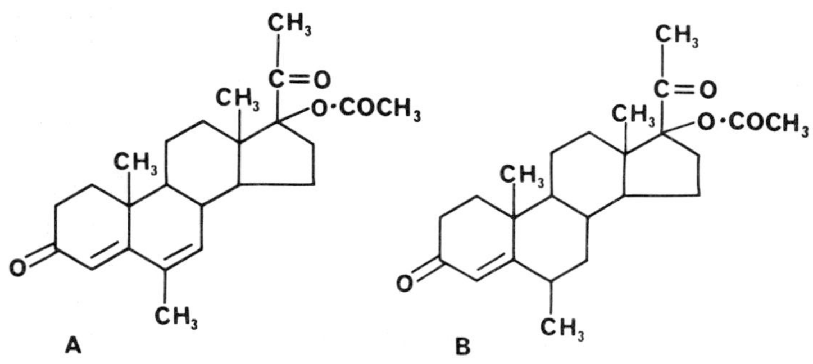

8.5 Endometrial Cancer

The relation between oestrogen and endometrial cancer dates back to the early 1950s when the long-term administration of oestrogens to rabbits and guinea pigs was shown to induce endometrial cancer. Furthermore, the oestrogenic action could be reversed by administering progesterone. Kelley and Baker in 1961 were the first to report the beneficial effects of progestins on patients with endometrial cancer.[18] By 1975 epidemiological surveys on oestrogen replacement therapy for postmenopausal women had demonstrated an increased incidence of endometrial cancer.

The majority of patients respond to the initial treatment of radiotherapy or surgery, but between 30 and 40 per cent develop recurrent disease and this group is treated with progestational agents. About one-third of patients with advanced disease obtain remissions, from this therapy, for periods from one to three years, but some cases are in remission for longer periods. The progestins are well tolerated and very few side-effects are observed. The drugs are commercially available in either oral preparations, e.g. megestrol acetate (Megace), medroxyprogesterone acetate (Provera) (see Figure 8.5) and norethindrone or as injections, e.g. medroxyprogesterone acetate (Depo-Provera). With the latter, abscesses sometimes occur at the site of injection, but generally the drugs have little adverse toxicity.

Until recently, such treatment had been empirical with little understanding of the underlying mechanisms of hormone action. These drugs have a number of biological effects, not all of which are understood. Consequently, there is no simple explanation for the anti-tumour actions of these agents. Progestins bind to the progesterone receptor in target tissues and this impairs the regeneration of the oestrogen receptor and stimulates the 17β-oestradiol dehydrogenase enzyme resulting in an increased intracellular breakdown of oestradiol. These actions lead to an antagonistic effect on the oestradiol stimulation of target tissues. Increased dosages lead to direct interactions with oestrogen, androgen and glucocorticoid receptors. In addition, the pituitary gonadotropins (LH and FSH) are decreased as well as plasma levels of cortisol and ACTH, but no adrenal insufficiency is apparent. At high doses, the progetins have a direct cytotoxic action. A beneficial side-effect of progestin therapy is that a significant number of patients obtain pain relief and experience a feeling of well-being. The mechanism of these effects is unknown[19]

The uterine endometrium is a target tissue for oestrogen and progesterone. Therefore the appropriate steroid receptors can be isolated from this tissue. The measurement of oestrogen-receptor levels in breast cancer has proved to be clinically useful in the management of the disease. The initial results suggest that determination of steroid receptors may be of importance in the treatment of endometrial cancers.[20]

8.6 Endocrine Therapy for Other Cancers

Progestational agents and oestrogen therapy are used to treat ovarian cancers, but the results have been disappointing. Surgery with adjuvant chemotherapy remains the standard treatment. However, the importance of oestrogen-receptor determinations to the management of breast cancer and the intensive research into hormone receptors in endometrial cancer has prompted an investigation into the receptor status of ovarian tumours. Both oestrogen and progesterone receptors have been detected in ovarian-cancer tissue and several clinical trials are in progress with tamoxifen to see if it has any therapeutic benefit.

In recent years, a number of non-gynaecological cancers have been shown to be hormone-dependent. In 1950 Kirkman and Bacon induced the formation of hormone-dependent renal cancer in hamsters by the administration of oestrogens.[21] Tumour formation was inhibited by the addition of progesterone or androgens. Receptor proteins for oestradiol, progesterone and dihydrotestosterone were detected in the animal renal cancers. Recently, all three receptors have been found in tissue samples from human renal cancers. This provides a molecular basis for the hormone dependency. Unfortunately, clinical results with progestins and androgens have been disappointing, with objective responses around 15 per cent in cases of advanced disease.[22] Chemotherapy and radiotherapy yield similar results. More information is required before the role of hormone therapy in renal cancer can be elucidated.

Another non-gynaecological tumour that possesses characteristics of hormone dependency is that of the pancreas. High levels of oestrogen receptors have been demonstrated in pancreatic-carcinoma tissue.[23] In addition, the activity of the aromatase enzyme, that converts androgens into oestrogens, and the 5α-reductase enzyme, responsible for converting testosterone into dihydrotestosterone, the more active androgen, are elevated in cancerous tissue.[24] This evidence suggests that pancreatic carcinoma may respond to endocrine therapy.

In the late 1930s glucocorticoids were shown to cause cell death to lymphoid tissue. Now, glucocorticoids are widely used in the treatment of many leukaemias and lymphomas. For some of these diseases very successful clinical results have been achieved. Lymphoid cancers are sensitive to cytotoxic agents, so the glucocorticoid (usually prednisone) is administered with a cytotoxic drug or combination of such agents. Unlike the sex steroids, glucocorticoids act on a wide variety of tissues so that a large number of complications may arise with such therapy. One advantage is that no bone-marrow depression is produced. However, the main side-effects are immunosuppression, water retention and hypertension. At present there is no general agreement on the value of glucocorticoid-receptor determinations for predicting the clinical response.[25]

8.7 Developments in Hormonal Agents

Most of the hormonal agents at present in clinical use were selected on an empirical basis. However, the elucidation of the role of the steroids in tumour growth and the discovery of hormone receptors has prompted the design of more potent specific hormonal drugs.

The aromatase enzyme complex has a key role in oestrogen production, so that the development of aromatase inhibitors offers a new approach to breast-cancer treatment. Aminoglutethimide is a relatively weak reversible inhibitor of aromatase activity, with additional inhibitory action on other steroidogenic enzymes. Intensive research has gone into the design of potent irreversible inactivators of aromatase activity. The most effective inhibitors are analogues of androstenedione, the natural substrate for the enzyme. In particular, 4-hydroxyandrostenedione is a specific irreversible inhibitor of the aromatase enzyme, three times as potent as aminoglutethimide, which causes an 80 per cent rate of tumour regression in the dimethylbenzanthracene-induced rat mammary tumour.[26] This drug is undergoing clinical evaluation in the Royal Marsden Hospital, Surrey, on postmenopausal patients with advanced breast cancer.

Unlike breast cancer, little improvement has been made in the endocrine therapy of prostatic cancer. Orchidectomy and oestrogen therapy remain the mainstays of the treatment regimens. Limited clinical studies have been made on a number of steroidal and non-steroidal anti-androgens which offer advantages over the existing treatments as regards side-effects and therapeutic benefit. Flutamide is a non-steroidal agent (see Figure 8.6), that competes with androgens for binding to the

Figure 8.6 Structural Formula of Anti-androgenic Agents (A = cyproterone acetate; B = flutamide)

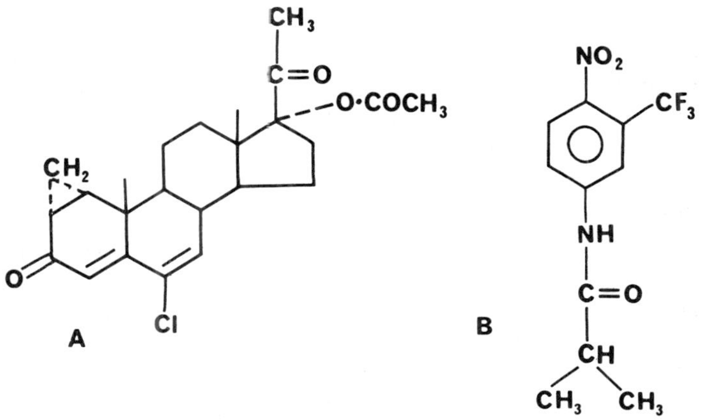

cytoplasmic androgen receptor, and suppresses all androgen-dependent functions. However, this also blocks the negative-feedback effect of androgens on hypothalamic-pituitary function, so LH secretion is stimulated which leads to an elevation of testosterone synthesis. Consequently the anti-androgen action of the drug decreases with time. However, no cardiovascular toxicity is associated with this agent and gynaecomastia was rarely observed.

The steroid, cyproterone acetate (see Figure 8.6), also binds to the androgen receptor and, in addition, it displays progestational and thus antigonadotropic properties which prevent the stimulation of testosterone synthesis associated with flutamide. Clinically, cyproterone acetate appears to be as effective as oestrogen therapy for objective responses, but lacks the adverse toxicity of oestrogen administration.[27]

Progestin therapy has long been established as an effective treatment for advanced endometrial cancer. Trials with conventional doses of progestins, such as medroxyprogesterone acetate, for other cancers gave unsatisfactory results. In the last decade high-dose progestin therapy has been demonstrated to give comparable rates of remission as the conventional endocrine treatments for advanced breast cancer. This is associated with little toxicity and the benefits of considerable pain relief and the feeling of well-being associated with administration of progestins. Trials have extended the use of high-dose progestins to ovarian, prostate and renal cancers with encouraging results.

The antigonadotropins are drugs that act primarily by suppressing the synthesis of the pituitary gonadotropins LH and FSH and this in turn leads to suppression of gonadal steroidogenesis. Danazol (see Figure 8.4) is a synthetic steroid with an antigonadotropic action, as well as having the ability to bind to androgen and progesterone receptors and inhibit several steroidogenic enzymes. It has been used for treating benign breast disease and endometriosis. Evaluation at the Royal Marsden Hospital, Surrey, showed that responses were less frequent than responses observed to tamoxifen and aminoglutethimide when used as first-line treatment.[28] At the time of writing, attention has focused on super-agonists of luteinising-hormone releasing hormone (LHRH), a decapeptide secreted from the hypothalamus. Pulsatile secretion of LHRH controls steroid-hormone synthesis from the gonads. Agonists of LHRH when given in high concentrations, and in a continuous manner, decrease concentrations of LH and FSH and therefore decrease gonadal steroid synthesis.[29] The compounds evaluated are buserelin and ICI 118630, polypeptides with modified residues. Both give responses in prostate cancer comparable with either orchidectomy or oestrogen therapy, but with far less morbidity than either treatment.[30,31] LHRH agonists are being evaluated in breast and endometrial cancers.

8.8 Future Trends

Over the last 40 years, endocrine therapy has been established as providing a 30–40 per cent objective response rate for a transient period in a number of gynaecological tumours. The indications are that in the future both the response rate and duration of response may be improved. The temporary nature of the remissions obtained by endocrine therapy was assumed to be due to a clone of cells within the tumour that are hormone-independent and resistant to such therapy. With the regression of the hormone-dependent cells the tumour takes on the characteristic of hormone independency. There is some evidence that this view is incorrect. For a long time, adrenalectomy was used as second-line treatment following relapse after oophorectomy, suggesting that the tumour cells retain some degree of hormone dependency. Patients with disseminated breast cancer who relapse on treatment with tamoxifen, grow an oestrogen-receptor-positive tumour that may respond to aminoglutethimide. Therefore sequential endocrine therapy may extend the duration of response. The use of drugs with little or no toxicity is advantageous for patients from the older age-group.

A trial at the Royal Marsden Hospital, Surrey, of tamoxifen, aminoglutethimide and danazol in combination endocrine therapy for advanced postmenopausal breast cancer patients has produced an objective response rate of 50 per cent[32] so that variations of these three drugs, together with high-dose progestin therapy either in sequence or combination, may offer the way forward for endocrine therapy of advanced breast cancer. The use of tamoxifen and the LHRH agonist for premenopausal patients appears promising and several trials are underway to explore the relevance of endocrine therapy as an adjuvant for surgery on patients with primary breast cancer.

Tamoxifen, aminoglutethimide and danazol were not designed for the treatment of breast cancer and in no case is the complete metabolism profile known. The precise mode of action of danazol is unclear, although tamoxifen and aminoglutethimide have relatively weak affinities for their target sites. Several research groups are studying the metabolism and mode of action of these three drugs with a view to the development of new improved analogues with enhanced anti-tumour activity.

Most of the advances in endocrine therapy have been made in the treatment of breast cancer. The last decade has seen the identification of a number of other tumours such as endometrial, ovarian and prostate that may respond to the established anti-endocrine drugs or the newer derivatives. Clinical trials are underway to discover their potency against these tumours and whether they can replace the traditional ablative or additive therapies.

Most of the previous work on steroid-hormone receptors has

concentrated on the role of oestrogen and progesterone receptors in breast cancer. Intensive research is going on into the study of these receptors in other gynaecological tumours such as ovarian and cervical. Other receptors remain to be explored, especially the androgen and glucocorticoid receptors. New methods are being devised to make receptor assays more accurate and reliable, in particular the development of monoclonal antibodies to steroid receptors has opened the possibility of utilising immunochemical techniques.

Advances in the study of hormone and receptor interactions will add greatly to the understanding of hormone-dependent tumours and the treatment of such disease processes.

References

1. White, J.W. (1895) The results of double castration in hypertrophy of the prostate. *Ann. Surg.*, *22*, 1-6
2. Beatson, G.T. (1896) On the treatment of inoperable cases of carcinoma of the mamma: suggestions for a new method of treatment with illustrative cases. *Lancet*, *ii*, 104-7
3. Jensen, E.V. & Jacobson, H.I. (1962) Basic guides to the mechanism of oestrogen action. *Rec. Prog. Horm. Res.*, *18*, 387-408
4. Folca, P.J., Glascock, R.F. & Irvine, W.T. (1961) Studies with tritium labelled hexoestrol in advanced breast cancer. *Lancet*, *ii*, 796-8
5. King, R.J.B., Cowan, D.M. & Inman, D.R. (1965) The uptake of [^{3}H]oestradiol by dimethylbenzanthracene induced rat mammary tumours. *J. Endocrinol.*, *32*, 83-90
6. O'Malley, B.W. & Mears, A.R. (1974) Female steroid hormones and target nuclei. *Science*, *183*, 610-20
7. Katzenellenbogen, B.S., Miller, M.A., Eckert, R.L. & Sudo, K. (1983) Antioestrogen pharmacology and mechanism of action. *J. Steroid Biochem.*, *19*, 59-68
8. Sledge, G.W. & McGuire, W.L. (1983) Steroid hormone receptors in human breast cancer. *Adv. Cancer Res.*, *38*, 61-75
9. Huggins, C., Stevens, R.E., Jr & Hodges, C.V. (1941) Studies on prostatic cancer. II. The effects of castration on advanced carcinoma of the prostate gland. *Arch. Surg.*, *43*, 209-14
10. Haddow, A., Watkinson, J.M., & Patterson, E. (1944) Influence of synthetic oestrogens upon advanced malignant disease. *Br. Med. J.*, *ii*, 393-8
11. McGuire, W.L., Pearson, O.H. & Segaloff, A. (1975) Predicting hormone responsiveness in human breast cancer. In W.L. McGuire, P.P. Carbone, E.P. Vollmer (eds) *Estrogen Receptors in Human Breast Cancer*, Raven Press, New York, pp. 17-29
12. Thompson, E.A., Jr & Siiteri, P.K. (1974) The involvement of human placental microsomal cytochrome P450 in aromatization. *J. Biol. Chem.*, *249*, 5373-8
13. Santen, R.J., Badder, E., Lerman, S., Harvey, H., Lipton, A., Boucher, A.E., Manni, A., Rosen, H. & Wells, S.A. (1982) Pharmacologic suppression of oestrogens with aminoglutethimide as treatment of advanced breast carcinoma. *Breast Cancer Res. Treat.*, *2*, 375-83
14. Santen, R.J. & Misbin, R.I. (1981) Aminoglutethimide: review of pharmacology and clinical use. *Pharmacotherapy*, *1*, 95-117
15. Jordan, V.C. (1982) Metabolites of tamoxifen in animals and man: identification, pharmacology and significance. *Breast Cancer Res. Treat.*, *2*, 123-38
16. Patterson, J.S. (1981) Clinical aspects and development of antioestrogen therapy: review of the endocrine effects of tamoxifen in animals and man. *J. Endocrinol.*, *89*, 67P-75P

17. Murray, R. & Pitt, P. (1983) Inhibition of adrenal steroid synthesis with aminoglutethimide in advanced prostatic cancer. *J. Steroid Biochem.*, *19*, 11s
18. Kelley, R.M. & Baker, W.H. (1961) Progestational agents in the treatment of carcinoma of the endometrium. *N. Engl. J. Med.*, *264*, 216-22
19. Kohorn, E.I. (1976) Oestrogens and endometrial carcinoma. *Gynecol. Oncol.*, *4*, 398-411
20. Vihko, R., Isotalo, H., Kuappila, A. & Vierikko, P. (1983) Female sex steroid receptors in gynaecological malignancies: clinical correlates. *J. Steroid Biochem.*, *19*, 827-32
21. Kirkman, H. & Bacon, R.L. (1950) Malignant renal tumours in male hamsters treated with oestrogens. *Cancer Res.*, *10*, 122-4
22. Bloom, H.J.G. (1973) Hormone-induced and spontaneous regression of metastatic renal cancer. *Cancer*, *32*, 1066-71
23. Greenway, B., Iqbal, M.J. Johnson, P.J. & Williams, R. (1981) Oestrogen receptor proteins in malignant and fetal pancreas. *Br. Med. J.*, *283*, 751-3
24. Iqbal, M.J., Greenway, B., Wilkinson, M.L., Johnson, P.J. & Williams, R. (1983) Sex-steroid enzymes, aromatase and 5α-reductase in the pancreas: a comparison of normal adult, foetal and malignant tissue. *Clin. Sci.*, *65*, 71-5
25. Bell, P.A. (1979) Assessment of the glucocorticoid responsiveness of human leukaemias and lymphomas. In B.A. Stoll (ed.) *Reviews on Endocrine-related Cancer*, vol. 4, ICI Pharmaceuticals, UK, pp. 5-12
26. Brodie, A.M.H., Garrett, W.M., Hendrickson, J.R.. Tsai-Morris, C.H. & Williams, J.G. (1983) Aromatase inhibitors: their pharmacology and application. *J. Steroid Biochem.*, *19*, 53-8
27. Neumann, F. & Jacobi, G.H. (1982) Antiandrogens in tumour therapy. In B.J.A. Furr (ed.) *Clinics in Oncology*, vol. 1, W.B. Saunders, Philadelphia, pp. 41-64
28. Coombes, R.C., Perez, D., Gazet, J.-C., Ford, H.-T. & Powles, T.J. (1983) Danazol treatment for advanced breast cancer. *Cancer Chemother. Pharmacol.*, *10*, 194-5
29. Nicholson, R.I. & Walker, K.J. (1983) Control of hormone dependent cancers of the breast and prostate by the LH-RH agonist ICI 118630. *J. Steroid Biochem.*, *19*, 83s
30. Allen, J.M. O'Shea, J.P., Mashito, K., Williams, G. & Bloom, S.R. (1983) Advanced carcinoma of the prostate: treatment with a gonadotrophin releasing hormone agonist. *Br. Med. J.*, *286*, 1607-9
31. Borgmann, W., Hardt, W., Schmidt-Gollwitzer, M., Adenauer, H. & Nagel, R. (1982) Sustained suppression of testosterone production by the luteinising-hormone agonist, buserelin in patients with advanced prostate carcinoma. *Lancet*, *i*, 1097-9
32. Powles, T.J., Gordon, C. Coombes, R.C. (1982) Clinical trial of multiple endocrine therapy for metastatic and locally advanced breast cancer with tamoxifen-aminoglutethimide-danazol compared to tamoxifen used alone. *Cancer Res.*, *42*, (Suppl.) 3458s-60s

Further Reading

Furr, B.J.A. (ed.) Hormone therapy. *Clinics in Oncology*, vol. 1, W.B. Saunders, Philadelphia

Jensen, E.V. (1981) Hormone therapy. In S.T. Crooke & A.W. Prestayko (eds) *Cancer and Chemotherapy*, vol. 3, Academic Press, New York, pp. 187-204

Stoll, B.A. (1981) *Hormonal Management of Endocrine Related Cancer*, Lloyd-Luke (Medical Books), London

9 RADIATION THERAPY

P. Workman

Contents

9.1 Introduction: the Clinical Problem

Radiation therapy involves the application of ionising radiation in the treatment of malignant disease. Its importance is illustrated by the fact that as many as 70 per cent of all patients with cancer will receive radiation therapy at some time during their treatment.

Treatment of cancer commonly requires eradication of both primary tumour and distant metastases. Because of serious damage to the bone marrow and intestinal mucosa, systemic radiotherapy (to the whole body or a major part of it) is not widely used today. There has, however, been some resurgence of interest in the technique, for example when it is combined with bone-marrow transplantation in the management of acute leukaemia. The main role of radiation therapy is in the treatment of localised tumour deposits. Although metastasis obviously presents a major problem which must be dealt with principally by systemic chemotherapy, it is often not realised that in many types of cancer, particularly tumours of the brain, head and neck and cervix, inability to control the growth of the localised primary tumour is in fact a major cause of treatment failure. Thus local control of tumours can lead to cure of the patient. A treatment designed to provide a high probability of local control is referred to as radical radiotherapy, and this must inevitably carry a small but finite probability of serious damage to normal tissues (see later). Where tumour cure cannot be obtained without unacceptable side-effects palliative radiotherapy may be administered. This might be used, for example, to delay the progress of more widespread disease or to reduce the size of the tumour deposits and so relieve symptoms such as pain. Depending on the circumstances, radiation therapy may be used alone or in combination with surgery, chemotherapy or hyperthermia.

Insufficient data are available to rank different types of human tumours of comparable size in terms of probability of local control for a given radiation dose. Radioresistance of human tumours is defined operationally on the basis of clinical experience: a tumour is termed resistant if it is not controlled locally by the highest radiation doses which can reasonably be given. The reasons for this radioresistance are complex (see Tubiana, Chapter 2 in Steel *et al.*, Further Reading) and some of these will be discussed in detail later. According to clinical experience Paterson in 1948 classified human tumours into three categories: radiosensitive ones such as lymphomas and seminomas; radioresistant ones, particularly sarcomas, melanomas and gastrointestinal tumours; those of intermediate sensitivity, such as adenocarcinomas and squamous carcinomas. The best example of a malignant tumour with a high probability of cure by radiation therapy would probably be a small basal-cell carcinoma of the skin.

To inhibit tumour growth it is necessary to destroy the reproductive potential (or 'sterilise') those cells capable of repopulating it ('stem cells'). Unfortunately, although it is true that differences in radiation sensitivity between different cell types do occur, there is no systematic difference in sensitivity between malignant cells and their normal counterparts.

It is important to make the distinction, however, between the radiosensitivity of individual cells and the radioresponsiveness of whole tissues or tumours. The latter is partly dependent on the former, but is also subject to complex factors such as cell proliferation kinetics, oxygenation, homeostatic repair processes and damage to blood vessels. Fortunately, therefore, the probability of achieving local tumour control can often exceed that of producing unwanted damage to normal tissues, the amount by which these probabilities differ being a measure of the 'therapeutic ratio'.

This important point is illustrated in Figure 9.1 which shows a plot of the radiation doses required to give different probabilities of tumour cure (0, 50, 100 per cent, etc.) as compared with the doses with the same probabilities of producing normal tissue necrosis. In this example, a cure rate of 85 per cent can be achieved with a dose of 6.5 Gy[a] if we accept a 10 per cent incidence of normal tissue damage. Necrosis rates of 3–5 per cent, or even 10 per cent in some sites, would normally be acceptable to the radiation therapist, these values being known as the normal tissue tolerance. Damage to normal tissues can become manifest over a very broad time-scale (weeks to years) and the types of damage are normally classified as 'early' or 'late' (see Chapter 3). Both types can limit the radiation dose; skin, lung, spinal cord and small bowel are the most common tissues concerned.

An additional point to note from Figure 9.1 is the steepness of the dose–response curves. This means that a small increment in dose can produce a considerable gain in tumour control or exacerbation of normal tissue necrosis.

In view of these considerations, it will be seen that the fundamental objective of radiation therapy, and indeed of therapy-orientated radio-

---

[a] The gray is the unit of absorbed dose; 1 gray (Gy) is defined as the absorption of 1 joule of energy per kilogram of material, e.g. tissue.

Figure 9.1 Schematic Diagram of Dose–response Curves for Tumour Cure and Normal-tissue Necrosis

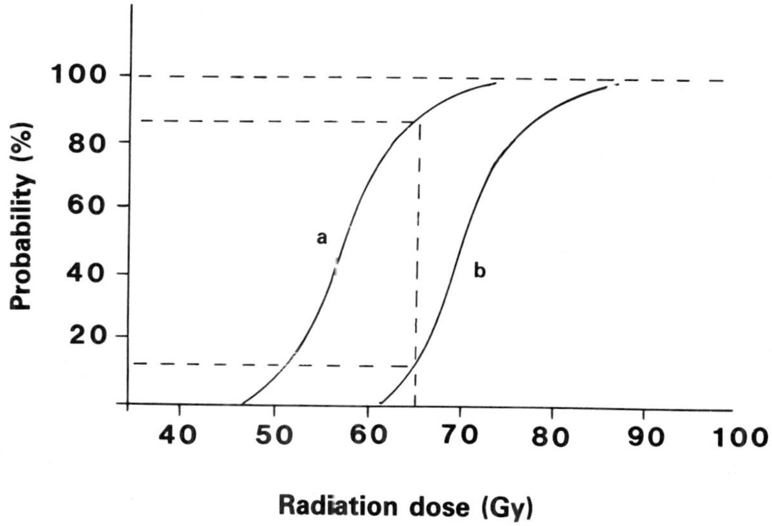

At a dose that gives an 85 per cent probability of tumour cure (**a**) the incidence of normal-tissue necrosis (**b**) will be 10 per cent. Redrawn from Duncan and Nias (see Further Reading)

biology research, is to deliver the maximum possible radiation dose to the tumour while at the same time minimising the dose to the surrounding normal tissues.

In the following pages the major effects of therapeutic radiation are highlighted, concentrating on biochemical events at the cellular and molecular levels.

9.2 Types of Radiation

Radiation is said to be ionising if it has sufficient energy to remove electrons from atoms or molecules. In therapy, X- and γ-rays are the most commonly used forms. These are electromagnetic radiations, being made up of streams of extremely high-energy photons.

Although produced by different methods, X-rays by accelerating fast electrons into a metal target (usually tungsten) and γ-rays from the decay of a radioactive isotope (usually cobalt-60), their biological properties are essentially identical. Most of this chapter will be concerned with them.

Other types of radiation used, or potentially usable, for radiation therapy are the particulate radiations, namely electrons, protons, α-particles, neutrons, negative π-mesons and heavy charged ions. These are known as particulate radiations, because of their particulate rather than electromagnetic nature, and have very different properties.

Radiations are classified as 'directly' or 'indirectly' ionising. The particulate radiations can damage directly the atomic structure of the absorbing material. In contrast, damage from electromagnetic radiations is mainly caused by the fast electrons released on absorption.

The more densely ionising particulate radiations are said to have a higher 'linear energy transfer' (LET) (see Chapter 3). The high LET radiations produce comparatively more biological damage than low LET radiations, and the ratio of doses to give the same effect is known as the 'relative biological effectiveness' (RBE).

9.3 Radiation Chemistry

Ionisation occurs randomly in irradiated material. Since cells are about 80 per cent water most of the ionisations involve water molecules. For this reason the radiation chemistry of water is very important. It is also surprisingly complex.[1,2] Breakdown products resulting from the interaction of ionising radiation and water include various ions and free radicals[b] that are short-lived and very reactive. Particularly important species for the production of biological damage are fast electrons (e^-) that can become hydrated by water molecules (e^-aq), the hydrogen atom $H^\bullet$ and the hydroxyl radical $OH^\bullet$. These can be generated as follows:

$$\begin{aligned}
H_2O &\xrightarrow{\text{Radiation}} H_2O^{+\bullet} + e^- \\
e^- + n\,H_2O &\longrightarrow e^-aq \\
H_2O^+ + H_2O &\longrightarrow OH^\bullet + H_3O^+ \\
H_2O &\xrightarrow{\text{Radiation}} H_2O^* \longrightarrow H^\bullet + OH^\bullet \\
e^-aq + H_3O^+ &\longrightarrow H^\bullet + H_2O
\end{aligned}$$

(H_2O^* represents the excited or energetic water molecule: this does not have sufficient energy to result in ionisation but can release the excitation energy by bond breakage to generate $H^\bullet$ and $OH^\bullet$.)

The two major radicals resulting from the radiolysis of water are e^-aq and $OH^\bullet$; these occur in roughly equal amounts, whereas the yield of $H^\bullet$ is about five times lower. The reactivity of these radicals is illustrated by

---

b. A free radical is a species with an unpaired electron in its outer shell; this is indicated by a bold superscript, e.g. the hydroxyl radical $OH^\bullet$ and $H^\bullet$ atom.

the fact that under conditions frequently used in clinical radiotherapy the steady-state concentration of free radicals in the treatment volume will be in the order of 10^{-15}M (see Wardman, Chapter 4 in Steel *et al.*, Further Reading.)

What happens when we consider organic solutes such as biomolecules in the cell? Chemical changes are produced in biomolecules by two separate mechanisms. The so-called 'direct effect' results from ionisation and excitation of biomolecules by the radiation impinging on them directly. The 'indirect effect', which predominates for X- and γ-rays, is due to effects of the reactive water species, described above. The $OH^\bullet$ radical and the $H^\bullet$ atom can react with organic solutes by abstraction (removal of an $H^\bullet$ atom):

$$RH + OH^\bullet\ (H^\bullet) \longrightarrow R^\bullet + H_2O\ (H_2)$$

or by addition

$$R + OH^\bullet\ (H^\bullet) \longrightarrow ROH^\bullet\ (RH^\bullet)$$

Abstraction is usually from a C–H bond and addition at olefinic and aromatic centres, both forming organic free radicals. The hydrated electron reacts by nucleophilic addition, e.g. to carbonyl groups, producing radical anions which then protonate:

$$R + e^{-}aq \longrightarrow R^{\bullet -} \xrightarrow{H^+} RH^\bullet$$

The fate of these short-lived organic free radicals is crucial to the ultimate fate of the biomolecule. The main reactions are dimerisation, dismutation, oxidation and reduction. Reaction with molecular oxygen produces organic peroxy radicals, e.g.

$$R^\bullet + O_2 \longrightarrow RO_2^\bullet$$

This prevents restitution which might otherwise occur through reduction by a thiol compound, e.g.

$$R^\bullet + R'SH \longrightarrow RH + R'S^\bullet$$

These reactions have important consequences for the 'oxygen effect' and radiosensitisation and protection (see later).

9.4 Radiation Biochemistry

So far we have dealt with the types of reactions that can lead to damaged biomolecules without specifying which cellular biomolecules constitute the 'target' for radiation killing. As in the development of modern molecular biology, early emphasis was on proteins, but in the last 20 years attention has switched to nucleic acids and DNA in particular (see

Dertinger and Jung, Further Reading). There has also been some interest in the role of membrane damage.

9.4.1 Proteins

Radiation can alter both the physicochemical and functional properties of proteins. Effects on enzymes have been studied in particular detail (see, e.g. Ref. 3). Physicochemical changes include fragmentation of the polypeptide chain, altered folding properties, cross-linking, aggregation and the destruction of individual amino acids. Side-chain groups are particularly labile. Functional changes include loss of enzyme activity.

While there is no doubt that radiation can produce severe damage to protein molecules, it is important to point out that the doses required are generally about ten times greater than those needed to sterilise cells. For example, doses of 1–2 Gy will sterilise two-thirds of a mammalian cell population, while tens of a gray, at least, are needed to produce appreciable inactivation of enzymes, particularly in concentrated solutions (see Figure 9.2).

Figure 9.2 Dose–response Curves for Inactivation of Purified DNAase by X-Rays, at Three Different Enzyme Concentrations (**a**, 5, **b**, 0.5 and **c**, 0.05 mg/ml)

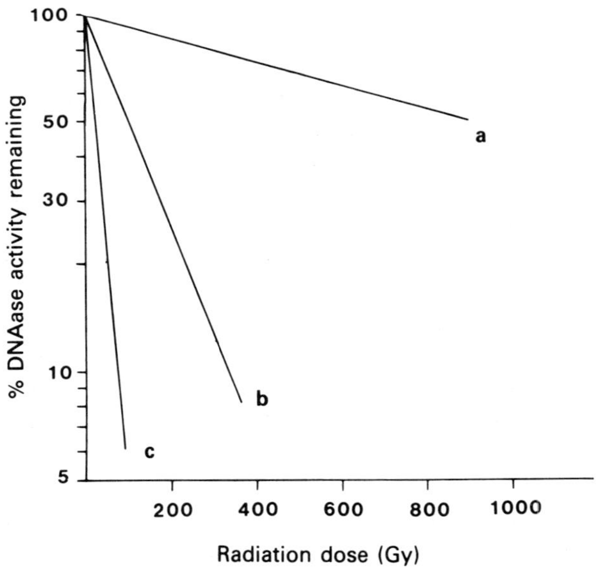

Note the semilogarithmic scale. Inactivation is greater at lower concentrations because of the increasing indirect effect due to reactive water species. Redrawn from Okada, S. (1957) *Arch. Biochem. Biophys.*, *67*, 102-112.

9.4.2 *Nucleic Acids*

The complex reactions associated with the irradiation of nucleic acids, particularly DNA and its constituents, have been studied in great detail.[2,4] All three primary reactive water species (e^-aq, $OH^\bullet$ and $H^\bullet$) can damage DNA components by the mechanisms described earlier. However, the $OH^\bullet$ radical reacts four times faster than the other two. Both base and sugar moieties are affected, the former being three times more susceptible to $OH^\bullet$ radical damage. In addition pyrimidines are damaged rather more efficiently than purines. For each 100 eV (1.6×10^{-17}J) of radiation energy absorbed by DNA the number of bases destroyed will be roughly two (this is the so-called 'G-value'; see Dertinger and Jung, Further Reading).

As an illustration the reaction of the $OH^\bullet$ radical with the pyrimidine thymine will be described.[2,4,5] The principle mode of attack is addition at the 5,6-ethylenic bond forming the 5-yl or 6-yl radical adducts. Reaction with oxygen leads to the formation of the corresponding peroxy radicals which on reduction give the hydroperoxides, predominantly the 5-hydroxy-6-hydroperoxy compound:

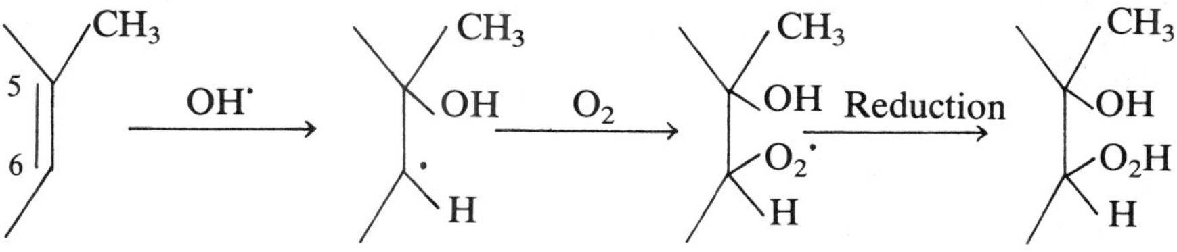

DNA thymine peroxides are unstable and undergo further reactions including the formation of thymine glycols and the opening of the pyrimidine ring. A putative key intermediate is thought to be N^1-formyl-N^2-pyruvylurea DNA.[5]

The $OH^\bullet$ radical appears to abstract hydrogen from all available positions on deoxyribose. Attack on this sugar component can cause elimination of bases from the DNA. Damage to the sugar or sugar-phosphate bond is thought to be the principal cause of DNA-strand breaks, which are of major importance. These can lead to local unzipping of the DNA strand and loss of hydrogen bonding. Moreover, if another break is formed close by in the opposite strand, this will generate a double-strand break (DSB). DSBs result in the production of smaller pieces of DNA with lower molecular weight. These breaks can be quantified by measuring sedimentation velocity or retention on a filter. A single-strand break (SSB) can be measured in a similar way after treatment with alkali to separate the DNA strands (see Figure 9.3)[6].

It can be seen from Figure 9.3 that strand breaks can be produced at relatively low radiation doses, e.g. 2.5 Gy, and assays with greater sensitivity are now being developed. For example, a DNA-unwinding

Figure 9.3 Effect of X-rays on the Alkaline Elution Patterns of DNA from L1210 Mouse Tumour Cells (———, Control ---, 2.5 Gy; –·–·–, 5 Gy)

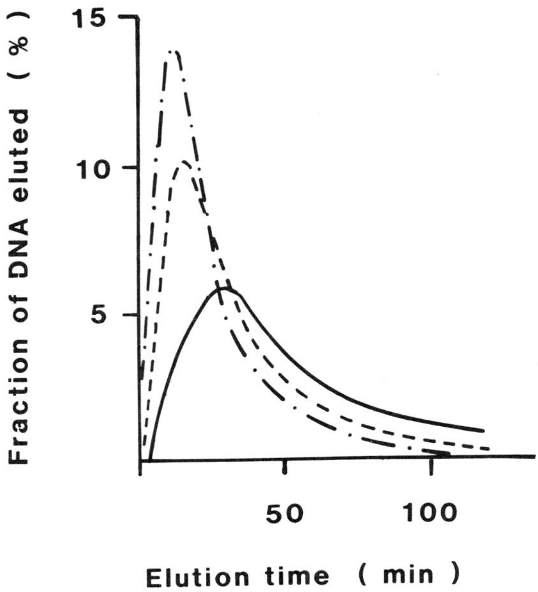

Irradiated DNA is eluted from the cellulose triacetate filter more rapidly than control DNA, indicating the presence of single-strand breaks. Redrawn from Kohn, K.W. & Grimek-Ewig, R.A. (1973) *Cancer Res., 33,* 1849-1853.

assay allows quantitation of strand breaks and base damage at low doses and the use of monoclonal antibodies will allow the identification of very specific lesions. The three most important types of DNA lesion, SSBs, DSBs and base damage, are illustrated in Figure 9.4. DNA cross-links can also be formed. To give an idea of the scale of the damage which can be produced, doses of 1–2 Gy, which sterilise two-thirds of a mammalian cell population, will produce about 50 DSBs, 1000 SSBs and an even greater number of damaged bases per cell.

The conformation of intact DNA appears to protect bases from radiation damage. For example, partial opening ('nicking') of the double-helical structure with DNAase leads to increased base damage.[4] Association of DNA with histones and other chromatin proteins would be expected to provide further protection from free radical attack.

Until recently there has been a tacit assumption that radiation damage is distributed randomly throughout the genome. However, evidence is accumulating to suggest that this is not the case. One might expect, for example, that replicating DNA would be particularly sensitive to radiation and this is indicated by the finding that nascent DNA, with no

Figure 9.4 Schematic Diagram of the Principal Types of Damage to the DNA Molecule

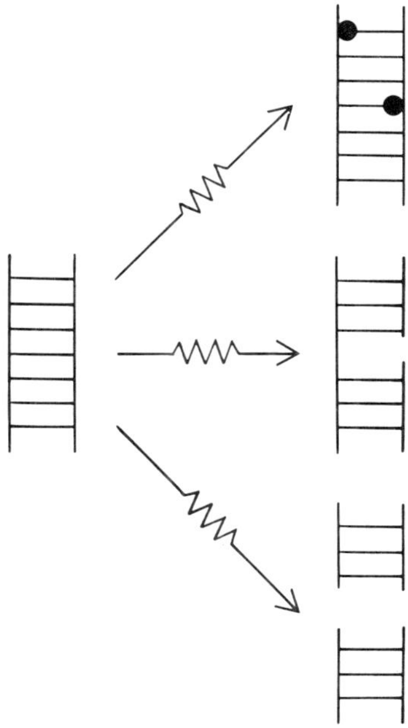

associated protein, receives three times more damage than mature chromatin DNA[7] which has associated protein. In addition, work with DNA-hybridisation probes has demonstrated the existence of mammalian DNA sequences with five to six times more SSBs than bulk DNA;[8] these are taken to be transcriptionally active genes where the chromatin structure is relaxed.

Target theory predicts that molecules of high molecular weight will be inactivated more readily than those of lower molecular weight (see Dertinger and Jung, Further Reading). It would follow from this that the processes of replication and transcription, involving the largest nucleic acid DNA, would be more radiosensitive than translation, which involves the various smaller ribonucleic acids. This is indeed the case. Likewise the greater radiosensitivity of DNA synthesis compared with RNA synthesis would be explained by the fact that the former requires the whole DNA molecule, whereas the latter involves only small sections of it.

On a larger scale, damage to DNA leads to the production of gross chromosome abnormalities, mutation and cancer induction.

It is interesting to consider the time-scale of events after irradiation of cells. In broad terms ionisation of atoms occurs within about 10^{-18}–10^{-12} s and free radical damage to biomolecules within 10^{-12}–1 s, whereas the production of the ultimate biochemical lesions, which may or may not be repaired, takes seconds to hours.

9.5 DNA Repair (see also Chapter 5)

An important feature of DNA damage is that even quite major lesions can be repaired. Many of the enzymes involved in DNA repair have now been identified[9] and mechanisms proposed for repair of SSBs, DSBs and base damage.[10] Nevertheless, virtually nothing is known about the removal of specific DNA base changes induced by ionising radiation.

SSBs are probably repaired by 'excision repair'. First, the damage is recognised, presumably as a result of a distortion in the chromatin structure. An 'incision' is then made and an area around the break is excised by endonucleases. Next, the gap is filled by insertion of the correct nucleotides through the action of DNA polymerase, with the undamaged strand as a template ('repair replication'). Finally, the newly synthesised material is joined to the strand under repair by ligase enzymes. For ionising radiation the inserted 'patch' is quite small, three to four nucleotides, and repair is complete within 1–2 h. This has been referred to as 'short-patch' repair.

Bases damaged by ionising radiation are probably removed by DNA glycosylases, a group of highly specific enzymes which cleave the base–sugar bonds in altered or damaged nucleotides. A DNA glycosylase activity is present in *Escherichia coli* which catalyses the release of thymine glycols from DNA.[9] Removal of a damaged base generates an apurinic or apyrimidinic site which can be rapidly corrected by excision repair.

If DSBs are formed without base or sugar damage in at least one strand, then strand-rejoining could occur through direct action of a ligase so long as the ends are held together in the meantime by the chromatin structure. Where base or sugar damage occurs in both strands, there will be no authentic template available for excision repair. Nevertheless, DSBs are repaired, and the most likely hypothesis proposes the involvement of recombination with homologous undamaged DNA.

A recent finding of interest has been that DNA excision repair requires poly(ADP-ribose) polymerase, an enzyme which adds successive ADP-ribose groups to chromatin proteins.[11] Increased enzyme activity is seen after radiation treatment and the rejoining of SSBs is prevented when the activity is inhibited.

The importance of DNA repair as a means of protection against radiation damage is illustrated by a number of inheritable human disorders in which the patients' cells exhibit hypersensitivity to radiation and deficiency in DNA repair.[12] The best example is ataxia telangiectasia (AT) in which repair of SSBs is apparently normal, but there is defective repair at sites of damaged bases. Cells from patients suffering from Fanconi's anaemia exhibit defects in the repair of both interstrand cross-links and base damage. Patients with repair defects develop extreme normal-tissue reactions to therapeutic doses of X- and γ-rays. Elucidation of the molecular basis of AT and other repair-deficiency syndromes may reveal common mechanisms for the development of associated symptoms such as cancer proneness, neurological defects and immunological deficiencies.

9.6 Is DNA the Biological Target?

As with cytotoxic drugs, it is difficult to prove that radiation damage to DNA represents the lethal event in mammalian cells, and the topic remains a controversial one. Nevertheless, there is considerable evidence to implicate DNA. The following reasons are perhaps the most convincing (for more details see Coggle, Further Reading):

(1) Various elegant techniques have been used to preferentially irradiate the nucleus or the cytoplasm. These have shown the nucleus to be perhaps 100 times more sensitive, and implicate the chromosomes in particular
(2) The pyrimidine analogue 5-bromodeoxyuridine is incorporated specifically into DNA and acts as an efficient radiosensitiser. This is attributed to an altered conformation of the DNA
(3) There is an excellent correlation between cell death at first mitosis and the presence of chromosome aberrations
(4) Functional DNA damage is detectable in the required dose range, whereas functional damage to proteins is only seen at much higher doses
(5) Increased radiation sensitivity is seen in cells from patients with DNA repair deficiencies.

Most radiobiologists would probably agree that DNA is the most likely target molecule, while pointing out that other forms of damage, for example to lysosomes, mitochondria and in particular cell membranes, may also be important. There is in addition considerable support for DNA–membrane complexes as important targets.[13]

Of course, DNA has considerable intuitive appeal since damage to critical sites in DNA would be likely to have profound consequences for the cell, more so than with proteins and RNA, for example, which are smaller molecules present in multiple copies. It is probably fair to say that DNA provides the most obvious link between radiation chemical events and the sterilisation of cells. If a unique damaging lesion exists it remains for the moment undefined. Among the more popular candidates are base damage, DSBs and sites of 'misrepair', i.e. where repair has been carried out incorrectly.

9.7 Cellular Radiobiology

Very high radiation doses indeed (tens of gray) are required to completely inhibit metabolism and cause rapid cell disintegration. This is known as 'interphase' or 'non-mitotic' death. Lower doses (1–2 Gy) are able to sterilise mammalian cells by inhibiting their potential for unlimited division. The cells may remain morphologically and biochemically normal, and even undergo a few cell divisions to produce sterile progeny, but will be unable to repopulate the tissue. Thus this type of 'reproductive' death is of most relevance to radiation therapy. The fact that damage is not usually expressed until cells attempt to undergo mitosis explains why more rapidly proliferating tissues are the most responsive to radiation. An important law in radiobiology, that of Bergonié and Tribandeau, states that the radiation response of a tissue is proportional to its mitotic activity. Thus, although the constituent cells may have similar intrinsic radiosensitivities, organs with many dividing cells (e.g. skin, bone marrow, intestinal epithelium, gonads) suffer more damage than those with few (e.g. brain, muscle, liver).

Reproductive death is assayed by measuring the ability of cells to form colonies of 32–64 cells (five or six doublings). The original demonstration that radiation effects on mammalian cells could be assayed by 'cloning' *in vitro*[14] was of major significance for radiobiology.

When the fraction of cells surviving (i.e.forming colonies) is plotted on a logarithmic scale against radiation dose on a linear scale the result is called a 'survival curve'. Although there is some controversy about the precise shape of this curve, the majority of curves for X- and γ-rays are of the kind shown in Figure 9.5. There are two components: a 'shoulder' region at low doses and an exponential region at higher doses. The shape of the curve is defined by the parameters D_o and n. The slope of the exponential region is described by D_o, the dose required to reduce the surviving fraction f to fe^{-1} (or $0.37f$), e.g. from 1 to 0.37 per cent. The 'extrapolation number' n is obtained by extrapolating the exponential region to the survival axis, and gives an indication of the size of the shoulder. The shoulder is also described by the 'quasi-threshold dose' D_q, the dose at which the exponential region crosses the 100 per cent survival value. For well-oxygenated mammalian cells exposed acutely to X- and γ-rays n is normally between 1 and 10, D_q between 0.5 and 2.5 Gy and D_o between 1 and 2 Gy. Doses of 10–15 Gy are required to sterilise 99.99 per cent of cells. It is worth emphasising that the doses of radiation used

Figure 9.5 Schematic Survival Curve for Mammalian Cells Exposed to X- and γ-Rays ($n = 3$; $D_o = 1.5$ Gy; $D_q = 1.6$ Gy)

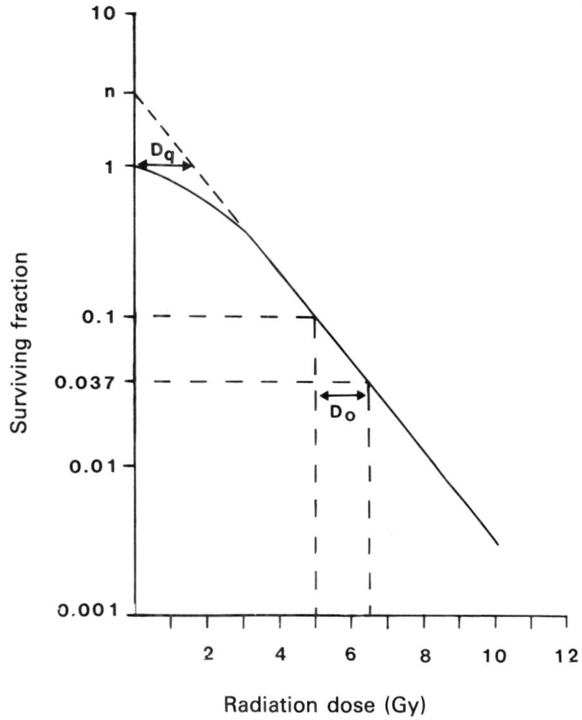

clinically are usually in the shoulder region.

The interpretation of radiation survival curves is also controversial (see Coggle, Further Reading). Mathematical modelling has been used to analyse the data in terms of the infliction of one or more pieces of damage, or 'hits', in one or more critical sites, or 'targets'. The most popular model is the multi-target, single-hit model where a single hit occurs in two or more targets. Attempts have been made to interpret these models mechanistically, particularly with respect to DSBs, but these are subject to criticism. One school of thought favours the idea that mammalian survival curves may really be simple single-hit exponential curves, with the precise shapes actually reflecting individual repair characteristics.

Two types of repair phenomenon have been studied in great detail at the cell-survival level. The shoulder or the radiation survival curve has been attributed to the repair of 'sublethal damage' (SLD), which prevents damage accumulating to a lethal level. This type of damage repair is measured by the size of D_q. It is normally demonstrated in a 'split-dose' experiment where the effect of a single radiation dose is compared with that of the same total dose divided into equal fractions given at different time intervals.[13] As the time between fractions increases the amount of cell-killing decreases. SLD repair requires oxygen and is metabolism-dependent.

The second phenomenon is the repair of 'potentially lethal damage' (PLD).[14] PLD is normally lethal, but can be repaired if postirradiation conditions are suboptimal for growth, e.g. if cells are held in density-inhibited cultures *in vitro* or in a solid tumour before a cloning assay is performed.

Both types of repair are understood only at a phenomenological level; the molecular mechanisms and their relation to the specific types of DNA repair described earlier are completely unknown.

9.8 Factors Affecting Cell-killing by Radiation

The principal factors affecting cell-killing by radiation are summarised in Figure 9.6.

9.8.1 Recovery

As far as radiation therapy is concerned, the significance of recovery from sublethal damage is that higher doses are required to give the same sterilisation of tumour cells when the dose is fractionated. However, normal tissue tolerance also increases with fractionation and for cell kinetic reasons the sparing may be greater than in the tumour (Figure 9.6**a**).

9.8.2 Oxygenation

The sensitivity of cells to X- and γ-rays is strongly dependent on the oxygen tension during irradiation (see Coggle, Further Reading). Well-oxygenated or 'oxic' cells with oxygen tensions above 30 mmHg are two to three times more sensitive than 'hypoxic' cells at very low oxygen tensions. This is known as the 'oxygen effect'. Oxygen is usually said to be a 'dose-modifying agent' in that the ratio of hypoxic to oxic doses needed to produce an equal level of survival is constant. There is, however, increasing evidence that the oxygen effect is actually reduced in the shoulder region.

Cells in normal mammalian tissues are generally considered to have oxygen tensions between those of venous and arterial blood (40 and 70 mmHg respectively). Tumours tend to outgrow their blood supply and, because oxygen can diffuse only 100–200 μm into tissue before it is consumed, those cells at greater distances from blood vessels will become hypoxic and therefore radioresistant.[16] Experimental animal tumours usually have 10–15 per cent hypoxic cells. Cell-survival curves for tumours irradiated *in vivo* therefore have two components: an initial steep slope

Figure 9.6 Factors Affecting Cell-killing by Radiation (**a**, Recovery; **b**, oxygenation; **c**, cell cycle; **d**, repopulation; **e**, dose rate; **f**, radiation quality)

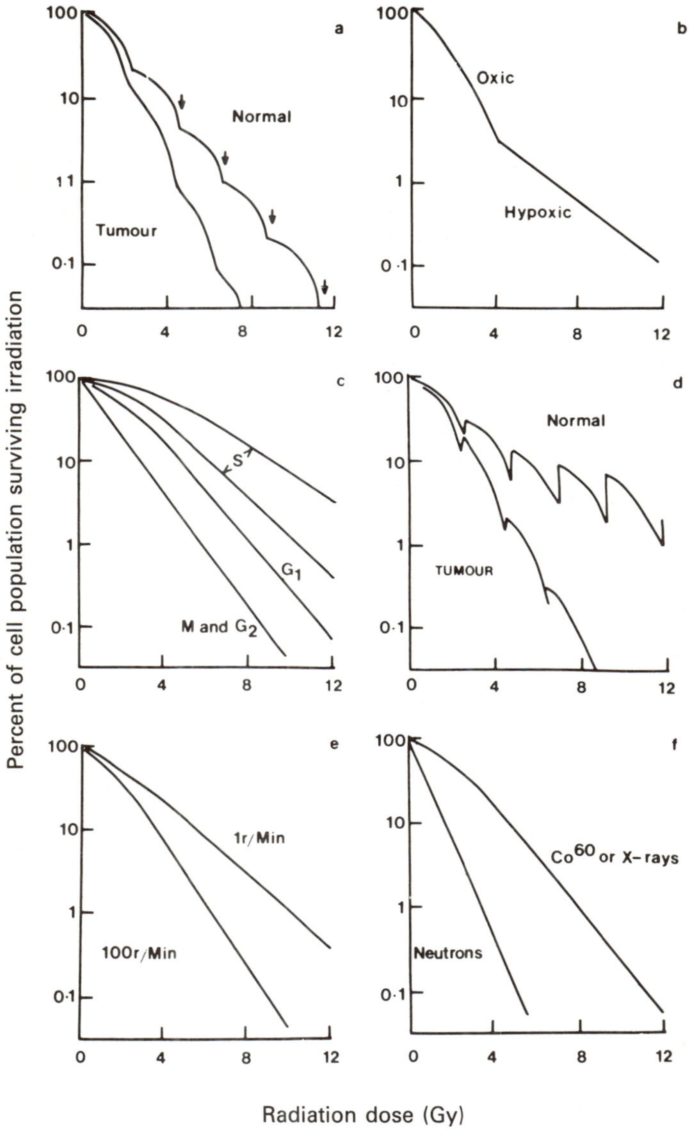

Source: Dixon, B. (1982) In *Cancer Topics and Radiobiology*, Pitman, London, pp. 17–30; reproduced with permission.

ıg the sensitivy of the oxic cells, giving way to a shallow slope ıing the resistance of the hypoxic population (Figure 9.6**b**).

wo mechanisms have been proposed for the oxygen effect at the ɔlecular level, both involving fast free radical processes. One is the ormation of damaging species such as hydroperoxy radicals. The second, most popular one, is the reaction of the oxygen with free radicals formed in target biomolecules to produce organic peroxy radicals (see p. 245). This reaction prevents restitution via hydrogen donation, by thiols for example, and is said to 'fix' the biological damage into a more severe form.

The oxygen effect is considered to have profound consequences for radiation therapy and we shall return to these in the final section.

9.8.3 Cell Kinetics

Radiosensitivity varies by a factor of about two over the different phases of the cell cycle.[17] Cells with a short G_1 are usually most sensitive in mitosis and G_2, least sensitive in S and of intermediate sensitivity in G_1 (Figure 6.9**c**). Those with a long G_1 show a similar pattern, but also have a resistant phase in early G_1 and a sensitive phase in late G_1. The overall response of an asynchronous population will depend on the proportion of cells in the different phases of the cycle.

There are no clear differences in the cycle times of tumour vs normal tissues, though normal tissues may have a greater proportion of stem cells in the resting phase G_0, whereas tumours may have a higher proportion of clonogenic cells in cycle.[18] Tumours may also have very high cell-loss factors. It has been suggested that gross parameters such as labelling index, growth fraction and cell-cycle time are unlikely to predict well the comparative response of normal and malignant tissues, whereas the number, proliferation characteristics and recruitment properties of the actual stem cells would be more helpful.[18] In some circumstances these cell kinetic factors can lead to greater repopulation in normal tissues than in tumours (Figure 9.6**d**).

9.8.4 Dose Rate

With X- and γ-rays cell-killing is reduced when the radiation dose is given over an extended period (see Hall, Further Reading). For example, Figure 9.6**e** shows a twofold reduction in biological effectiveness when the dose rate is reduced from 1 to 0.01 Gy/min (100 to 1 rad/min). This dose-rate effect is due to two factors already discussed: repair of SLD and repopulation resulting from cell division during the exposure.

9.8.5 Radiation Quality

For a given dose, greater cell-killing is achieved with high LET radiations such as neutrons than with low LET radiation such as X- and γ-rays (Figure 9.6**f** and see Hall, Further Reading).

9.9 Improvements in Radiation Therapy

Improvements in radiotherapy have stemmed principally from the empirical observations and experience of radiation therapists over the last 50 years. Developments in the sophistication of treatment machines, radiation physics and techniques for tumour localisation such as computerised tomography (CT) scanning have all played their part. Compared with traditional low-energy X-rays sets, high-energy treatment machines (cobalt-60 units and linear accelerators) allow higher doses to be delivered deep within tissues, while sparing the sensitive cells of the skin. CT scanning, which detects minor differences in tissue density with X-rays by the use of computer techniques to construct a series of cross-sectional images, has greatly improved the accuracy of tumour localisation. Computers are also used to improve treatment planning. Early diagnosis is also important. Large tumours respond less well than small tumours since there are more cells to eradicate; however, the dose must actually be reduced to avoid exceeding normal tissue tolerance with the larger treatment field.

Research into radiation chemistry, biochemistry and radiobiology have greatly improved our understanding but have not yet contributed directly to improved therapy. Nevertheless, a number of experimental treatments are now being evaluated. The following is a summary of some current approaches being used to improve radiation therapy (see also Ref. 20).

9.9.1 Higher Radiation Doses

Higher doses to the tumour are being achieved by more accurate localisation and treatment planning, particularly with the application of CT. As well as using external beam therapy clinicians use radiation sources implanted in the body to deliver very high local doses of radiation to a small volume.

9.9.2 Hypoxic Cells

Over the years radiation therapists have found that the best way to avoid excessive normal-tissue damage is to divide or 'fractionate' the total dose into a number of smaller ones given on different days. Also the death of oxic cells causes tumour shrinkage and allows the more radioresistant hypoxic cells to reoxygenate so that they can be eradicated by subsequent doses. Despite the use of fractionation many workers feel that hypoxic cells remain a major obstacle in local control and a number of approaches are being tried to deal with them (see Hall, Further Reading).

One is to further optimise fractionation by giving two or three doses each day ('hyperfractionation'). Another is to have patients breath pure oxygen in a chamber at a pressure of 2–3 atm (203–304 kPa; 1 atm = 101.325 kPa) ('hyperbaric oxygen') to try to oxygenate the hypoxic cells.

This is a difficult and somewhat hazardous procedure and, in general, the expected gains in local control have not been seen. For this reason there has been much interest in the use of drugs to sensitise hypoxic cells.

The most studied have been the 'electron-affinic' chemicals, especially the nitroimidazoles such as metronidazole and misonidazole. Their mechanism of action is complex and probably involves several factors. One involves reaction with free radicals produced in biological molecules, including DNA, resulting in 'fixation' of the damage in an unrepairable form. This is analogous to the fixation of radical damage by oxygen described previously, but in this case results in the formation of DNA-sensitiser adducts. Another mechanism involves reaction with radioprotective SH groups. Compounds such as misonidazole are able to diffuse further than oxygen to reach hypoxic cells and are extremely effective radiosensitisers in animal models. In recent clinical trials, neurotoxicity has proved a problem, and analogues less toxic are now being introduced. Nitroimidazoles are also being used to detect hypoxic cells and to increase response to some cytotoxic drugs ('chemosensitisation').

Many other agents have been investigated as radiosensitisers, such as bromodeoxyuridine. This will sensitise both oxic and hypoxic cells, but relies on greater incorporation into rapidly proliferating tumour cells. Additional types of experimental drugs include thiol depletors (e.g. diethylmaleate and buthionine sulphoxime) and inhibitors of PLD repair (e.g. nucleoside analogues) and of poly(ADP-ribose)polymerase (e.g. benzamide derivatives).

Another experimental approach to combat hypoxic cells is the use of high LET radiations, such as neutrons, for which the oxygen effect is smaller. There is also less repair of sublethal damage.

9.9.3 Radioprotectors

Thiols act as radioprotective agents (see Hall, Further Reading). They probably do this by repairing free-radical damage, thus preventing its fixation by oxygen (the 'radical-scavenger' hypothesis). Thus, as already discussed, donation of a hydrogen atom by a thiol compound can restore the organic free radical to its original form. Thiophosphate derivatives have been developed which in animal models can confer greater radioprotection on normal tissues compared with that on tumours. Many factors are involved and the mechanisms are not fully understood, but clinical studies are in progress.

9.9.4 Combined Modality Treatment

Radiation therapy is often combined with chemotherapy or surgery. Either of these can be used to reduce the bulk of the tumour to a size more readily curable by radiation. Combination with chemotherapy is particularly attractive because this allows simultaneous treatment of primary tumour and distant metastases. Good examples of the combined use of chemotherapy and radiotherapy are the treatment of childhood tumours and Hodgkin's disease. However, the value of this type of combination, as opposed to chemotherapy alone, has been demonstrated in only a few clinical studies, and enhanced toxicity can be a problem.

There is a developing interest in the use of heat or 'hyperthermia' in cancer therapy.[19] Under certain circumstances heat and radiation can be combined to advantage. The mechanism of the interaction is complex and not well understood, but an effect on the repair of DNA damage may well be important. Difficulties with the use of hyperthermia include technical ones concerned with heat delivery and dosimetry and an effect known as 'thermal tolerance' whereby tumour cells develop resistance to further heating. Nevertheless clinical trials are underway combining heat with either radiation therapy or chemotherapy.

9.10 Concluding Remarks

It is hoped that these experimental approaches resulting from radiobiology research will improve further on the empirically based practice of radiation oncology. Moreover, a new era lies ahead: that of the application of the powerful techniques of modern molecular biology to study the molecular mechanism of radiation damage which remains poorly understood at the mechanistic level. For example, one area that should reap immediate benefit is the field of DNA repair. A molecular understanding of such phenomena as 'sublethal' and 'potentially lethal' damage is urgently required and may lead to the development of a new generation of modifying drugs for the radiation therapist.

References

1. Schwarz, H.A. (1981) Free radicals generated by radiolysis of aqueous solutions. *J. Chem. Educ.*, *58*, 101-5
2. Scholes, G. (1983) Radiation effects on DNA. *Br. J. Radiol.*, *56*, 221-31
3. Okada, S. (1957) Inactivation of deoxyribonuclease by X-rays. II. The kinetics of inactivation in aqueous solutions. *Arch. Biochem. Biophys.*, *67*, 102-12
4. Ward, J.F. (1975) Molecular mechanism of radiation-induced damage to nucleic acids. *Adv. Radiat. Biol.*, *5*, 181-239
5. Teoule, R., Bert, C. & Bonicel, A. (1977) Thymine fragment damage retained in the DNA polynucleotide chain after gamma irradiation in aerated solutions. II. *Radiat. Res.*, *72*, 190-200
6. Kohn, K.W. & Grimek-Ewig, R.A. (1973) Alkaline elution analysis, a new approach to the study of DNA single-strand interruptions in cells. *Cancer Res.*, *33*, 1849-53
7. Warters, R.L. & Childers, T.J. (1982) Radiation induced thymine base damage in replicating chromatin. *Radiat. Res.*, *90*, 564-74
8. Chiu, S.M., Oleinick, N.L., Friedman, L.R. & Stambrook, P.J. (1982) Hypersensitivity of DNA in transcriptionally active chromatin to ionizing radiation. *Biochem. Biophys Acta*, *699*, 15-21

9. Lindahl, T. (1982) DNA repair enzymes. *Ann. Rev. Biochem.*, *51*, 61-87
10. Hanawalt, P.C., Cooper, P.K., Ganesan, A.K. & Smith, C.A. (1979) DNA repair in bacteria and mammalian cells. *Ann. Rev. Biochem.*, *48*, 783-836
11. Durakacz, B.W., Omidiji, O., Gray, D.A. & Shall, S. (1980). (ADP-ribose)$_n$ participates in DNA excision repair. *Nature (Lond.)*, *283*, 593-6
12. Friedberg, E.C., Ehmann, U.K. & Williams, J.I. (1979) Human diseases associated with defective DNA repair. *Adv. Radiat. Biol.*, *8*, 85-174
13. Elkind, M.M., Sutton-Gilbert, H., Moses, W.B., Alescio, T. & Swain, R.W. (1965) Radiation response of mammalian cells grown in culture. *Radiat. Res.*, *25*, 359-76
14. Phillips, R.A. & Tolmach, L.J. (1969) Repair of potentially lethal damage in X-irradiated HeLa cells. *Radiat. Res.*, *39*, 317-18
15. Dixon, B. (1982) Therapeutic radiobiology. In C.A.F. Joslin (ed.) *Cancer Topics and Radiobiology*, Pitman, London, pp. 17-30
16. Thomlinson, R.H. & Gray, L.H. (1955) The histological structure of some human lung cancers and the possible implications for radiotherapy. *Br. J. Cancer*, *9*, 539-49
17. Terasima, T.T. & Tolmach, L.J. (1963) Variations in several responses of HeLa cells to X-irradiation during the division cycle. *Biophys. J.*, *3*, 11
18. Steel, G.G. (1977) *Growth Kinetics of Tumours*, chapter 8, Clarendon Press, Oxford
19. Storm, F.K. (ed.) (1983) *Hyperthermia in Cancer Therapy*, G.K. Hall, Boston
20. Kaplan, H.S. (1979) Experimental frontiers in radiation therapy. In S. Okada, M. Imamura, T. Terashima & H. Yamaguchi (eds) *Proc. 6th Int. Congress of Radiation Research*, Toppan, Tokyo

Further Reading

Chadwick, K.H. & Leenhouts, H.P. (1981) *Molecular Theory of Radiation Biology*, Springer-Verlag, New York

Coggle, J.E. (1983) *Biological Effects of Radiation*, Taylor and Francis, London

Dertinger, H. & Jung, H. (1970) *Molecular Radiation Biology*, vol. 12, Heidelberg Science Library, Longman, New York

Duncan, W. & Nias, A.H.W. (1977) *Clinical Radiobiology*, Churchill Livingstone, Edinburgh

Hall, E.J. (1978) *Radiobiology for the Radiologist*, 2nd edn, Harper and Row, Hagerstown

Joslin, C.A.F. (1982) *Cancer Topics and Radiotherapy*, Pitman, London

Steel, G.G., Adams, G.E. & Peckham, M.J. (1983) *The Biological Basis of Radiotherapy*, Elsevier/North-Holland, Amsterdam

10 CANCER CHEMOTHERAPY I: DESIGN AND MECHANISM OF ACTION OF CYTOTOXIC DRUGS

P.B. Farmer

Contents

10.1 Introduction

The fundamental goal facing designers of anticancer drugs is the discovery of molecular features of cancer cells which are distinct from those of normal cells and which may thus be exploited as selective targets for a drug. In the drug treatment of other diseases advantage can be taken of the unique nature of the invading species. Thus, in bacterial infections β-lactam antibiotics (penicillins, cephalosporins) inhibit a transpeptidase reaction involved in the biosynthesis of a peptidoglycan component, characteristic of the bacterial cell wall. Similarly bacterial utilisation of *p*-aminobenzoic acid for the synthesis of folic acid (not carried out by mammalian cells) may be inhibited by sulphonamides. For cancer cells, however, unique features have proved difficult to identify and consequently the clinically used anti-tumour agents are, unlike antibacterial agents, not totally selective toward the target cells. Although cure can be achieved for a limited number of tumours (see Chapter 1) the majority of human cancers are refractory to chemotherapy. It is to be hoped that the current studies of oncogenes (see Chapter 4) may lead to the development of a new highly selective class of anti-tumour agents, although the realisation of this possibility is many years hence.

As will be discussed in Chapter 11 there are two approaches to treating cancer with drugs: cytotoxic chemotherapy which forms the basis of most present-day therapy and non-cytotoxic therapy which has not so far yielded dramatic clinical results but holds considerable promise. In this chapter the design and mechanism of action of cytotoxic drugs will be discussed. In the subsequent chapter the methods used for discovering these drugs, their development for clinical use and some aspects of human response to them will be considered, together with a description of the initial results from non-cytotoxic therapy. The cytotoxic drugs that have reached clinical practice (now more than 30) (Figure 11.1, Chapter 11) have been selected on the basis of animal-tumour screening tests. The nature of this selection process is such that compounds passing through the screen are frequently of the same chemical classes as those previously selected. This may be a true representation of the value of these classes of compounds or it may be a reflection that the test systems are particularly susceptible to these types of material. The major classes of synthetic compounds in clinical use are alkylating agents and antimetabolites. Natural products, derived from plant or bacterial sources, represent around one-quarter of anticancer agents in use. Hormones, and their antagonists, that are receiving rapidly increasing use are described in detail in Chapter 8.

10.2 Alkylating Agents

Alkylating agents (RX, R = alkyl group, X = leaving group) are characterised by their ability to react with nucleophilic (nucleus-loving) centres (Y) to form a covalently bound alkylated product (RY). The discovery that alkylating agents used as chemical-warfare agents (e.g. sulphur mustard) showed powerful cytotoxic properties against lymphoid tissue and blood-forming organs led in the 1940s to the clinical testing of these compounds against lymphoid tumours and leukaemias.[1] Subsequently several thousand analogues, containing a wide variety of alkyl leaving groups, have been synthesised and subjected to anti-tumour testing.

10.2.1 Leaving Group X: Mechanisms of Alkylation

Alkylating agents are generally classed into subgroups according to the nature of their leaving group (Table 10.1). The loss of the leaving group during an alkylation may either follow a unimolecular course, known as an S_N1 mechanism (*S*ubstitution, *N*ucleophilic, *1* molecule) or a bimolecular mechanism (S_N2) in which the nucleophile (Y^-) is involved in the rate-determining step:

$$S_N1 \quad RX \xrightarrow[\text{slow}]{} R^+ + X^- \xrightarrow[\text{fast}]{Y^-} RY$$

$$S_N2 \quad RX + Y^- \xrightarrow[\text{slow}]{} [X\text{---}R\text{---}Y]^- \xrightarrow[\text{fast}]{} X^- + RY$$

These mechanisms are extremes of the possible processes and intermediate mechanisms are frequently followed.

For S_N2 mechanisms second-order reaction kinetics with respect to the concentration of alkylating agent and nucleophile are followed. The nature of the nucleophile is of great importance to the fate of an S_N2-reacting alkylating agent, sulphur nucleophiles normally reacting faster than nitrogen and oxygen nucleophiles; an S_N1-reactor should theoretically be less selective in the nucleophile it alkylates.

For the nitrogen mustard class of alkylating agents (Table 10.1), which are widely used in chemotherapy, there may be intramolecular displacement of the leaving group (Cl^-) by the mustard nitrogen atom. The

Table 10.1 Alkylating Agents

Class	Structure	Example		Expected length of cross-link (atoms)
Nitrogen mustards	$R\text{-}N(CH_2CH_2Cl)_2$	Melphalan, **1**	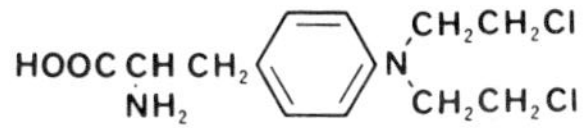	5
Alkyl sulphonates	$R\text{-}OSO_2R'$	Busulphan, **2**	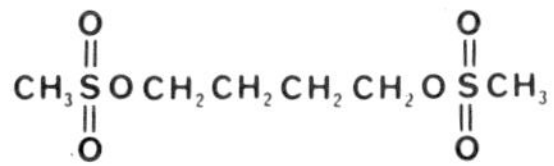	4
Aziridines	$R\text{-}N(CH_2)_2$ (aziridine ring)	Triethylenemelamine, **3**	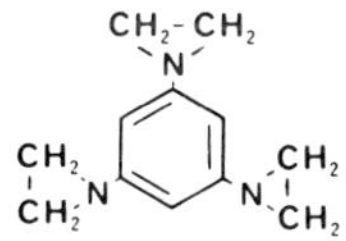	9
Epoxides	$R\text{-}CH(O)CH\text{-}R^1$ (epoxide)	Dianhydrogalactitol, **4**	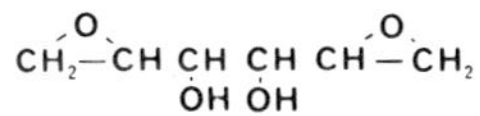	6[a]
Triazenes	$R\text{-}N{=}N\text{-}N(CH_3)_2$	5-(3,3-Dimethyltriazeno)imidazole-4-carboxamide; DTIC, **5**	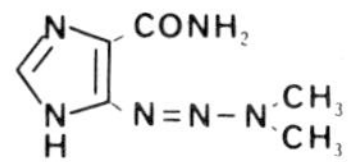	—[b]
Nitrosoureas	$R\text{-}N(NO)\text{-}C(=O)\text{-}NHR'$	1,3-Bis(2-chloroethyl)-1-nitrosourea; BCNU, **6**	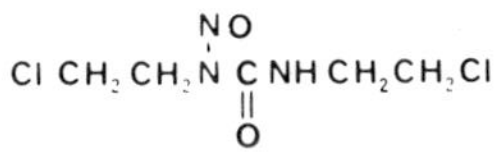	2

[a] Under S_N2 conditions epoxides are normally attacked by nucleophiles at the less highly substituted position, i.e. in this case the terminal position

[b] Formaldehyde, a metabolite of DTIC, may be able to cause one-carbon cross-links

resulting aziridinium ion may then react with nucleophiles by an S_N2 mechanism.

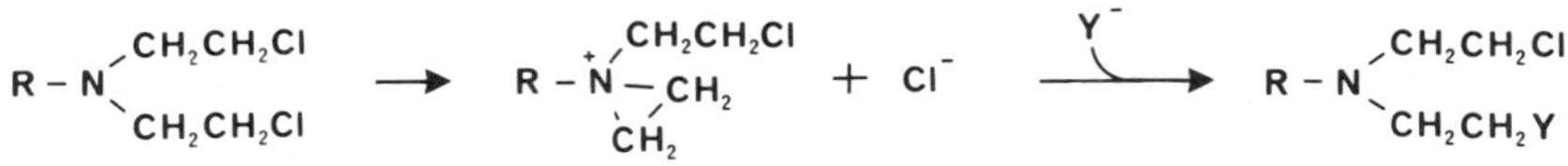

The existence of an aziridinium intermediate in nitrogen mustard reactions has been elegantly demonstrated by studies of mustards labelled with a stable isotope at one of the CH_2 groups in the chloroethyl side-chain. Nucleophilic opening of an aziridine ring, if formed, will occur equally at each of the CH_2 groups in its ring, resulting in one-half of the label being transferred to the adjacent CH_2 group in the alkylated product.[2]

If the R group in the nitrogen mustard is electron-withdrawing the nucleophilicity of the nitrogen will be reduced and aziridine formation less likely. Thus chemical modification of the R group may be used to vary the reactivity of mustards.

The mechanism of reaction of epoxides, aziridines and alkanesulphonates (Table 10.1) is generally S_N2 under physiological conditions. Triazenes and nitrosoureas, however, react by more complex mechanisms involving the production of an alkyl diazonium ion (see later).

10.2.2 Nucleophile Y: Mechanism of Biological Action

The nucleophilic centres with which alkylating agents interact are normally based on oxygen, sulphur or nitrogen atoms. The most abundant nucleophile present *in vivo* is water; reaction with water (hydrolysis) and consequent detoxification is thus a common fate for alkylating agents. Other susceptible intracellular oxygen centres include carboxylic acids, phosphates (including internucleosidic DNA phosphates), the phenolic group of tyrosine (present in proteins), and the extranuclear oxygen atoms on nucleic acid bases (e.g. O^6 of guanine, cf. Chapter 5). Thiol (SH) groups are much more powerful nucleophiles than hydroxyl (OH) groups and reaction of alkylating agents with sulphur-containing amino acids (e.g. cysteine), peptides (e.g. glutathione) and proteins is widespread. Reaction with nitrogen centres is exemplified by that with purine and pyrimidine bases in nucleic acids (especially *N*-7 of guanine) and with the amino group of lysine and the imidazole group of histidine in proteins.

The biological action of alkylating agents falls into four main categories, cytotoxic, mutagenic, carcinogenic and teratogenic, each of which is believed to be related to the interaction of the compounds with some of the nucleophiles mentioned above (see Ross, Further Reading, and Chapter 5 for discussion of mutagenic and carcinogenic mechanisms). Alkylating agent cytotoxicity is reflected by a rapid depression of DNA synthesis, a lack of cell division and the formation of giant cells (cf. review by Connors in Sartorelli and Johns, Further Reading). The cytotoxic affects of alkylating agents are most pronounced on rapidly proliferating tissues, which have a high proportion of cells in cycle.

Both cytotoxicity and anti-tumour effectiveness are generally associated with alkylating agents bearing more than one alkylating group, and it has therefore been suggested that these compounds cause these effects by cross-linking two nucleophilic centres in complementary strands of the DNA helix (Figure 10.1). The result of this would be an effective inhibition of cell replication. Monofunctional alkylation of DNA purines or pyrimidines may give rise to abnormal base-pairing, base loss or chain scission, resulting in mutagenesis and carcinogenesis (Chapter 5).

Figure 10.1 (a) DNA Interstrand Guanine-guanine Cross-link Produced by a Bifunctional Alkylating Agent; (b) Hypothesised Production of a Cross-link Between the *N*-7 Positions of Two Guanine Residues in DNA by a Nitrogen Mustard (Phosphoramide Mustard)

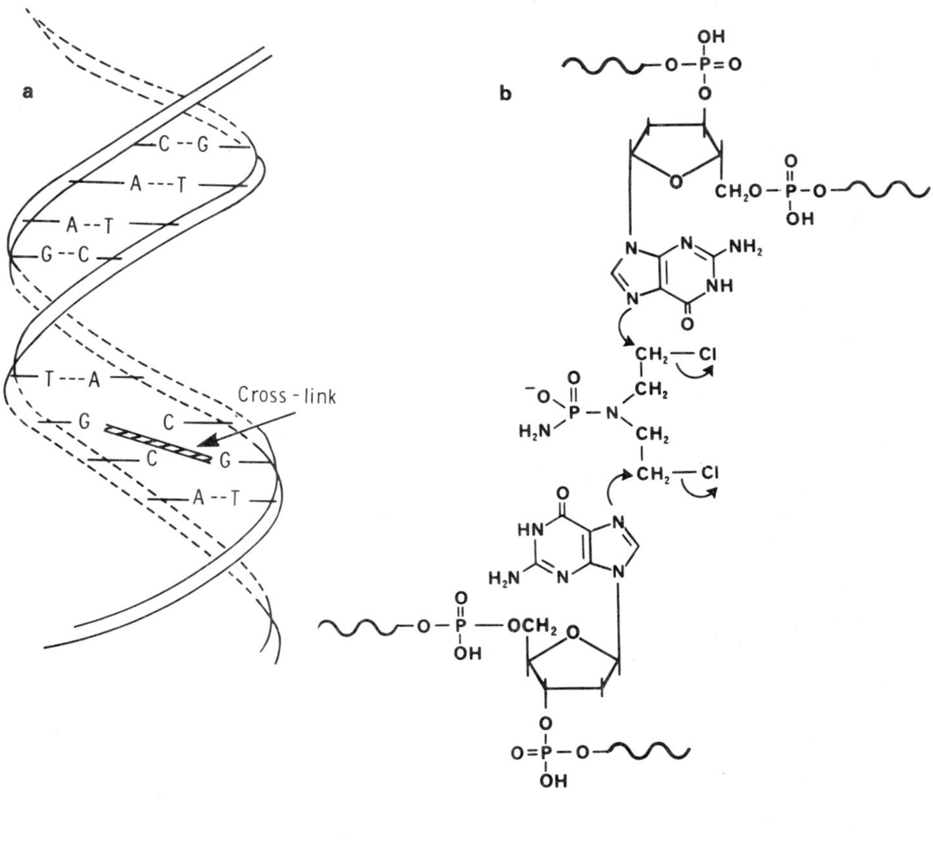

Inter- (or intra-)strand DNA base-pair cross-links have been observed experimentally in systems *in vitro* for most of the clinically used alkylating agents (e.g. chlorambucil, melphalan, phosphoramide mustard, busulphan, CCNU [1-(2-chloroethyl)-3-cyclohexyl-1-nitrosourea]). The length of the cross-link produced (summarised in Table 10.1) varies from agent to agent and presumably is of relevance to the relative biological effectiveness of the compounds. As the reaction of alkylating agents with DNA may occur at any stage of the cell cycle, the drugs are not cycle-phase specific in their action.

10.2.3 Alkyl Group R: Alkylating Agents used in the Clinic

The main preoccupation of those involved in synthesising alkylating agents for anti-tumour testing has been to vary the nature of the alkyl group R in such a way as to increase the concentration of the active drug at the tumour. (This could be achieved, for example, by using drugs that are activated by the tumour enzymes or drugs that are incorporated selectively into tumour cells by transport processes favoured by these cells.) The wide variation of R groups that have been employed is well illustrated by considering the alkylating agents that are currently used in the clinic.

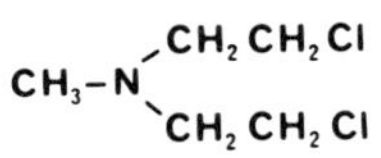

7

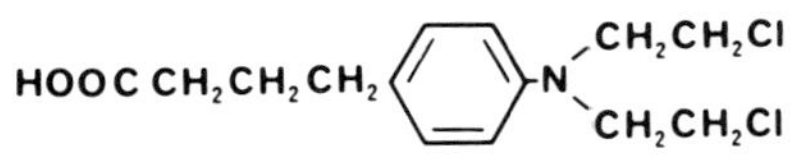

8

10.2.3.1 Nitrogen Mustards. Mechlorethamine (**7**) (nitrogen mustard) has the simplest chemical structure of the clinically used mustards and was the first alkylating agent to be used in cancer treatment.[1] It is a highly reactive chemical that is readily deactivated by hydrolysis. It is commonly used in combination therapy (Chapter 11). Melphalan (**1**, Table 10.1) is an example of a compound where the R group mimics an endogenous compound (phenylalanine).[3] Melphalan is also hydrolysed *in vivo* and no other metabolism is known. Chlorambucil (**8**) is the phenylbutyrate derivative of nitrogen mustard. This is extensively metabolised *in vivo* by β-oxidation to phenylacetic mustard which is also an active alkylating agent.[4]

Cyclophosphamide (**9**, Figure 10.2) (see reviews by Brock and Hohorst[5] and by Grochow and Colvin in Ames *et al.*, Further Reading) has been the most widely used nitrogen mustard. It is worth considering in some detail the mechanistic work that has been carried out for cyclophosphamide as it presents a good example of the value of fundamental biochemical and metabolism studies in the search for novel anticancer agents. Cyclophosphamide was first synthesised as a prodrug (a compound that is activated by metabolism *in vivo*), the hypothesis being that it would be a substrate for phosphatases or phosphamidases (thought to be at a high concentration in some tumours) and would liberate at the sites where the enzyme is present an active nitrogen mustard [e.g. $HN(CH_2CH_2Cl)_2$]. As predicted it was an inactive compound *in vitro* but showed powerful anti-tumour effects *in vivo*. Attempts to obtain evidence to support the original hypothesis have proved unsuccessful and it is now known that cyclophosphamide is activated by an alternative pathway involving metabolism by the cytochrome *P*-450 mixed-function oxidase system. The currently accepted metabolic scheme is shown in Figure 10.2.

Figure 10.2 Metabolism of Cyclophosphamide (**9**) [M represents the nitrogen mustard group, $N(CH_2CH_2Cl)_2$]

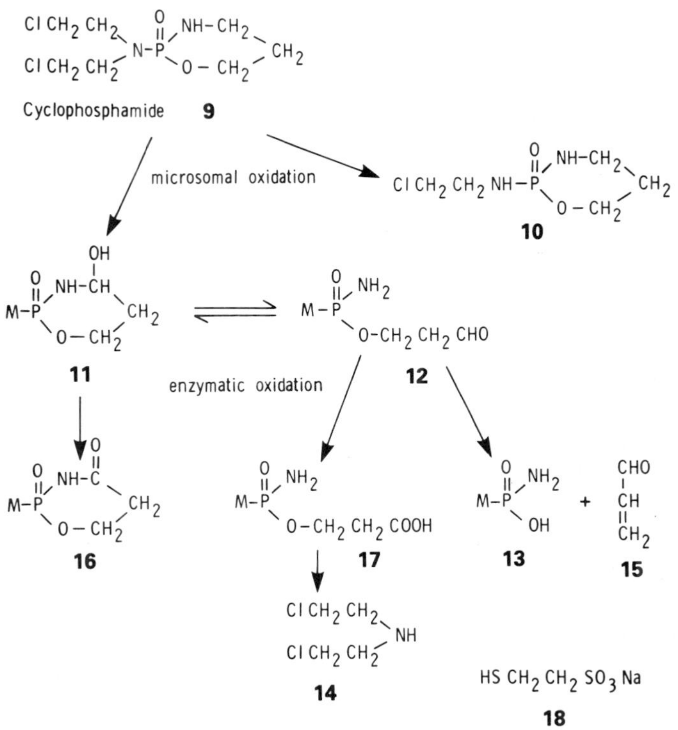

The drug is first hydroxylated at the CH_2 groups adjacent to its N atoms. Metabolism of the nitrogen mustard side-chain yields an inactive monochloroethyl analogue (**10**). Metabolism within the ring yields an unstable and active metabolite, 4-hydroxycyclophosphamide (**11**), which is in equilibrium with aldophosphamide (**12**). Aldophosphamide breaks down non-enzymically to phosphoramide mustard (**13**). Phosphoramide mustard is very much more active as an alkylating agent than cyclophosphamide (because of availability of electrons to promote intramolecular aziridine formation) and is likely to be responsible along with nornitrogen mustard (**14**) for the alkylating properties of cyclophosphamide *in vivo*. Concomitant with the production of phosphoramide mustard is that of acrolein (**15**) which is thought to be responsible for the dose-limiting bladder toxicity caused by cyclophosphamide (haemorrhagic cystitis). In competition with the spontaneous breakdown of 4-hydroxycyclophosphamide and aldophosphamide are enzymic oxidations, leading to 4-ketocyclophosphamide (**16**) and carboxyphosphamide (**17**) both of which are of low toxicity.

Knowledge of this metabolic pathway has led to improvements in the way the drug is used therapeutically and has led to the design of new analogues. The discovery that one of the metabolites (acrolein) was responsible for the toxic side-effects and did not contribute to anti-tumour activity led to the use of mesna (**18**) (sodium mercaptoethane sulphonate) as a uroprotectant.[6] Mesna reacts with acrolein and dramatically lowers the bladder toxicity caused by cyclophosphamide. It is also clear from the metabolic pathway that the requirement for metabolism could be avoided by administering 4-hydroxycyclophosphamide, or a labile precursor of it, to the patients. The experimental use of such 4-substituted cyclophosphamides has been started and should lead to better defined and more predictable pharmacokinetics for the active species.

10.2.3.2 Methane Sulphonates, Aziridines and Epoxides. None of these classes of compounds has had the same clinical impact as the nitrogen mustards. Only one methane sulphonate, busulphan (**2**, Table 10.1) is in widespread use. As an S_N2 reactor its metabolism is governed largely by reaction with SH groups although DNA cross-linking has also been shown to occur.[7] Its pharmacological properties are somewhat different from those of the nitrogen mustards in that it has a selective effect on the development of the granulocyte and other myeloid elements. Busulphan is consequently used in the treatment of chronic myelogenous leukaemia. Aziridines (ethyleneimines) that have reached clinical practice include triethylenemelamine (TEM, **3**, Table 10.1) and triethylenethiophosphoramide (thioTEPA, **19**), although the use of these is no longer extensive. Interest in aziridines has been revived by the introduction of aziridinylbenzoquinone (AZQ, **20**),[8] a compound of high lipid solubility which is capable of entering and treating tumours in the central nervous system. Similar remarks apply to the diepoxide dianhydrogalactitol (DAG, **4**, Table 10.1) which has also been undergoing clinical trial.[9]

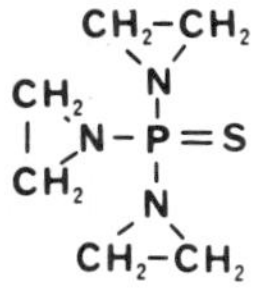

19

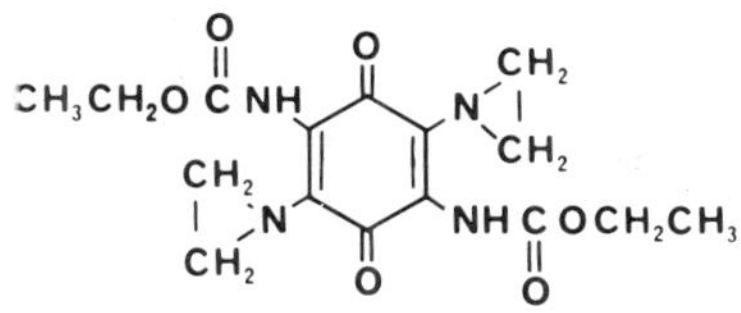

20

10.2.3.3 Triazenes. These have the unusual feature among anti-tumour alkylating agents in that they only contain one alkylating centre. This is liberated after oxidative demethylation of the dimethyltriazene (Figure 10.3). Formaldehyde is also produced in this metabolism although the significance of this to the anticancer activity is unknown. For DTIC [5-(3,3-dimethyltriazeno)imidazole-4-carboxamide: dacarbazine, **5**; Table 10.1], the only clinically used triazene, this metabolic route has been well established in animals; the hydroxymethyl intermediate (**21**) has been isolated and the production of methylated nucleic acids after DTIC administration has been demonstrated (see review by Farmer and Newell in Ames *et al.*, Further Reading). Although some temporary remissions have been achieved with DTIC in the treatment of malignant melanoma the drug is highly toxic. This may in part be due to its photolytic breakdown to a toxic diazonium compound (**22**); novel photostable analogues are being developed for clinical use.

Figure 10.3 Metabolism of 5-(3,3-Dimethyltriazeno)imidazole-4-carboxamide (DTIC, **5**)

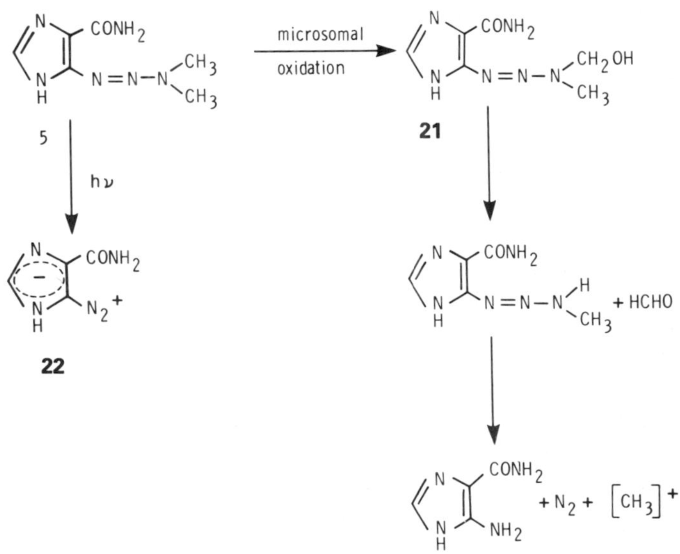

10.2.3.4 Nitrosoureas. These are converted into alkylating agents by a spontaneous chemical process in aqueous solution (Figure 10.4). For all clinically used nitrosoureas except streptozotocin the alkylating species is effectively a chloroethyl carbonium ion ($ClCH_2CH_2^+$) which is capable of producing cross-links of two carbon atoms between two nucleophilic centres. Inter- and intra-strand cross-links of this type have been demonstrated *in vitro* for BNCU (1,3-bis(2-chloroethyl)-1-nitrosourea, **6**; Table 10.1) and CCNU (**23**),[10,11] the main site of alkylation being transcriptionally active chromatin (see review by Tew *et al.*)[12]. Together with the production of the alkylating agent by the nitrosoureas is that of an isocyanate, which is also capable of reacting with (carbamoylating) biologically important sites, especially in protein. For example, CCNU liberates cyclohexyl isocyanate (**24**) which carbamoylates lysine (and other) residues in proteins. Such protein reactions may be responsible for some of the observed enzyme inhibitions caused by nitrosourea treatment.

Figure 10.4 Decomposition of 1-(2-Chloroethyl)-3-cyclohexyl-1-nitrosourea (CCNU, **23**)

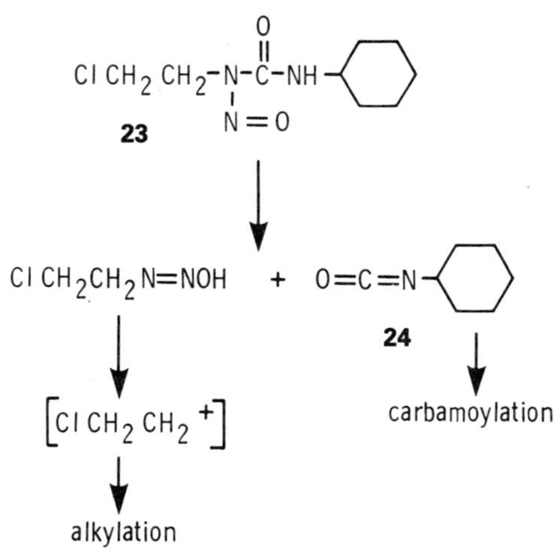

Nitrosoureas, like nitrogen mustards, are proliferation-dependent anticancer agents, but show no phase-specificity in their action. In other aspects they differ from the nitrogen mustards; for example, it is possible to derive cells that have acquired resistance to nitrosoureas but are still sensitive to nitrogen mustards,[13] which indicates that the biological mechanisms of action of the two groups are not identical. (Similar remarks apply to the mode of action of the triazenes.)

Clinically, the use of the nitrosoureas has been beset by problems caused by the high myelosuppression which they cause. Consequently the recent development of nitrosoureas has concentrated on ways of minimising this toxicity. Streptozotocin (**25**) which is a naturally occurring methylnitrosourea, and the synthetic analogue of this, chlorozotocin (**26**), are two interesting analogues in that the isocyanates produced in their breakdown are detoxified by internal reactions (i.e. they carbamoylate the sugar rings within their structures).[12,14] Both these drugs show lower myelosuppression than BCNU and CCNU and it is to be hoped that further developments with this mechanistic approach may lead to nitrosoureas of significant clinical value.

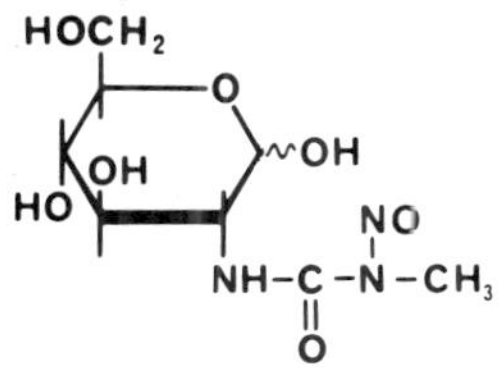

25

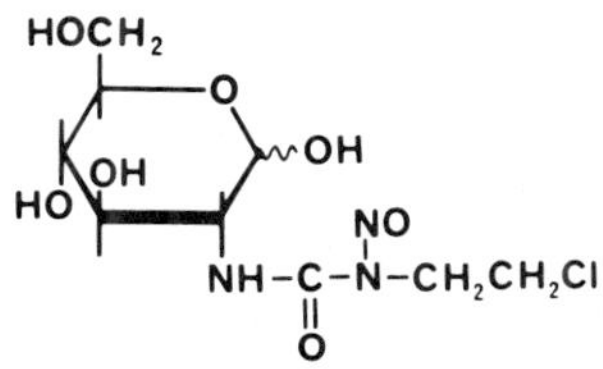

26

10.3 Antimetabolites

An antimetabolite is a compound with a chemical structure related to that of an endogenous metabolite, which interferes with the function of the latter by competing with it for an enzyme (cf. *p*-aminobenzoic acid analogues as antibacterial agents; see section 10.1). Anticancer activity is shown by antimetabolites for folic acid, purines and pyrimidines (reviews in Ames *et al.*, and in Sartorelli and Johns, Further Reading).

10.3.1 Antifolates

The observation by Farber in 1948[15] that folic acid (**27**) conjugates administered to children with acute leukaemia made their condition worse led to the use of the antifolate aminopterin (**28**) (4-aminofolic acid) in cancer treatment. The use of aminopterin has now been superseded by that of amethopterin (**29**), more commonly known as methotrexate.

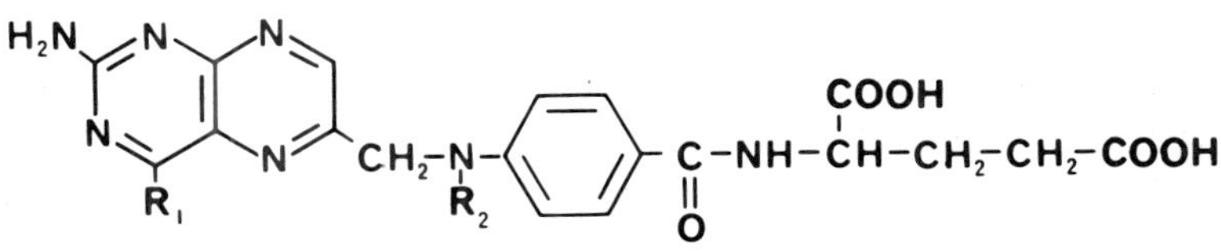

27	$R_1 = OH$	$R_2 = H$
28	$R_1 = NH_2$	$R_2 = H$
29	$R = NH_2$	$R_2 = CH_3$

Methotrexate is a potent inhibitor of the enzyme dihydrofolate reductase (Figure 10.5**a**) and thus blocks tetrahydrofolate production. Tetrahydrofolate is a precursor for N^5, N^{10}-methylene tetrahydrofolate, which is involved in the catalytic transfer of one-carbon unit from deoxyuridine monophosphate (dUMP) to thymidine monophosphate (TMP) by thymidylate synthetase. During this process methylene tetrahydrofolate is converted back into dihydrofolate. As this cannot be reduced again to give a further supply of active cofactors (because of the presence of the inhibitor methotrexate), DNA synthesis will be interrupted. Reduced folates are also required for the formation of one-carbon-transferring coenzymes involved in the biosynthesis of inosinic acid (IMP) which is the precursor for purines (Figure 5**b**). Additionally interference with protein synthesis and the biosynthesis of some amino acids (e.g. serine, methionine) are also caused by lack of folate coenzymes.

Figure 10.5 Sites of Action of Methotrexate (MTX)

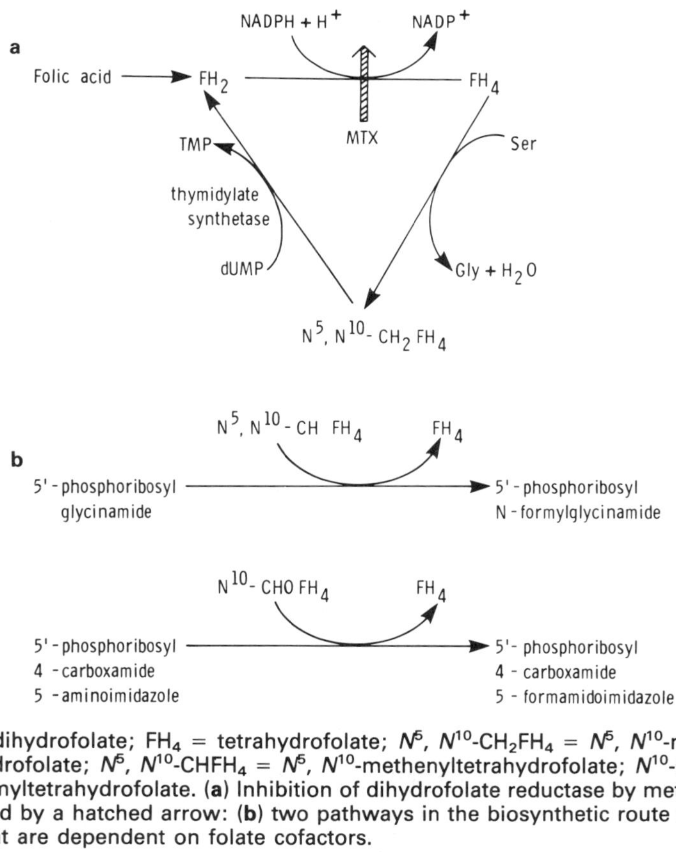

FH_2 = dihydrofolate; FH_4 = tetrahydrofolate; N^5, N^{10}-CH_2FH_4 = N^5, N^{10}-methylene-tetrahydrofolate; N^5, N^{10}-$CHFH_4$ = N^5, N^{10}-methenyltetrahydrofolate; N^{10}-$CHOFH_4$ = N^{10}-formyltetrahydrofolate. (**a**) Inhibition of dihydrofolate reductase by methotrexate, indicated by a hatched arrow: (**b**) two pathways in the biosynthetic route to inosinic acid that are dependent on folate cofactors.

Cultured cells treated with methotrexate become purineless and/or pyrimidineless (the term 'thymineless death' is often used to describe cells in which DNA synthesis has stopped but RNA and protein synthesis have continued in an unbalanced manner). Animals treated with toxic doses of methotrexate may be rescued by treatment with thymidine or mixtures of purines and pyrimidines. In the clinic rescue techniques are frequently used after high-dose methotrexate treatment. The rescuing agent is normally leucovorin (5-formyltetrahydrofolate) (**30**) that replenishes the reduced folate pool in the cell. The use of this compound is described in more detail in Chapter 11.

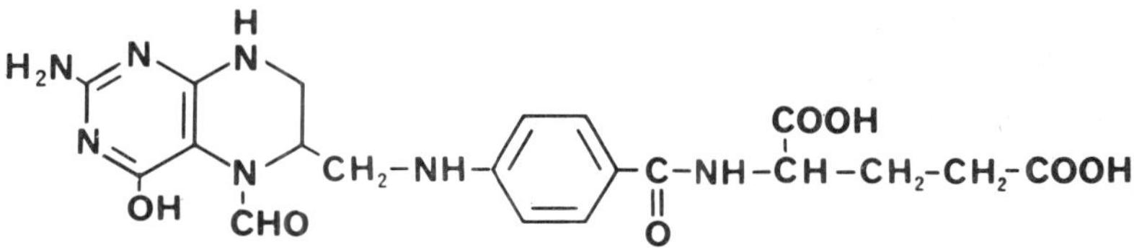

30

Methotrexate shows no selectivity in its action on dihydrofolate reductase in tumour cells as compared with the enzyme in normal cells. The cytotoxic action is, however, dependant on cellular proliferation and is phase-dependent, the drug acting at the DNA-synthesis stage (S-phase). Many other features may govern a cell's susceptibility to methotrexate, including transport of the drug into the cells, levels of dihydrofolate reductase (and its rate of synthesis) and the presence of 'salvage' pathways for thymidylate synthesis. (Salvage pathways are biosynthetic routes that bypass the metabolic block, e.g. in this case the formation of thymidylate from preformed thymidine by thymidine kinase.)

10.3.2 Pyrimidine Antimetabolites

10.3.2.1 5-Fluorouracil, 5-Fluoro-2'-deoxyuridine. 5-Fluorouracil (**31**) provides an example of rational synthesis resulting in a clinically useful anticancer agent. The initial observation was that chemically induced rat hepatoma cells utilised uracil for nucleic acid pyrimidine biosynthesis more rapidly than normal liver. Charles Heidelberger and his colleagues[16] attempted to exploit this difference by synthesising a compound that would be mistaken by living cells for uracil but which would act as an antimetabolite. Because of increased uptake it was hoped the compound would be selectively taken up by tumour cells. Since a remarkable increase in toxicity was known to occur when fluorine was substituted for hydrogen in acetate it was decided to synthesise a uracil derivative with substitution of a fluorine atom for a hydrogen atom at the C-5 position. In

5-fluorouracil (**31**) the fluorine atom, which is slightly bulkier than the hydrogen of uracil, is not recognised as foreign by enzymes which convert 5-fluorouracil into nucleotides. However, once converted by the cell into 5-fluorodeoxyuridine monophosphate (5-FdUMP) this compound binds tightly to the enzyme thymidylate synthetase competing with the natural substrate deoxyuridine monophosphate and inhibits formation of thymidine monophosphate, an essential precursor of thymidine triphosphate required for DNA synthesis. The production *in vivo* of 5-FdUMP may also be achieved by administration of 5-fluoro-2′-deoxyuridine (**32**) which is phosphorylated by thymidine kinase. 5-Fluorouracil is incorporated into RNA and affects its processing and function, which may also be of significance with regard to the cytotoxicity of the drug.

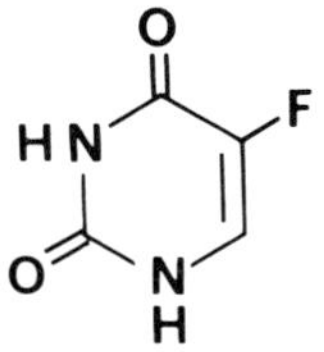

31

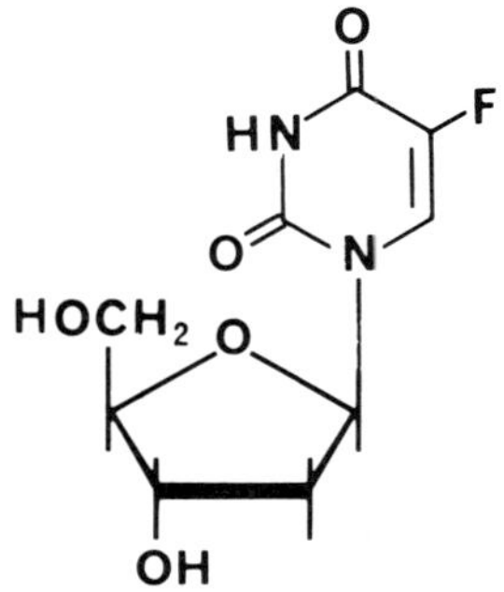

32

10.3.2.2 Cytosine arabinoside (1-ß-D-Arabinofuranosylcytosine, araC) (**33**) This is an analogue of cytidine, in which the hydroxyl group at the 2-position of the sugar is inverted, to be *trans* to that in the 3-position. Cytosine arabinoside is a S-phase-specific agent and inhibits DNA synthesis. Deoxycytidine kinase catalyses the phosphorylation of Cytosine arabinoside to its 5′-monophosphate, which is then converted into the triphosphate (araCTP). Although the mechanism of action of Cytosine arabinoside is not clear, araCTP inhibits DNA polymerase. AraC is also incorporated into DNA and, to a lesser extent, into RNA, and its nucleotides inhibit ribonucleoside diphosphate reductase, which is involved in the production of deoxyribonucleotides for DNA synthesis. The quantitative relations of these mechanisms to the anti-tumour activity of Cytosine arabinoside has not been established.

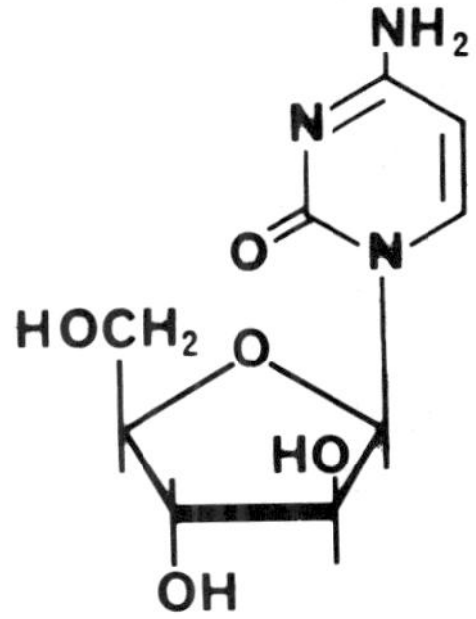

33

10.3.3 Purine Antimetabolites

10.3.3.1 6-Mercaptopurine (**34**). This is an analogue of hypoxanthine in which the extracyclic oxygen atom is replaced by a sulphur atom. 6-Mercaptopurine exerts a number of inhibitory effects on purine biosynthesis and purine nucleotide interconversion. It is activated by phosphorylation catalysed by hypoxanthine-guanine phosphoribosyltransferase, which yields thionosinic acid (Figure 10.6). (6-Mercaptopurine is a competitive inhibitor for the normal process catalysed by this enzyme, the conversion of hypoxanthine into inosinic acid.) Thioinosinic acid inhibits the conversion of inosine monophosphate (IMP) to xanthine monophosphate (XMP) by IMP dehydrogenase, and both the conversion of IMP

Figure 10.6 Biosynthetic Pathways Inhibited by Thioinosinic Acid (thio-MP)

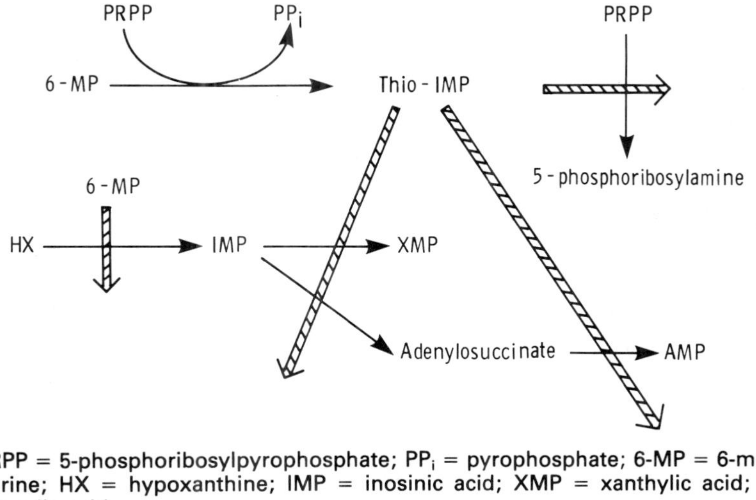

PRPP = 5-phosphoribosylpyrophosphate; PP_i = pyrophosphate; 6-MP = 6-mercaptopurine; HX = hypoxanthine; IMP = inosinic acid; XMP = xanthylic acid; AMP = adenylic acid.

into adenylosuccinate and of adenylosuccinate into adenosine monophosphate (AMP). An additional potentially important site of inhibition of thioinosinic acid is the conversion of 5-phosphoribosylpyrophosphate (PRPP) to 5-phosphoribosylamine, which is an early stage in the synthesis *de novo* of purines. After conversion into its triphosphate, thioinosinic acid can inhibit the synthesis of the coenzyme NAD. DNA incorporation of metabolised 6-mercaptopurine (as 2′-deoxythioguanosine; see later) also occurs, which raises the possibility of further inhibition of DNA replication.

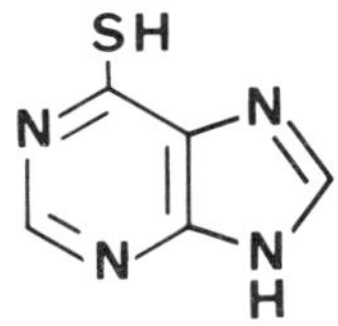

34

As can be seen from the above, 6-mercaptopurine has a complex effect on the purine pathways. Despite extensive study of these effects the mechanism responsible for cell death is not (and, in view of the complexity of the cellular biochemistry involved, may never be) understood. Anti-tumour activity of the drug and its toxicity toward normal tissues is greatest toward rapidly proliferating cells.

10.3.3.2 6-Thioguanine (**35**). This is also converted into its ribotide (thioGMP) by the action of hypoxanthine-guanine phosphoribosyltransferase. Like thioinosinic acid, thioGMP acts as a feedback inhibitor of the PRPP to 5-phosphoribosylamine conversion. It also inhibits the formation of XMP from IMP and of GDP and GMP. There is extensive incorporation of 6-thioguanine into nucleic acids, the extent of which is correlated, in experimental systems, with the cytotoxicity of the drug.

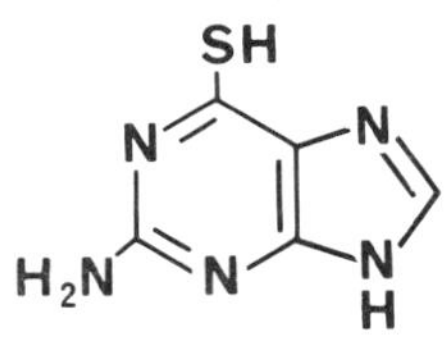

35

10.4 Natural Products

Natural materials have been the source of a large proportion of the clinically used anticancer agents (see also Chapter 11). The biochemical effects caused within cells by these compounds are generally not understood in as great a detail as those from synthetically derived alkylating agents and antimetabolites. Natural-product drugs are also normally of great chemical complexity which limits the potential for analogue synthesis and the development of structure–activity relations.

10.4.1 Mitotic Inhibitors

After the discovery of anti-tumour activity in extracts from the periwinkle plant (*Vinca rosea*) two alkaloids vincristine (**36**) and vinblastine (**37**) were isolated and subjected to clinical testing.[17] These compounds are cell-cycle-specific agents and inhibit cells in metaphase. The mitotic spindle that is formed in cells during their division is partially composed of 'microtubules' (structures of around 25 nm width, related to fibrillar subunits). In cultured cells, vincristine and vinblastine cause the disappearance of these microtubules, probably because of their binding to the protein tubulin that is involved in the formation of these spindle structures. As a result the mitotic spindle does not form the structure required for orderly segregation of the chromosomes in metaphase. Interaction of vinca alkaloids with the microtubule system also leads to the formation of aggregates known as microtubule crystals, which again will upset the function of the mitotic spindle. The chemical association of the vinca alkaloids with these systems is extremely structure-specific. Even small structural modifications, such as the removal of the acetyl group at position 4, cause loss of activity, presumably by altering the binding of the alkaloid to tubulin.

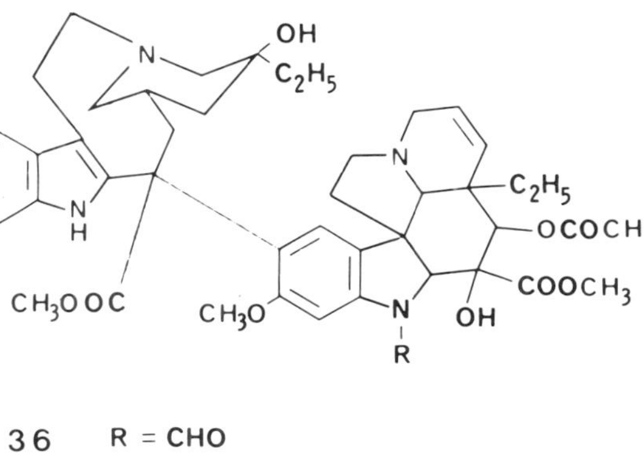

36 R = CHO

37 R = CH_3

The resin of the May apple (*Podophyllum peltatum*) has also been used as a source of antimitotic agents. The main active constituent is podophyllotoxin that is used in the treatment of warts. Chemical modification of this structure has led to two analogues VM 26 (teniposide) and VP 16-213 (etoposide) which act on the cell in an uncertain manner at the premitotic stage and have no effect on the microtubule assembly. They thus differ slightly from the vinca spindle poisons. Both of these agents are at the early stage of clinical use.

10.4.2 Antibiotics

Streptomyces culture fluids have been the source for a wide variety of antibiotics having anticancer properties (see Sartorelli and Johns, Further Reading). These include actinomycin D (dactinomycin) (**38**), whose mode of action is believed to be due to its strong binding to double-stranded DNA, the glycopeptide, bleomycin, which causes breakage of DNA strands, the anthracyclines, doxorubicin (adriamycin) (**39**) and daunorubicin (daunomycin) (**40**), which interchelate between DNA base pairs and produce reactive free radicals, mithramycin (whose structure contains an aglycone, chromomycinone, linked to three sugars, olivose, oliose and mycarose), which binds to DNA and inhibits RNA synthesis and mitomycin (**41**), which is believed to be reduced *in vivo* to a bifunctional alkylating agent capable of cross-linking DNA.

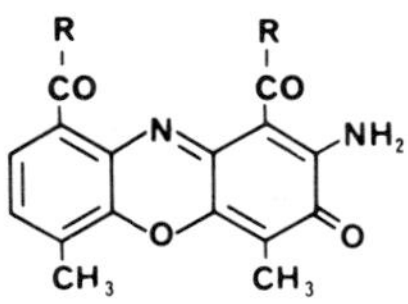

38 R =

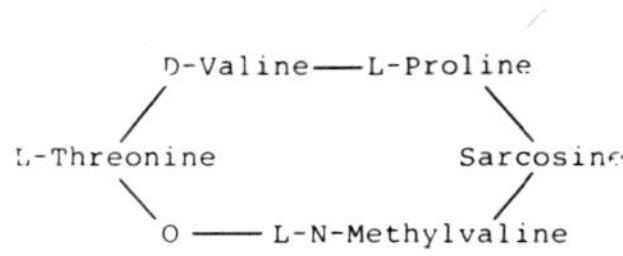

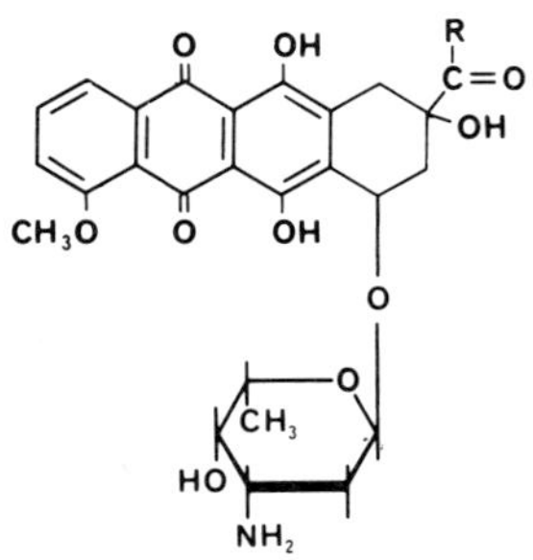

39 R = CH_2OH

40 R = CH_3

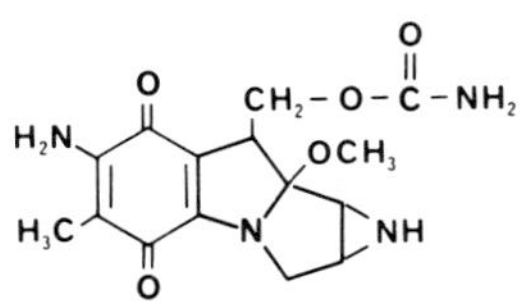

41

10.4.3 L-Asparaginase

The discovery that a variety of tumour cells have a nutritional requirement for the amino acid asparagine (e.g. see Ref. 18) led to a great deal of excitement about the use of the enzyme L-asparaginase (L-asparagine amidohydrolase) as an anticancer tool. Administration of this enzyme should remove extracellular asparagine and only cells that are capable of producing their own stocks of the amino acid should survive. The anti-tumour effect of L-asparaginase was first suspected when a mouse lymphoma was shown to regress permanently after administration of guinea-pig serum.[19] The effect was not due to immunological factors but was caused by the presence of L-asparaginase which is present in this serum. Although the enzyme can be isolated from this source it is normally prepared from *Escherichia coli*.

In principle the use of L-asparaginase for therapy should be close to the ideal situation of inhibiting a pathway specific for tumour cells, with no toxic effects on normal cells. In practice the compound has not lived up to its promise; it is quite toxic (liver, pancreas, central nervous system, etc.) and has had only partially successful use on the treatment of acute lymphoblastic leukaemia. Immunologic sensitisation to the drug, which may be severe, is also seen in some treated patients. Asparaginase is, however, not toxic to the bone marrow which should make it an attractive component for combination therapy (Chapter 11).

10.5 Other Drugs

10.5.1 Platinum Complexes

The discovery by Rosenberg *et al.* (Chapter 11)[20] that *cis*-dichloroplatinum complexes show anticancer activity has led to widespread investigation of organoplatinum compounds. Cisplatin [*cis*-diamminedichloroplatinum (II)] (**42**) was the first platinum compound to reach the clinic; several more analogues are now at the clinical-trial stage. Cisplatin behaves like an alkylating agent in that it reacts with nucleophiles, including the bases in DNA where inter- and intra-strand cross-links are formed. Reaction with sulphur nucleophiles also occurs extensively. Cisplatin is a highly nephrotoxic compound and it has been suggested that this is caused by the binding of the drug to SH groups in kidney tubules. As kidney damage is often dose-limiting in the use of cisplatin, diuretics are frequently administered with the drug. Rescue techniques [e.g. the administration of a competing nucleophile like sodium diethyldithiocarbamate (DDTC)] have also been used with considerable success in animals in alleviating kidney toxicity without affecting anti-tumour activity.[21] The new analogues of cisplatin now being tested [e.g. JM8, *cis*-diammine-1, 1-cyclobutane dicarboxylate platinum (**II**) (**43**)] should exhibit lower nephrotoxicity than the parent drug.

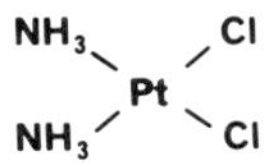

42

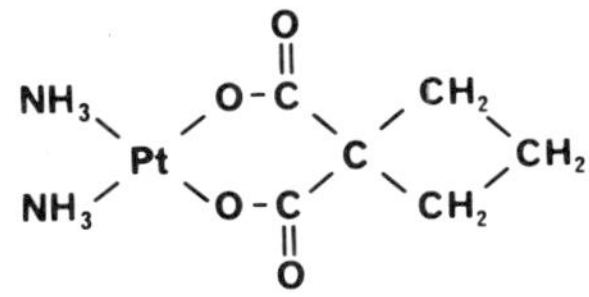

43

10.5.2 Miscellaneous

A challenging puzzle to those interested in the mechanism of action of anticancer drugs is the cause for the activity of several *N*-methyl compounds. Examples of these are procarbazine (**44**) (widely used in combination therapy, e.g. of Hodgkin's disease), hexamethylmelamine (**45**), pentamethylmelamine and *N*-methylformamide (**46**). The *N*-methyl groups in these compounds are essential for their anticancer action; replacement by an ethyl group decreases their activity. Extensive metabolic *N*-demethylation occurs for the melamine derivatives yielding formaldehyde (cf. triazenes, section 10.2.3.3), although the relevance of this to the anti-tumour activity is unknown.

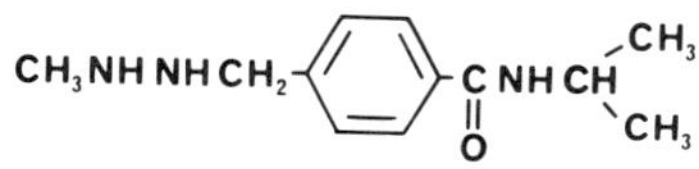

44

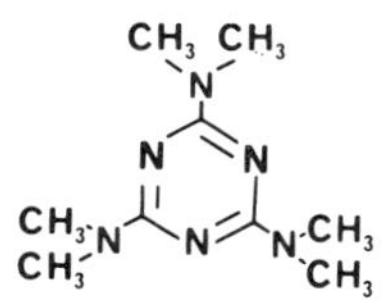

45

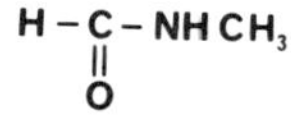

46

The three other main areas of drug therapy of cancer are the use of hormones (and antagonists), of radiation sensitisers and of non-cytotoxic drugs. These are discussed in Chapters 8, 9 and 11 respectively.

10.6 Search for Increased Selectivity

Increased selectivity of a drug to its target may be achieved either by increasing its transport to, or by increasing its effectiveness at, its site of action. Selectivity of transport to a particular organ may be relatively easy to establish; for example, by local infusion of the drug or by taking advantage of partition-coefficient differences (e.g. lipid-soluble drugs may enter the central nervous system). Achievement of selectivity of transport toward tumour *cells*, however, presents another order of difficulty. Several attempts are being made to design drugs which will home in preferentially on to tumour cells, and which will thus break away from the slightly unsatisfactory concept, characteristic of most of the compounds described above, of selecting their target cells on the basis of their growth rate.

Extramustine and prednimustine represent two of these attempts in which a cytotoxic alkylating function is bound to a hormonal residue, the hope being that the tumour would have receptor sites for the hormones and would thus receive a high local concentration of active drug (see Ames *et al.*, further Reading). Similar approaches are being made using drugs linked to tumour-specific antibodies (immunotargeting). A recent review[22] likens the procedure to the actions of Retiarius, a gladiator in ancient Rome who cast a net to entangle and draw his opponent to be killed with a trident. The authors thus refer to immunotargeting as Retiarian therapy, as it relies on recognition, interaction and subsequent kill. This approach is fraught with many difficulties. The chemistry to combine drugs to antibodies while retaining antigen recognition has been worked out but the major difficulty occurs in directing enough drug to the target. In general only a very small proportion of the antibody dose finds its way to the tumour. While this may be sufficient for localisation it is often not sufficient for therapeutic use. Similarly once the antibody–drug combination finds its target there is no guarantee that the complex will enter the cell for therapy to take place. Use of antibodies conjugated with ribosome-inactivating proteins such as ricin, abrin or the A chain of diptheria toxin have been used in attempts to overcome this difficulty. These proteins penetrate membranes and only a few molecules are needed to kill a cell. The major problem is therefore with antibody specificity. The possibility that better targeting can be achieved by using membrane-fixed antigens rather than secreted materials (see Chapter 7) is currently being investigated. The production of new antibodies to tumours and particularly the use of monoclonals should result in a rapid expansion of this work and further knowledge of oncogene-derived products may lead the way to highly specific target sites to which the drug may be bound.

Other aspects of modifying drug-delivery methods, and their relevance to improving anticancer therapy, are discussed in Chapter 11.

10.7 Problems Associated with Cytotoxic Therapy

10.7.1 Toxicity

Because anti-tumour drugs are not totally selective for tumour cells, they show many toxic effects to other organs. For proliferation-dependent agents these include the bone marrow (haematopoietic suppression), bladder (cystitis) and hair follicles (loss of hair). Effects on the gastrointestinal tract, the reproductive tissues and the central nervous system are also observed; nausea and vomiting are very common side-effects of anticancer agent therapy. Many anticancer drugs are also carcinogenic (see Chapter 5 for carcinogenicity of alkylating agents), mutagenic and teratogenic. As anticancer treatment becomes more successful, with greater survival time for treated patients, the incidence of drug-related (iatrogenic) cancers is becoming more apparent. The drugs that have been associated with subsequent cancer developments include cyclophosphamide, nitrogen mustard, chlorambucil and melphalan.[23]

Many of these toxic side-effects may be minimised by clinical intervention. For example, bladder toxicity may be minimised by hydration or administration of diuretics or nucleophilic compounds (see sections 10.2.3.1 and 10.5.1). It is possible also that the increase in bladder cancer that is caused by cyclophosphamide may be reduced by the use of the uroprotectant mesna.[24] Doses of alkylating agent that would normally result in bone-marrow death may successfully be given to patients if autologous marrow transplantation is carried out.[25] For this a sample of healthy marrow is removed before treatment and reimplanted, subsequently, when the plasma level of active drug has fallen below danger level. The development of these and other rescue techniques has meant that higher and higher doses of drugs are being given to patients.

Many anticancer agents (e.g. cyclophosphamide, cytosine arabinoside, methotrexate) are powerful immunosuppressants. Although this is normally a disadvantageous property for patients receiving therapy (due to the lack of resistance caused to infections and possibly to iatrogenic cancer), these drugs may be used profitably in the treatment of immunological conditions, and as immunosuppressants after organ transplant.

10.7.2 Resistance

Resistance to anticancer agents frequently limits their use. Some of the mechanisms by which resistance may arise are as follows:

(1) decreased transport of the drug into the cell (e.g. alkylating agents, vinca alkaloids, methotrexate)
(2) gene amplification, in which the messenger RNA for an enzyme that is inhibited by the drug is produced in greater quantities (e.g.

dihydrofolate reductase for methotrexate)
(3) increased DNA repair (alkylating agents)
(4) production of modified enzymes to avoid an inhibition (e.g. modified dihydrofolate reductase for methotrexate)
(5) increased thiol concentration (alkylating agents)
(6) increased level of deactivating enzyme (e.g. cytidine deaminase which inactivates cytosine arabinoside)
(7) decreased level of activating enzyme (e.g. deoxycytidine kinase for cytosine arabinoside, hypoxanthine-guanine phosphoribosyltransferase for 6-mercaptopurine and 6-thioguanine).

Generally there is little one can do to overcome resistance apart from increasing the dose of the drug. Normally alternative therapy is used when resistance arises.

10.8 Conclusion

It is not possible in a single chapter to give more than a minimal overview of the mechanisms of action of the anticancer drugs in use and development, and the reader is recommended to consult Further Reading references for more information. It should be apparent from the above, however, that the concept of administering highly toxic compounds, coupled with extensive (and frequently very expensive) rescue techniques, is not ideal, either for the patient or for the clinician. Only an increase in selectivity of the drugs, which is of especial importance for the treatment of the slowly dividing tumour cells which are abundant within solid tumours, will resolve the situation. Although our existing drugs do not possess this selectivity they can, when used correctly, produce valuable clinical results. Studies of their metabolism, cell-cycle specificity, pharmacokinetics and interactions with other drugs are all essential prerequisites for the successful use of these anticancer agents, and are discussed in more detail in the subsequent chapter.

References

1. Gilman, A. & Philips, F.S. (1946) The biological actions and therapeutic applications of the β-chloroethylamines and sulfides. *Science*, *103*, 409-15
2. Colvin, M., Brundrett, R.B., Kan, M.-N.N., Jardine, I. & Fenselau, C. (1976) Alkylating properties of phosphoramide mustard. *Cancer Res.*, *36*, 1121-6
3. Bergel, F. & Stock, J.A. (1954) Cytoactive amino acid and peptide derivatives. Part. 1. Substituted phenylalanines. *J. Chem. Soc.*, 2409-17
4. McLean, A., Woods., R.L., Catovsky, D. & Farmer, P. (1979) Pharmacokinetics and metabolism of chlorambucil in patients with malignant disease. *Cancer Treat. Rev.*, *6* (Suppl.), 33-42
5. Brock, N. & Hohorst, H.-J. (1977) The problem of specificity and selectivity of

alkylating cytostatics: studies on *N*-2-chloroethylamido-oxazaphosphorines. *Z. Krebsforsch.*, *88*, 185-215

6. Brock, N., Pohl, J. & Stekar, J. (1981) Detoxification of urotoxic oxazaphosphorines by sulphydryl compounds. *J. Cancer Clin. Oncol.*, *100*, 311-20
7. Tong, W.P. & Ludlum, D.B. (1980) Cross-linking of DNA by busulfan. Formation of diguanyl derivatives. *Biochim. Biophys. Acta*, *608*, 174-81
8. Schilsky, R.L., Kelley, J.A., Ihde, D.C., Howser, D.M., Cordes, R.S. & Young, R.C. (1982) Phase I trial and pharmacokinetics of aziridinylbenzoquinone (NSC 182986) in humans. *Cancer Res.*, *42*, 1582-6
9. Chiuten, D.F., Rozencweig, M., Van Hoff, D.D. & Muggia, F.M. (1981) Clinical trials with the hexitol derivatives in the US. *Cancer*, *47*, 442-51
10. Kohn, K.W. (1977) Interstrand cross-linking of DNA by 1,3-bis(2-chloroethyl)-1-nitrosourea and other 1-(2-haloethyl)-1-nitrosoureas. *Cancer Res.*, *37*, 1450-4
11. Gombar, C.T., Tong, W.P. & Ludlum, D.B. (1981) Mechanisms of actions of the nitrosoureas – IV. Reactions of bis-chloroethylnitrosourea and chloroethylcyclohexylnitrosourea with deoxyribonucleic acid. *Biochem. Pharmacol.*, *29*, 2639-43
12. Tew, K.D., Smulson, M.E. & Schein, P.S. (1981) Molecular pharmacology of nitrosoureas. *Recent Results Cancer Res.*, *76*, 130-40
13. Kline, I., Woodman, R.J., Gang, M. & Venditti, J.M. (1971) Effectiveness of antileukemic agents in mice inoculated with leukemia L1210 variants resistant to 5-(3,3-dimethyl-1-triazeno)-imidazole-4-carboxamide (NSC-45388) or 5-[3,3-bis(2-chloroethyl)-1-triazeno]imidazole-4-carboxamide (NSC-82196). *Cancer Chemother. Rep.*, *55*, 9-28
14. Hammer, C.F., Loranger, R.A. & Schein, P.S. (1981) Structures of the decomposition products of chlorozotocin: new intramolecular carbamates of 2-amino-2-deoxyhexoses. *J. Org. Chem.*, *46*, 1521-31
15. Farber, S., Diamond, L.K., Mercer, R.D., Sylvester, R.F. & Wolff, J.A. (1948) Temporary remissions in acute leukemia in children produced by folic acid antagonists, 4-aminopteroylglutamic acid (aminopterin). *N. Engl. J. Med.*, *238*, 787-93
16. Heidelberger, C., Chaudhari, N.K., Danneberg, P., Mooren, D., Griesbach, L., Duschinsky, R., Schnitzer, R.J., Pleven, E. & Scheiner, J. (1957) Fluorinated pyrimidines. A new class of tumour-inhibitory compounds. *Nature (Lond.)*, *179*, 663-6
17. Johnson, I.S., Armstrong, J.G., Gorman, E. & Burnett, J.P. (1963) The vinca alkaloids: a new class of oncolytic agents. *Cancer Res.*, *23*, 1390-1427
18. Neuman, R.E. & McCoy, T.A. (1956) Dual requirement of Walker carcinosarcoma 256 *in vitro* for asparagine and glutamine. *Science*, *124*, 124-5
19. Kidd, J.G. (1953) Regression of transplanted lymphomas induced *in vivo* by means of normal guinea pig serum. I. Course of transplanted cancers of various kinds in mice and rats given guinea pig serum, horse serum or rabbit serum. *J. Exp. Med.*, *98*, 565-82
20. Rosenberg, B., Van Camp, L. & Krigas, T. (1965) Inhibition of cell division in *Escherichia coli* by electrolysis products from a platinum electrode. *Nature (Lond.)*, *205*, 698-9
21. Borch, R.F., Katz, J.C., Lieder, P.H. & Pleasants, M.E. (1980) Effect of diethyldithiocarbamate rescue on tumor response to *cis*-platinum in a rat model. *Proc. Natl. Acad. Sci. USA*, *77*, 5441-4
22. Edwards, C.D. & Thorpe, P.E. (1981) Targetting toxins – the retiarian approach to chemotherapy. *Trends Biochem. Sci.*, *6*, 313-16
23. Schmahl, D., Habs, M., Lorenz, M. & Wagner, I. (1982) Occurence of second tumors in man after anticancer drug treatment. *Cancer Treat. Rev.*, *9*, 167-94
24. Habs, M.R. & Schmahl, D. (1983) Prevention of urinary bladder tumours in cyclophosphamide treated rats by additional medication with the uroprotectant sodium 2-mercaptoethane sulphonate (Mesna) and disodium 2,2-dithio-bis-ethane sulphonate (Dimesna). *Cancer*, *51*, 606-9
25. McElwain, T.J., Hedley, D.W., Burton, G., Clink, H.M., Gordon, M.Y., Jarman, M., Juttner, C.A., Millar, J.L., Milsted, R.A.V., Prentice, G., Smith, I.E., Spence, D. & Woods, M. (1979) Marrow autotransplantation accelerates haematological recovery in patients with malignant melanoma treated with high dose melphalan. *Br. J. Cancer*, *40*, 72-80.

Further Reading

Ames, M.M., Powis, G. & Kovach, J.S. (eds) (1983) *Pharmacokinetics of Anticancer Agents in Humans*, Elsevier/North-Holland, Amsterdam

Chabner, B. (1982) *Pharmacological Principles of Cancer Treatment*, W.B. Saunders, Philadelphia

Ross, W.C.J. (1962) *Biological Alkylating Agents*, Butterworth, London

Sartorelli, A.C. & Johns, D.G. (1975) *Handbook of Experimental Pharmacology*, vol. 38, *Antineoplastic and Immunosuppressive Agents*, part II, Springer-Verlag, Berlin

11 CANCER CHEMOTHERAPY II: LABORATORY TO CLINIC

G. Powis

Contents

11.1 Where Do Anticancer Drugs Come From?

There are currently over 30 approved drugs for the treatment of cancer in the USA and 36 in the UK. This does not include hormonal agents such as adrenal cortical steroids (cortisone, prednisone), androgens (testosterone), oestrogens (diethystilboestrol), progesterone and adrenocorticotropic hormone (ACTH) which are used in treatment of some hormonally dependent tumours (see Chapter 8). Since the first anticancer drug, nitrogen mustard, was introduced in 1949 there has been a steady increase in the number of anticancer drugs available (Figure 11.1). Approximately two-thirds of the drugs are synthetic while the remainder are of natural origin, being plant or fermentation products. These drugs represent just a very few of an enormous number of compounds that have been tested in animals for anti-tumour activity. Over 275 000 compounds with known structure and an equal number of natural-product extracts have been tested in the USA by the National Cancer Institute in its screening programme, which started in 1955. Screening programmes, although generally on a more limited scale, exist in countries outside the USA and testing is also carried out by the drug industry. Screening for anti-tumour activity relies on the assumption, which remains to be proven, that a correlation exists between a compound's activity against transplantable tumours in animals and its therapeutic effectiveness in human cancer. In the early years of screening for anti-tumour activity compounds were selected by an empirical random process. In recent years more attention has been paid to a rational selection. As is the case with many drugs used in medicine most anticancer drugs or their precursors were found serendipitously or in laboratory screens and attempts at drug design have been most successful when modifying the structure of a compound with established activity, so-called 'analog' development. Random selection and rational synthesis have both played a part in the development of some of the most widely used anticancer drugs and some examples are given below.

Natural products are an excellent source of complex chemicals with a wide range of biological activities. A large number of fermentation broths from exotic micro-organisms collected, often in soil samples, from all over

Figure 11.1 Anticancer Drugs Approved for Use in the USA (not including hormonal agents)

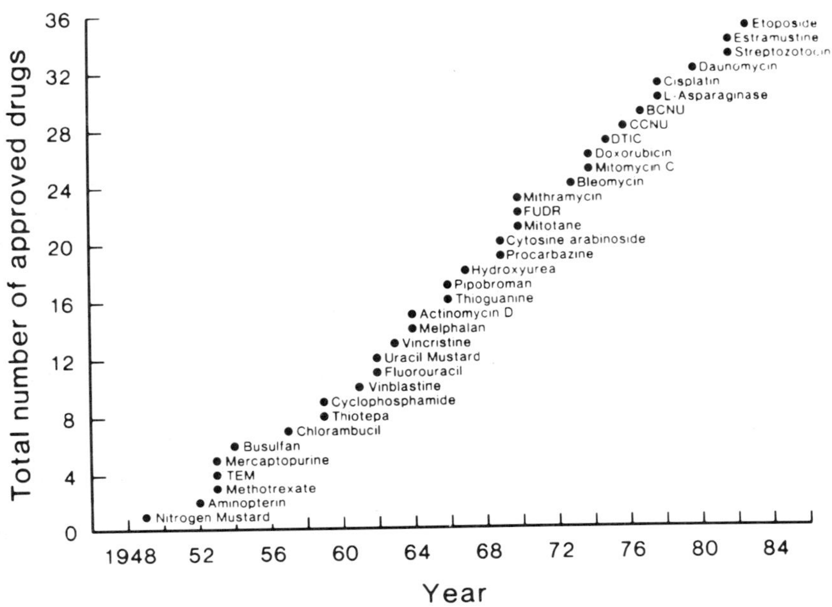

Adapted from DeVita **et al.** (1979) *Cancer Clin. Trials, 2,* 195-216; with permission, Masson Publishing USA, Inc., New York.

the world have been tested for anti-tumour activity. Initial screening is often random but can later focus on strains related to micro-organisms producing active anti-tumour agents. Daunomycin is an anthracycline antibiotic with anti-tumour activity isolated in 1963 by researchers at the Farmitalia Research Laboratories in Milan, Italy, from cultures of *Streptomyces peucetius*.[2] The *Streptomyces* cultures came from soil samples from Castel del Monte, in Puglia, Italy, hence the name of the species *S. peucetius*. Later a closely related compound, doxorubicin, was isolated from a mutant strain of *S. peucetius*. Doxorubicin is now preferred clinically to daunomycin because of its broader spectrum of activity against solid tumours.

The anti-tumour activity of *cis*-diamminedichloroplatinum (cisplatin) was discovered in 1965 through an astute observation by Barnett Rosenberg and his colleagues[3] that alternating electrical current delivered through platinum electrodes inhibited the growth of bacterial cells in culture and caused the bacteria to grow in long filaments. This pattern of unbalanced bacterial growth was recognised as being similar to that produced by a variety of alkylating agents and by radiation. It was soon found that platinum released from the electrodes in the form of hexachloroplatinate anions reacted, in the presence of light, with ammonium ions in the medium to form a complex containing both chloride and ammonia. Examination of several complexes of this type showed that the *cis*-isomer of $Pt(NH_3)_2Cl_2$, *cis*-diamminedichloroplatinum, produced characteristic inhibition of bacterial growth and more importantly had potent anti-tumour activity in animal models (see Chapter 10).

5-Fluorouracil provides an example of rational synthesis resulting in a clinically useful anticancer agent (see Chapter 10). The synthesis of cyclophosphamide was also rationally conceived to exploit a reported enzymic difference between malignant and normal cells, namely higher levels of phosphoramidase activity in tumour cells. Although the original biochemical hypothesis proved untenable cyclophosphamide has proven to be one of the most useful anticancer drugs, having a broad spectrum of activity and a relatively wide margin of safety (see Chapter 10).

Stages in the typical development of a new anticancer drug are shown in Figure 11.2. Compounds to be tested for anti-tumour activity can come from non-industrial suppliers, including universities and research institutes, from world-wide collection programmes for plant and animal material and fermentation products and from the industrial community. The time taken from acquisition of a drug to its introduction into the clinic can take anywhere from 5 to 15 years and is a costly process.[1] A typical cost for

Figure 11.2 Stages of Development of an Anticancer Drug

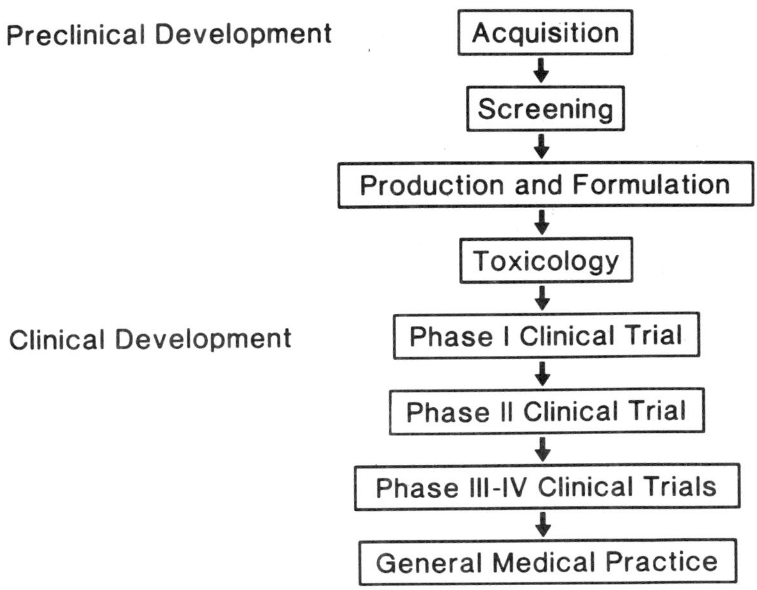

preclinical development of an anticancer drug is over $0.5 million. Clinical testing through phase II trial adds a further $1 million.[4] The cost of running a drug-development programme is, however, much greater than this because of the number of compounds which have to be discarded because of toxicity, difficulty of formulation or lack of clinical effectiveness. In an attempt to improve the cost effectiveness of the drug-discovery process a computerised statistical–heuristic model has been used by the National Cancer Institute in the USA since 1980 to aid in preselection of compounds for screening. The model uses molecular structure features as predictors of biological activity based on the total experience of structure–activity relations in the murine P-388 leukaemia screening system. An unknown compound is scored for novelty and potential activity by adding the weights of a dozen or so structural features. Of more than 20 000 potential compounds available in 1982, 10 000 were selected for screening. The different stages of anticancer drug development after acquisition are dealt with below.

11.2 Screening Methods

11.2.1 Systems in vitro

A wide range of test systems *in vitro* have been employed in the past to screen for anti-tumour activity, generally measuring inhibition of a single parameter of response, and have included slime, mould, bacteria, fungi, viruses, frog and chick embryos, drosophila as well as tissue-culture systems.[5] There is little evidence that non-tumour systems offer a substantial advantage over tumour systems for the screening of cytotoxic agents and most work *in vitro* now employs tumour cells growing in tissue culture. Systems *in vitro* currently find their greatest use for preliminary screening where only small amounts of compound are available and in the process of isolating the active principal from crude natural-product extracts.

A recent development that could greatly expand the role of systems *in vitro* in screening for anti-tumour drug activity has been the colony-formation assay. It is believed by many investigators that only a small fraction of cells ('stem' cells) within a given tumour are capable of self-perpetuation. 'Stem' cells are thought to be responsible for tumour growth and metastasis and regrowth of the tumour after incomplete surgical or therapeutic intervention. Therefore eradication of the stem-cell population is thought by many to be a key approach to successful therapeutic management of cancer. In culture stem cells form colonies in a semisolid medium since, unlike normal cells, they do not require attachment to plastic surfaces. Several methods have been developed *in vitro* which permit assessment of the growth and drug sensitivity of the proliferative fraction of tumour cells.[6] One such method is outlined in Figure 11.3. Solid-tumour specimens are minced into fine fragments. Single cells are liberated from the connective-tissue matrix with the aid of hydrolytic enzymes such as trypsin, proteinase and/or collagenase. The cells are suspended in aliquots of culture media to which drugs and drug vehicle are added. Tumour-cell growth in various drug environments is monitored in terms of colony formation over the course of 5–21 days incubation. A potential use of this assay is to determine the drug sensitivity (or resistance) profile of a given patient's tumour and then to select single- or multiple-agent therapy on an individual-patient basis. Another use of the assay is to screen new agents for cytotoxic activity. Conceptually such an assay differs from other screening methods in that each potentially useful agent can be tested against a large number and variety of human tumours. While these applications are highly attractive a number of problems with human stem-cell colony-formation assays remain to be resolved. These include assessment of cultures exhibiting poor cloning efficiency, appropriate characterisation of growth and drug sensitivity in heterogeneous populations of tumour cells within a given specimen, or derived from the sarcomatous-type tumour, preparation of cell suspensions lacking debris and cellular aggregates, development of an efficient drug-activating system for drugs which require hepatic activation and the fact that some antimetabolites, for example methotrexate, may appear to be inactive because the growth medium is rich in added cofactors.

Figure 11.3 Soft-agar Colony-formation Assay of Human Solid Tumours

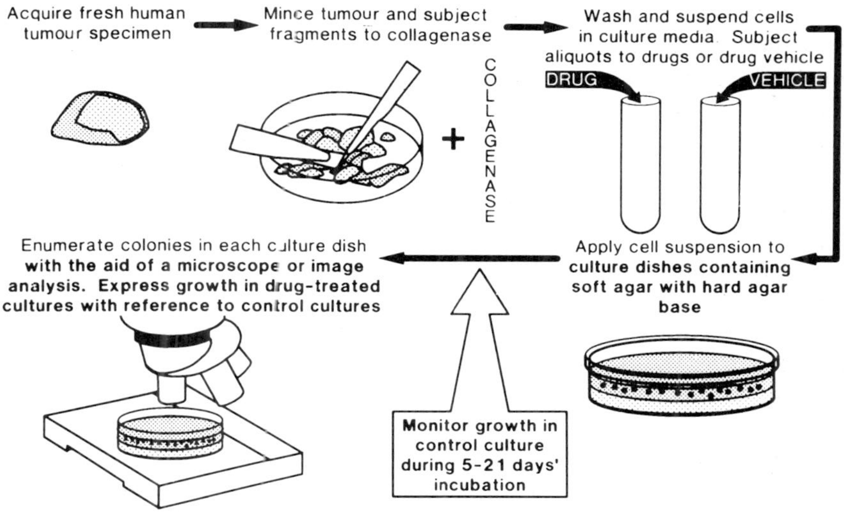

A basic limitation of all systems *in vitro* is that while they can measure inhibition of a specific enzymic or developmental process, or inhibition of cellular proliferation, they provide no indication of the specificity of a drug for tumour cells compared with that for normal cells. Companion assays *in vivo* are always required to determine if a compound has a favourable therapeutic index (section 11.2.2). It is conceivable, however, that advances in molecular biology could eventually lead to a system *in vitro* measuring drug specificity for a system unique to a tumour cell.

11.2.2 Animal Screening

Screening and evaluation of anti-tumour drugs in animals is usually conducted with transplantable tumours in the mouse and rat, although chicken, guinea pig, hamster and rabbit have also been used. The animal is inoculated systemically or parenterally with a defined number of tumour cells, usually 10^5 or 10^6, and the tumour allowed time to establish itself, usually for one to several days (in the latter case called an 'advanced' experimental tumour model). The animal is then administered the test compound at a range of doses up to the LD_{50} (the lethal dose for 50 per cent of the animals). Response to chemotherapy is assessed by measuring tumour growth by palpating or caliper measurements of tumour size, measurement of tumour weight or volume after excision of the tumour, the extent of haemorrhage and necrosis or histological, cytological or biochemical alteration of the tumour cells. Perhaps the most widely used factor for assessing response to chemotherapy is survival time of the animal. An important aspect of any animal model is that it should involve an examination of the effect of an agent being tested on normal cells of the host as well as on tumour cells. Clearly, there is little advantage to an agent that destroys most, or even all, of the tumour cells if the host itself is killed in the process. Figure 11.4 shows the effect of anti-tumour drugs on the survival time of mice tumoured with L1210 leukaemia. Mouse survival times increase to a maximum at an optimal drug dose and then decrease at higher doses as toxicity of the drug to the host becomes dominant. A way frequently used to express drug safety is the *therapeutic index*, that is the ratio of the dose of a drug producing a defined toxic effect to the dose of a drug producing a defined therapeutic effect. This assumes that the toxic effect of the drug in animals will be the same as the dose-limiting toxicity in man. If dose-limiting toxicity in man cannot be adequately assessed in animals, for example nausea or vomiting or headache, a *therapeutic index* measured in animals may have little relevance for the clinical use of the drug. A modification of the therapeutic index sometimes used with anti-tumour drugs is the *therapeutic ratio*, that is the ratio of the dose of drug producing the maximum increase in survival time of tumour-bearing animals (optimal dose) divided by the dose producing a 40 per cent increase in survival time. For two drugs that give the same or a similar increase in survival time the drug with the higher *therapeutic index* or *therapeutic ratio* (and hence the greater degree of safety) may be more desirable for use in man.

Figure 11.4 Dose–response Plot of Anti-tumour Drugs Against Advanced Experimental L1210 Leukaemia

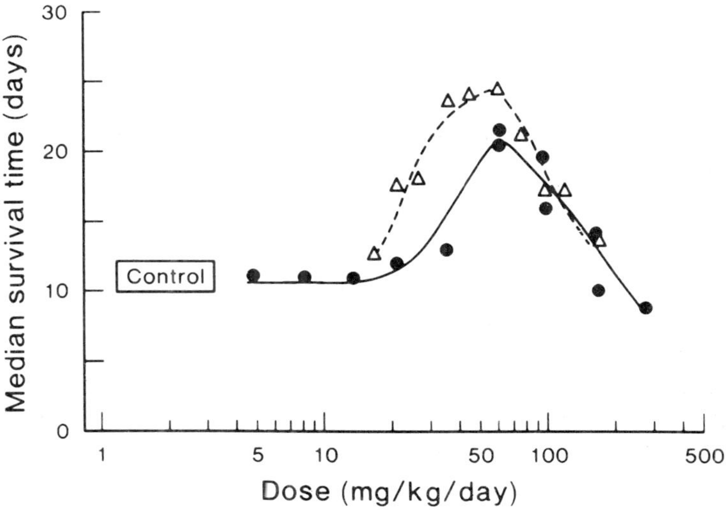

Mice were inoculated subcutaneously in a hind limb with a suspension of leukaemic cells from the spleen of a stock mouse with tumour. Drug treatment was begun when the tumour at the site of inoculation measured 7–12 mm in diameter (that is at seven to eight days). Cyclophosphamide or methyl-GAG (methylglyoxal bisguanylhydrazone) were administered subcutaneously in the axillary area daily until death of the animal.

Cyclophosphamide (△)

$$\text{Therapeutic ratio} = \frac{\text{optimal dose}}{\text{ILS}_{40}} = \frac{60\ \text{mg/kg}}{20\ \text{mg/kg}} = 3.0$$

$$\text{Increase in life-span (ILS) at optimal dose} = \frac{(24.5\ \text{day} - 10\ \text{day})}{10\ \text{day}} \times 100 = 145\%\ \text{(maximum effectiveness)}$$

Methyl-GAG (●)

$$\text{Therapeutic ratio} = \frac{\text{optimal dose}}{\text{ILS}_{40}} = \frac{60\ \text{mg/kg}}{36\ \text{mg/kg}} = 1.7$$

$$\text{ILS at optimal dose} = \frac{(21\ \text{day} - 10.5\ \text{day})}{10.5\ \text{day}} \times 100 = 100\%\ \text{(maximum effectiveness)}$$

Cyclophosphamide is the preferred drug in this test system

Data from Goldin, A. & Venditti, J.M. (1962) *Cancer Chemother. Rep.*, *17*, 145-152.

If a drug shows anti-tumour activity in the primary-animal model, usually intraperitoneally inoculated murine L1210 leukaemia or P388 leukaemia, or biological activity in some other systems, it then passes to a tumour panel (Figure 11.5). This consists of tumours with special utility, usually a spectrum of transplantable mouse tumours matched by corresponding human tumour xenografts growing in nude mice. Nude mice derive their name because the primary phenotypic alteration is the absence of a haircoat. However, the biomedically important trait in nude mice is thymic dysgenesis and the almost complete absence of T-lymphocytes so that tissue grafts from divergent species (xenografts) will grow in nude mice.[8] A problem with human tumour xenografts growing at the subcutaneous site in nude mice is the relatively slow growth rate necessitating approx. 60–90 days for each test, which is a hindrance in testing compounds at a favourable rate. A system which leads to more rapid screening with human tumour xenografts is the subrenal capsule assay where a small fragment of human tumour xenograft (approx. 1.0 mm^3) is inserted under the renal capsule of a nude mouse. Tumour growth is measured by using a stereoscope with a micrometer eyepiece over an eleven-day period. An alternative recently developed rapid screening system is the tumoured ear assay where a human tumour xenograft is grown in the external ear of the nude mouse and response to chemotherapy assessed by measuring tumour weight after ten days. Animal models that predict for a particular type of tumour are reserved for use when indicated specifically by prior drug information, for example brain tumours *in situ* where potential efficacy is indicated by the ability of a drug to cross the blood–brain barrier, which is usually associated with high lipid solubility, as well as anti-tumour activity.

11.2.3 Route- and Schedule-dependency Studies

Route-dependency studies are conducted in animals to see if an agent retains its anti-tumour activity when given by mouth to determine whether the agent might be a candidate for this route of administration in man. Clinically, most anticancer drugs are administered intravenously but some drugs, for example methotrexate, 5-fluorouracil and 6-thioguanine, can be given by mouth. For some anticancer drugs, for example methyl-CCNU and hexamethylmelamine, the oral route may be the only route for administration usually because of limited solubility which precludes parenteral administration.

The possibilities for scheduling a single drug by varying the duration of administration, the number of days the drug is given and the time between courses of therapy are almost limitless. Clinical instincts still play a large role in designing therapeutic regimes but schedule-dependency studies in animals can provide important information on the optimal conditions of drug use to aid in designing clinical trials. The optimal drug

Figure 11.5 Typical Anticancer-drug Screening Programme, Employed by the Division of Cancer Treatment, National Cancer Institute

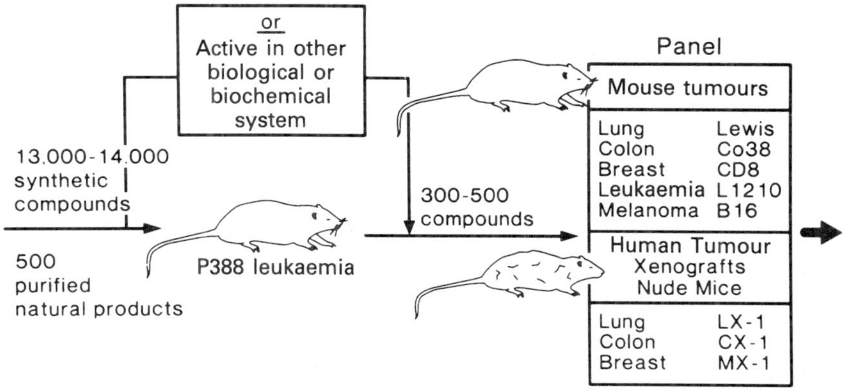

The number of compounds passing through the screening programme in 1982 are indicated. Mouse leukaemia P388 was chosen as a prescreen because accumulated data suggested it was sensitive to most classes of clinically active drugs but was sufficiently restrictive to avoid overloading the system. A number of agents (approx. 10 per cent) are tested in the tumour panel regardless of their P388 activity because of reported activity in other screening programmes or special chemical or biological features. The subcutaneous implantation site was chosen when the human xenograft system was first adopted in the panel in 1976 but was subsequently replaced by the subrenal capsule assay. Positivity in any one of the panel models qualifies a drug for further preclinical investigation

schedule will depend on the interplay of toxicological, pharmacological and biochemical actions of the drug and on the kinetics of normal and malignant cell growth. In 1965, Howard Skipper[9] reported studies on the therapy of experimental L1210 leukaemia in mice. An important feature of the experimental model he proposed, shown in Figure 11.6, was that each dose of drug killed a constant fraction of the cells, that is cell-killing was a first-order process. To affect a cure Skipper found that it was necessary for the treatment to kill every leukaemia cell in the host since even a single L1210 cell can grow, proliferate and eventually kill the mouse.

Use can be made in scheduling chemotherapy of a feature found in most anticancer drugs apart from simple alkylating agents, which is that they kill only dividing cells, in other words they are cell-cycle specific. In the mouse approx. 20 per cent of bone-marrow stem cells are not in cell cycle, but nearly all the leukaemia cells are in cycle. The most effective therapy is therefore to allow sufficient time between courses of drug therapy for bone-marrow function to recover before initiating another course of therapy. In mice with L1210 leukaemia the most effective treatment with the anticancer drug cytosine arabinoside consists of

Figure 11.6 Importance of Drug Scheduling in Treatment Regimens for Experimental Leukaemia

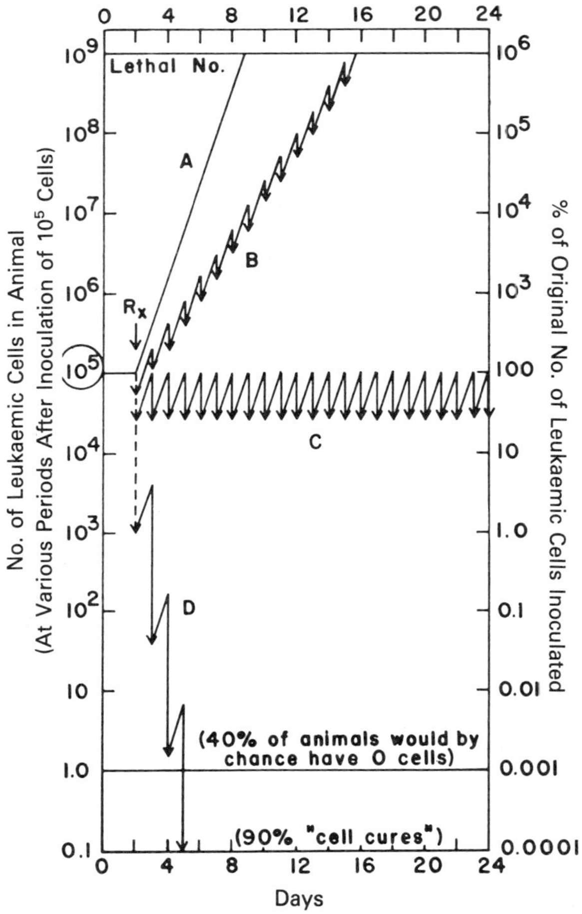

The doubling time of L1210 leukaemia is 13 h. Mice are inoculated with 10^5 leukaemia cells and, in the absence of any treatment, after a two-day lag associated with transplantation the number of leukaemic cells in the body increases logarithmically to reach 10^9, which is approximately the lethal tumour burden for the mouse, in eight days (line A). Drug treatment (R_x) is initiated on day 2. Daily treatment with a drug at a dose that produces a 50 per cent cell kill (line B) provides only a moderate increase in life-span of the animal. Daily treatment at a dose that produces a 75 per cent cell kill (line C) would allow the host to live out a normal life-span but only if the leukaemic cells do not become resistant to the drug, do not metastasise to sites in the body where the drug cannot reach and if cumulative drug toxicity does not occur. The likelihood of one or more of these events occurring is quite high and would in all probability lead to the failure of therapy. If the mice can tolerate daily treatment with a dose of drug that produces 99 per cent cell kill (line D) then 90 per cent of animals would be cancer-free after four doses. Unfortunately, a dose of drug producing 99 per cent cell kill is likely to kill many rapidly dividing normal cells, usually bone-marrow or gastrointestinal cells, as well as leukaemia cells and be unacceptably toxic to the mouse.

Source: Skipper, H.E. (1965) *Cancer Res., 25,* 1544-50; with permission.

continuous exposure of tumour cells to the drug for twice their generation time (2×13 h = 26 h), by administering drug every 3 h, and interruption between courses of treatment sufficient to allow complete recovery of bone marrow which in mice takes three days. Extrapolation of these findings to patients with acute myelogenous leukaemia (AML), where the generation time of AML cells is two to four days, has led to five-day continuous infusion of cytosine arabinoside given every two to three weeks. Compared with daily administration of cytosine arabinoside intermittent courses provide an increased remission rate and duration of remission in AML. Success in the treatment of cancer with drugs can only be achieved if the time required to restore critical normal functions is significantly shorter than the time required for the tumour to re-establish its original mass. This may be a problem with rapidly dividing leukaemia cells but should be less of a problem with solid tumours which divide more slowly. The doubling time of the average human solid tumour is between 50 and 60 days. However, most solid tumours are quite heterogeneous with respect to replicating cell populations. Cell growth is usually most rapid at the periphery of the neoplasm with necrotic regions in the centre and an intermediate zone in which cells are viable but non-dividing. These non-dividing cells are, of course, most resistant to cell-cycle-specific drugs. Solid neoplasms may also contain areas difficult for drug to penetrate producing 'sanctuaries' for the tumour cells. The problem of destroying every single tumour cell within the host is clearly much more difficult in solid tumours than in leukaemias; cancer chemotherapy against human solid tumours has generally been less successful than against leukaemias.

11.3 Preclinical Testing

11.3.1 Toxicology

Once a drug has been shown to exhibit anti-tumour activity in experimental models and has been formulated in a suitable form for intravenous administration it is subject to toxicity testing in animals.[10] In general, toxicity testing of anti-tumour drugs is less extensive than for other types of drugs intended for human use. The aim of animal toxicity studies is to establish the nature, the extent and reversibility of toxicity that might be encountered when the drug is given to man and to establish a safe dose at which to start clinical trials. In general it has been found that if a given toxicity occurs in both rodent and dog it will occur in man. If the toxicity is absent in both rodents and dog, it is unlikely to occur in man. It has been found that there is no safety advantage in the use of monkeys in anti-tumour-drug toxicity testing in addition to dogs. In most cases the dog has proved to be more sensitive than rodents to anti-tumour-drug toxicity, particularly in the major area of gastrointestinal and haematological toxicities, and provides most of the useful toxicity data. A rodent, in the past the mouse but now more frequently the rat, is, however, included as a second species. The correlation of toxic effects on rapidly dividing normal tissues, such as the gastrointestinal tract and bone marrow, between rodents, dogs, monkeys and man is good. The correlation of other toxic effects is not as good and routine pathological examination of animal tissue is usually not performed. The major clinical organ toxicities of anticancer drugs (and their frequencies) are haematological (80 per cent), gastrointestinal (73 per cent), nervous system (24 per cent), hepatic (18 per cent), urinary (11 per cent) and respiratory (4 per cent). If a drug is suspected of possessing a particular type of toxicity, perhaps because of its relation to other drugs, special tests may be performed. An example is the cardiotoxicity frequently encountered among anthracycline derivatives. It is not possible to predict from animal studies all potential types of dose-limiting toxicity of a drug in man. Nausea and vomiting are frequent and unpleasant side-effects of many anticancer drugs in man and for some drugs, for example pentamethylmelamine and *cis*-diamminedichloroplatinum, can be dose-limiting. There has been no convenient test for detecting this type of toxicity in animals although recent studies suggest that the ferret may offer a useful model.

Anticancer drugs are usually administered on the basis of surface area (square metre) rather than on the basis of body weight. A surface-area relation allows a better comparison of doses between species and between humans of grossly different sizes. The maximally tolerated doses for anticancer drugs are approximately the same for mouse, rat, dog, monkey and human when expressed on a milligram per square metre basis but vary inversely with absolute body size when expressed on a milligram per kilogram basis.[11] This is due, in part, to the fact that body surface area is more closely related than body weight to cardiac output which determines blood flow to the principal organs of drug elimination, the liver and kidneys.

11.3.2 Biochemistry and Pharmacology

Biochemical studies, although not essential to the preclinical development of an anticancer drug, are often conducted to determine the basic mechanism of antineoplastic activity of the drug, whether it is an inhibitor of DNA, RNA or protein synthesis, and whether the drug or a metabolite is the active agent. The primary mechanism of action of anticancer drugs is illustrated in Table 11.1 and has been discussed further in Chapter 10.

Preclinical pharmacological studies of drug and metabolite disposition in animals combined with information from schedule-dependency and toxicology studies can be of help in choosing an appropriate dosing schedule for clinical trials. For example, long-acting drugs should not be

Table 11.1 Primary Mechanism of Action of Some Anticancer Drugs

Site of action	Drug	Mechanism of action
DNA (direct interaction)	Busulphan Cisplatin Cyclophosphamide DTIC Mitomycin C Nitrogen mustards Nitrosoureas Procarbazine Thiotepa	Covalent reactions with DNA often leading to cross-linking of DNA strands
	Bleomycin	Causes DNA cleavage
	Daunomycin Doxorubicin	Intercalate between base pairs and inhibit nucleic acid synthesis, also bind covalently to DNA
	Actinomycin D	Binds to minor groove of DNA inhibiting nucleic acid synthesis
DNA synthesis	Cytosine arabinoside	Inhibits DNA polymerase, also incorporates into DNA
	Hydroxyurea	Inhibits ribonucleoside reductase
Purine biosynthesis	Methotrexate	Inhibits one-carbon transfer
	Mercaptopurine Thioguanine	Inhibit thymidilate synthetase
Pyrimidine biosynthesis	Methotrexate	Inhibits one-carbon transfer
	Fluorouracil FUdR	Inhibit thymidilate synthetase
Microtubular proteins	Vinca alkaloids	Bind to tubulin, inhibit mitotic spindle formation
Protein synthesis	L-Asparaginase	Hydrolyses L-asparagine, depleting circulating pools of this amino acid

administered as frequently as short-acting drugs while some drugs are so rapidly eliminated that they may have to be given at frequent intervals or by continuous infusion to maintain appreciable therapeutic concentrations in blood. Knowing which excretory pathway, hepatic or renal, predominates for a particular drug can guide the clinician in deciding whether to avoid patients with impaired hepatic or renal function. For example, care has to be exercised in giving methotrexate, a drug which is primarily eliminated by the kidneys, to patients with poor renal function. Distribution studies of drug into the cerebrospinal fluid of animals can show whether a drug crosses the blood–brain barrier and might have activity against tumours in the central nervous system in man. Antitumour drugs have been specifically synthesised with high lipid solubility for their ability to penetrate the brain, for example aziridinylbenzoquinone. It cannot be assumed, however, that a drug will distribute and be metabolised and eliminated in man in the same way as in animals. For example, the half-life of cytosine arabinoside is relatively long in rodents and treatment every 3 h is sufficient to provide continuous blood levels. In contrast cytosine arabinoside is rapidly deaminated in man and may have to be given by continuous intravenous infusion.

11.4 Clinical Testing

11.4.1 Phase I, II, III and IV Studies

Most chemotherapy regimens employ cycles of intensive therapy repeated as frequently as normal-tissue tolerance will allow and are designed on the principle that the response to treatment is a direct function of the dose of drug administered and the frequency of repetition of treatment. The validity of these principles has been demonstrated in animal models (see for example, the work of Skipper with leukaemia L1210 cited previously; section 11.2.3) and are assumed to apply in the human situation. Clinical experience suggests that for most anticancer drugs that kill rapidly dividing cells and cause gastrointestinal and bone-marrow toxicity a three-to-six-week cycle of treatment is optimal.

There are four phases of clinical testing for an anticancer drug before it can enter general medical practice; these are listed in Figure 11.2. The main purpose of phase I trials is to identify a maximally tolerated dose of the drug on one or more schedules suggested by the preclinical data, while also looking for some evidence of therapeutic effect. Phase I trials are conducted in terminally ill cancer patients instead of healthy volunteer subjects, thus offering the phase I patients the possibility of therapeutic benefit. Studies are conducted on small groups of patients, usually no more than 15–30, having a wide variety of tumours, for whom other forms of therapy such as surgery, chemotherapy or radiotherapy have proven ineffective or offer no hope of improvement. The trial is initiated at a low dose of drug usually in three patients and then, after a suitable period of time has elapsed in which to observe any toxicity which occurs, another group of patients is treated at a higher dose. The initial dose of drug should be sufficiently high to make the trial efficient and yet low enough to avoid unacceptable toxicity. Experience with numerous anticancer drugs in man over the last three decades has shown that a relatively safe dose at which to start human trials is 10 per cent of the LD_{10} (the LD_{10} is the dose of drug which produces death in 10 per cent of animals) in rodent or alternatively, one-third the maximum tolerated dose in the most susceptible species, usually the dog. Escalation of doses in phase I trials is usually by a system which provides for large initial escalations followed by smaller later increments, until a maximal tolerated dose is achieved. As has been previously discussed, clinical trials may reveal toxicities not suspected from animal testing, for example the cardiac toxicity of daunomycin, the ototoxicity of *cis*-diamminedichloroplatinum and the pancreatitis associated with L-asparaginase.

The decision to proceed with phase II testing is made on the basis of toxicity seen in the phase I trial and not on an absence of a positive clinical effect. Phase II testing is designed to evaluate the therapeutic effectiveness of the drug, namely the types of tumour that respond, dose–response relations and the extent of toxicity in relation to therapeutic effect. Groups of up to 30 patients with specific tumour types, ideally previously untreated, are studied. All the tumour types represented in the preclinical panel, breast, colon, lung, melanoma, acute leukaemia, lymphoma and Hodgkin's disease, are studied in addition to other tumour types. A complete phase II trial can require 600 patients or more; even so with such small numbers within each tumour type it is sometimes difficult to determine if a drug has activity sufficient to dictate expansion of the trial. In general, partial-response rates (that is, some objective regression of the tumour) place an agent in the category of potential clinical usefulness. A basic assumption in phase II testing and the use of a maximal tolerated dose of drug is that toxicity is a prerequisite for optimal anti-tumour activity. This concept arose in the early days of cancer chemotherapy with the clinical empiricism that toxicity indicated that a biologically active dose of drug had been administered. The concept has not been subject to critical appraisal with solid tumours. The increasing availability of comparative pharmacological data in animals and man and the introduction of newer agents, particularly biological-response modifiers (see later), may mean that the concept of maximal tolerated dose will have to be re-evaluated.

If a drug is effective in phase II trials, phase III trials and phase IV testing are initiated. In phase III trials the effectiveness of the drug is compared with existing standard therapy usually in a randomised clinical trial, although occasionally against historical controls, on a large group of patients and a specific tumour type. Therapeutic and toxic effects are compared with standard therapy. In the final phase IV of clinical testings the drug is integrated into primary-treatment regimes in combination with surgery or radiation therapy and compared with therapeutic programmes without added drugs. Decisions have to be made as to whether the toxic effects are minimal enough to risk giving the drug to a patient whose tumour will not necessarily reoccur while long-term toxicities of the drug, for example, sterility, bone-marrow aplasia and the occurrence of secondary tumours, have to be monitored.

11.4.2 Clinical Pharmacology Studies

The use of preclinical pharmacological and cytokinetic data in the rational design of treatment regimens has been discussed previously. Pharmacokinetic studies in conjunction with phase I–IV trials constitute a major component of the clinical pharmacology studies of a new anticancer drug and provide data which, taken together with the animal data, can form the basis for the design of improved treatment regimens. Useful information that can be gained from clinical pharmacological studies include whether the drug accumulates in the body with daily administration and whether blood levels achieved are within a therapeutic range based on animal studies and studies *in vitro* of cytotoxicity. By defining the rate of drug elimination and the relation of drug elimination to successive doses, pharmacokinetic studies can help to determine optimal intervals between successive drug doses. If a drug is given by mouth an assessment can be made of the drug's bioavailability, that is the fraction of the dose that reaches the systemic circulation after an oral compared with an intravenous dose of drug. It has been shown, for example, that methotrexate given by mouth exhibits saturable absorption and that almost twice as much drug is absorbed when a 100 mg dose is administered by mouth, as when four 25 mg doses are taken by mouth at 2-h intervals. Pharmacokinetic studies may contribute to an understanding of the mechanism of increased drug toxicity in selected patients and can be of help in tailoring drug delivery to a particular patient's needs, for example, patients with poor liver function who exhibit increased bone-marrow toxicity when given doxorubicin. Doxorubicin is metabolised by the liver and excreted in the bile and increased toxicity has been found to be associated with increased plasma concentrations of the drug due to delayed elimination. Based on these observations dosage-reduction schemes for doxorubicin aimed at maintaining plasma concentrations of doxorubicin close to normal, have been proposed depending on the extent of elevation of serum bilirubin. Monitoring of plasma methotrexate concentrations is frequently employed to determine when leucovorin rescue of high-dose methotrexate therapy (see section 11.6.4) can be safely terminated. This occurs when the plasma methotrexate concentration falls below 10^{-7} M.

The major limitation of the greater application of pharmacokinetic principles to anticancer therapy at the present time is the limited knowledge of the pharmacodynamics of cytotoxic drugs. Pharmacodynamics is the relation between measured drug concentration and the therapeutic or toxic effect of the drug. Bolus administration of drug provides high peak drug levels in the blood which decline rapidly due to distribution and elimination of the drug. It has been proposed that this form of administration might be optimal for a drug that is non-cell-cycle-dependent and which does not have to be present during a specific phase of a cell cycle, for example, most alkylating agents, nitrosoureas and other drugs that interact chemically with DNA. For cell-cycle-specific drugs, such as S-phase-specific methotrexate or cytosine arabinoside and, particularly for a drug that is rapidly eliminated such as cytosine arabinoside, prolonged intravenous infusion may be necessary to encompass as much of the tumour cell-cycle as possible. It has not been possible to prove the validity in man of even these simple concepts, due in large part to the difficulty of obtaining and accurately measuring a significant therapeutic response to single-agent chemotherapy for most types of tumour. There are, however, a few instances where the toxic side-effects of an anticancer drug have been shown to be related to the time of exposure to a suprathreshold concentration of drug, as in the case of the gastrointestinal and bone-marrow toxicities of methotrexate. A major emphasis of clinical pharmacology studies of anticancer drugs in the future will be to define accurately the pharmacodynamics of these agents.

11.5 Predictability of Animal Screening Methods

A crucial question to ask, once it has been demonstrated that it is possible to create animal-model systems for screening and evaluation of anti-tumour drugs and, once these drugs have been tested in the clinic, is how relevant are animal-model systems to the clinical situation? Do the test systems identify compounds that are active against tumours in man? The continued asking of this question has directed the development of the anticancer-drug screening programmes in use today. In the early years of anti-tumour drug testing a limited number of screening systems were used, usually leukaemia L1210, sarcoma 180 and adenocarcinoma 755. Because of a concern that this limited screen might not be detecting compounds of potential clinical value, a much wider screening programme that employed over 100 test systems was initiated. However, after ten years evaluation and over 100 000 compounds tested it was determined that most established clinically active drugs could have been detected by utilisation of a limited number of animal models and broad-spectrum screening was discontinued. In 1975 the current testing programme with

P388 leukaemia as the prescreen, followed by a panel of eight tumour types including human xenografts was initiated. Tables 11.2 and 11.3 show a comparison of the activity of some clinically established anticancer drugs in the current National Cancer Institute screening panel and against individual human tumours.[12] In general, drugs show greater activity against transplantable murine tumours than against human tumour xenografts growing in nude mice. The decreased sensitivity of the human tumour xenografts might reflect an intrinsic resistance of human tumours to therapy and could provide an important advantage in drug selection if it is also accompanied by the identification of new types of anti-tumour drugs. The degree of correlation between responses of mouse and human tumours of similar tissue origin is no greater than the correlation when human and mouse tumours of different origin are compared. Clinically active drugs show a broad spectrum activity in the screening panel. It

Table 11.2 Activity of Clinically Established Drugs in the National Cancer Institute Screening Panel

	Cyclophosphamide	Methotrexate	Melphalan	5-Fluorouracil	6-Mercaptopurine	Procarbazine	Actinomycin D	Vincristine	Chlorambucil
Murine Tumours									
L1210 leukaemia	+	+	+	+	+	+	+	+	+
P388 leukaemia	+	+	+	+	+	+	+	+	+
B16 melanoma	+	−	+	+	−	+	+	+	+
Lewis lung	+	−	+	+	−	−	−	−	−
Colon 26	+	−	+	+	+	−	+	+	+
Colon 38	+	−	+	+	+	+	+	+	−
CD8F1 mammary	+	+	+	+	+	−	+	+	+
Xenografts[a]									
Colon CX-1	−	−	−	+	ND	ND	ND	−	ND
Colon CX-2	+	−	+	+	−	ND	ND	ND	ND
Colon CX-5	−	ND	ND	−	ND	ND	ND	ND	ND
Mammary MX-1	+	+	+	−	−	+	+	+	+
Lung LX-1	−	−	+	−	−	+	+	−	−
Total active (%)	75	36	91	75	50	67	89	70	67

[a] All human tumour xenografts were implanted subcutaneously
ND = not done

Data from Goldin, A. & Venditti, J.M. (1981) *Recent Results Cancer Res., 76,* 176-91.

appears that the broader the spectrum of activity of a drug in the animal-tumour systems the broader the spectrum of activity observed against human tumours. CD8F1 and MX-1 mammary tumours predict activity against human mammary tumours relatively accurately, prediction for activity against human colon tumours is reasonably accurate while the predictability for activity against small-cell lung tumour appears to be least accurate. The majority of the clinically evaluated drugs were uncovered before the introduction of the current screening panel. A

Table 11.3 Activity of Clinically Established Drug in Human Tumours

	Cyclophosphamide	Methotrexate	Melphalan	5-Fluorouracil	6-Mercaptopurine	Procarbazine	Actinomycin D	Vincristine	Chlorambucil
Colon	⊕	⊕	–	+	–	–	–	–	–
Melanoma	⊕	–	⊕	–	–	–	⊕	–	–
Lung (small-cell)	+	+	–	–	–	⊕		⊕	
Breast	+	+	+	+				+	⊕
Ovary	+	⊕	+	+				–	+
Cervix	+	⊕	⊕	⊕				⊕	⊕
Prostate	⊕			⊕					
Choriocarcinoma		+			⊕				+
Pancreas				+					
Larynx									
Stomach				+					
Kidney									
Wilms	+							+	
Head and neck	+	+		+	⊕	+			⊕
Brain		⊕				⊕		⊕	
Bladder	⊕			⊕					
Neuroblastoma	+							+	+
Retinoblastoma	+								
Myeloma	+		+						+
Sarcoma	+	+	+				+	+	⊕
Testes	+	+	+				+		+
Leukoses	+	+			+	+	⊕	+	+
Lymphomas	+	+	+			+	+	+	+
Total active (%)	100	92	80	82	50	67	88	75	83

\+ Drug activity based on adequate evaluation: ⊕ evidence of activity but not established

Data from Goldin, A. & Venditti, J.M. (1981) *Recent Results Cancer Res., 76,* 176-91.

definitive analysis of the correlation between animal test data and clinical activity of the screening panel awaits the accumulation of a more extensive data base and clinical evaluation of new compounds selected by the current screening panel. A recent analysis of the total National Cancer Institute experience of experimental screening data for almost 2000 compounds and the comparison of results in humans and mice for nearly 70 drugs that have been clinically evaluated against solid tumours has shown a low correlation between screening and clinical results.[13] Based on these findings a modified approach with a P388 prescreen and an abbreviated tumour panel of L1210 leukaemias, B16 melanoma and the MX-1 mammary xenograft, which would uncover most drugs presently shown to be clinically active, has been suggested. In this scheme testing in other models would be used to establish priorities for development and clinical use of active agents.

11.6 Combination Chemotherapy (Polychemotherapy)

Drugs used to treat cancer are rarely given singly, more usually they are given in combinations to two, three or even more agents. Combinations of drugs are generally found to be more effective in treating cancer in man than single agents. There are several theoretical reasons why drug combinations should exhibit an increased therapeutic index. There can be additive anti-tumour activity but subadditive toxicity to the host, and biochemical or cytokinetic synergism of the drugs. Combinations of drugs may also provide a broader coverage of cells with intrinsic resistance to a particular drug within a heterogeneous tumour-cell population and may prevent the development of new resistant cell lines within a tumour. A number of overlapping rationales can often be employed to explain why a given combination chemotherapy programme has clinical activity.

11.6.1 Subadditive Host Toxicity

By choosing drugs whose spectrum of toxicity does not overlap but where all of the drugs are at least partially effective at killing tumour cells, a larger cytotoxic dose can be given to the patient than could be attained by giving a single agent alone. This approach, known as subadditive host toxicity, leads to a wider range of side-effects and greater discomfort to a patient but minimises the risk of a lethal effect. Drugs such as bleomycin, hexamethylmelamine, prednisone, vincristine and L-asparaginase that lack bone-marrow toxicity are particularly useful for combining with myelosuppressive drugs. Typical side-effects of the MOPP regimen, a combination of drugs used to treat Hodgkin's disease (a type of lymphoma) are shown in Table 11.4. In the era of single-agent chemotherapy Hodgkin's disease was an incurable disease, today it is curable. Combination chemotherapy for Hodgkin's disease is a careful exercise that employs doses of drugs with proven efficacy titrated against toxicity. The choice of many drug combinations is still empirical, based on clinical observation of activity of the drugs against certain tumour types and knowledge of expected toxicities. The following factors are considered desirable in developing a new drug combination in this manner: (1) each drug should be active when used alone against the tumour in question; (2) the drugs should have different mechanisms of action; (3) the toxic effect of the drugs should not overlap so that each drug can be given at or near its maximum tolerated dose.

Table 11.4 Drugs Used To Treat Hodgkin's Disease

Drug	Mechanism	Activity complete response (%)	Major toxicity
Nitrogen mustard	Alkylating agent	20	Marrow toxicity, nausea, vomiting
Vincristine (Oncovin)	Mitotic inhibitor	< 10	Neuropathy, alopecia, constipation
Procarbazine	Alkylating agent	< 10	Marrow toxicity, nausea, vomiting
Prednisone	Unknown	< 5	Osteoporosis, hypertension, diabetes, peptic ulcer
MOPP regimen		> 80	All of the above

In the era of single-agent chemotherapy responses were generally of short duration (two to six months) and Hodgkin's lymphoma was an incurable disease. The MOPP regimen, a combination of the single listed agents above, produces complete remission in 81 per cent of patients and in approximately one-half of all cases remissions persist for up to ten years. These patients can be considered cured.

Source: De Vita, V.T., Lewis, B.J., Rozencweig, M. & Muggia, F.M. (1978) *Cancer, 42,* 979-90.

11.6.2 Cytokinetic Rationale for Combination Chemotherapy

For log-phase cells in culture, many transplanted leukaemias and some human leukaemias the growth fraction (the fraction of cycling cells) approaches unity. However, for solid tumours the growth fraction tends to decrease with increasing tumour size and the number of non-cycling cells (G_o cells) increases. The growth fraction for very small solid tumours can approach unity but for large tumours can be as low as 0.05–0.1. As has already been discussed, anticancer drugs are generally more active against proliferating cells than against non-proliferating cells. Additionally some anticancer drugs are more effective against cells in one phase of the cell cycle than in other phases of the cell cycle, so-called cell-cycle phase-specific drugs, while other anticancer drugs are cell-cycle phase non-specific. Most combinations of chemotherapeutic agents employed in man involve cell-cycle phase-specific drugs, for example methotrexate, cytosine arabinoside or vincristine, combined with drugs that are cell-cycle phase non-specific, for example, 5-fluorouracil, anti-tumour antibiotics and alkylating agents. The MOPP regimen discussed previously contains both cell-cycle phase-specific and -non-specific drugs.

With few exceptions direct application of cytokinetic principles to the treatment of human cancer has met with little success. This demonstrates the large difference between human tumours and tumours in experimental animals where cytokinetic principles have successfully been applied to treatment. It may prove possible, however, when our knowledge of human tumour cytokinetics has advanced, successfully to apply cytokinetic principles to treatment of human cancer. Particularly promising approaches for the future are recruitment of non-dividing cells into cycle before chemotherapy, synchronisation of cells into various stages of the cell cycle and temporarily placing normal cells out of cycle while treating with cell-cycle phase-specific drugs.

11.6.3 Biochemical Rationale for Combination Therapy

Increasingly rational biochemical approaches are being adopted to the development of therapeutic drug combinations. Possible mechanisms of co-operative biochemical interaction between two active drugs are shown in Table 11.5. It must be pointed out that while many of these mechanisms of interaction have been demonstrated *in vitro*, they may not be valid for whole animals or the clinical situations. In *sequential metabolic blockade* drugs produce their effect on a metabolic pathway at two different sites by competing with normal substrates and thus may exert a greater cytotoxic effect by reducing the possibility that a metabolic block is bypassed. The combination of methotrexate plus 5-fluorouracil provides an example of sequential blockade in the synthesis of DNA. Both drugs inhibit synthesis of thymidilate, 5-fluorouracil by the binding of its nucleotide 5-FdUMP to thymidilate synthetase and methotrexate by depleting the intracellular reduced-folate pool. It might be expected that inhibition of thymidilate synthetase by 5-FdUMP would preserve the reduced-folate pool and negate the effect of inhibition of dihydrofolate reductase by methotrexate. This is the situation when 5-fluorouracil administration precedes methotrexate. When methotrexate precedes or is given together with 5-fluorouracil, consistently more beneficial results are seen. It is thought that this could be due to sequential blockade of dTMP synthesis. In *concurrent metabolic blockade* there are alternative pathways for formation of a common essential metabolic product so that inhibition of one pathway is insufficient to inhibit production of the essential metabolite. Only when both pathways are inhibited is production of the metabolite decreased. A general form of multiple biochemical blockade which might serve as a basis for increased anti-tumour effect is referred to as *complementary inhibition*. In this instance one of the inhibitors acts on a metabolic pathway involved in the synthesis of a cellular macromolecule (DNA, RNA or protein), while the second agent acts directly against the macromolecule. The combination of an antimetabolite which inhibits DNA synthesis by blocking purine or pyrimidine nucleotide synthesis in conjunction with an alkylating agent that directly attacks DNA is a frequently used drug combination, although such combinations have usually been arrived at empirically rather than on a basis of complementary inhibition.

Another mechanism that can lead to increased anti-tumour effectiveness is *single-enzyme inhibition* where binding of a drug to an enzyme forces the enzyme into an allosteric configuration which favours the binding of another drug. An example is the synergistic effects of guanosine and its analogues in combination with methylmercaptopurine riboside (MMPR) because of increased inhibition of purine biosynthesis *de novo* at the level of phosphoribosylpyrophosphate (PRPP) amidotransferase. *Concerted metabolic inhibition* has been proposed to occur where there is inhibition of two sites within a regulated metabolic network. Inhibition of a specific enzyme may produce changes at a site(s) distant from and, apparently unrelated to, the site of primary inhibition through feedback inhibition and sequential activation by mechanisms involving variations in metabolic pools controlling the system. Examples of concerted metabolic inhibition are found among combinations involving various inhibitors of DNA synthesis. Another form of interaction is exerted through an effect on the metabolism of a drug by another drug leading to *enhanced activation* or a *decreased inactivation* of the drug. An example of the former case is the synergistic interaction between 6-mercaptopurine and MMPR. After conversion into the 5′-monophosphate, MMPR inhibits PRPP amidotransferase and in this way inhibits utilisation of PRPP in the pathway *de novo* of purine nucleotide biosynthesis leaving more PRPP available for the activation of 6-mercaptopurine to 6-thioinosinic acid by hypoxanthine-guanine phosphoribosyltransferase. An example of the decreased metabolic activation of one active drug by another active drug is the inhibition of degradation of 6-thioguanylic acid to 6-thioguanine by MMPR. Finally, there may be *increased cellular uptake* of an active drug due to another active drug, for example the increased uptake of methotrexate in the presence of vincristine. In addition to combinations of two active agents as described above it is also possible to increase the activity of an active drug by combining it with an inactive agent. The metabolic activation of cytosine arabinoside at low concentrations by deoxycytidine kinase with UTP as

Table 11.5 Proposed Mechanisms of Biochemical Interaction Between Anticancer Drugs

Type of interaction	Description	Examples
1. Sequential inhibition 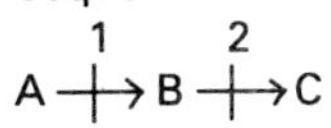	Inhibition of two sites along a linear sequence	Methotrexate + 5-fluorouracil Azaserine + 6-mercaptopurine
2. Concurrent inhibition 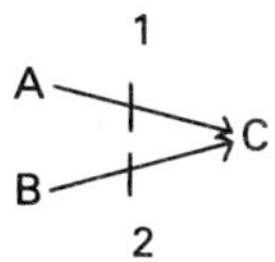	Inhibition of two sites on alternate pathways involved in the synthesis of the same product or macromolecules	Methotrexate + 6-mercaptopurine Cytosine arabinoside + 6-mercaptopurine
3. Complementary inhibition 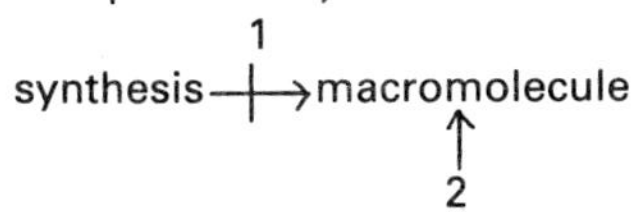	One drug causes macromolecular damage while the other inhibits synthesis and/or repair	Alkylating agent (nitrogen mustard) + 5-fluorouracil

4. Single-enzyme inhibition 1 + E′ ⟶ E′1 2 → ↑ E	Binding of the first drug to the enzyme increases the binding affinity of the other	Guanosine and MMPR
5. Concerted inhibition 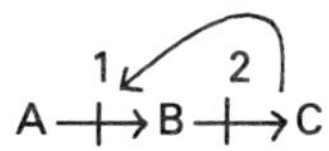	Inhibition of two sites within a regulated metabolic framework	Various inhibitors of DNA synthesis
6. Enhanced activation	The effects of one drug favour activation of another drug	Methotrexate + 5-fluorouracil 6-mercaptopurine + MMPR
7. Decreased inactivation	One drug inhibits the metabolic degradation of another	6-Thioguanylic acid + MMPR
8. Increased uptake	One drug increases the cellular uptake of another	Methotrexate and vincristine

Adapted from Mihich, E. & Grindey, G.B. (1977) *Cancer, 40*, 534-43 and Blum, R.H. & Frei, E. (1979) *Methods Cancer Res., 17*, 215-57.

the favoured phosphate donor is increased by concurrent administration of uridine which increases intracellular pools of UTP. Metabolic degradation of 6-mercaptopurine is prevented by concurrent administration of allopurinol which inhibits xanthine oxidase, the major pathway for 6-mercaptopurine degradation. Induction of hepatic enzymes responsible for activation of cyclophosphamide by phenobarbitone can lead to increased rate of activation of the cyclophosphamide.

11.6.4 Biochemical Basis of 'Rescue' Therapy

In some cases the biological effects of an anticancer drug can be selectively reversed and administration of a non-cytotoxic agent can be used to 'rescue' a patient from the toxic effects of the anticancer drug. An example of this approach to therapy is reversal of the toxicity of high-dose methotrexate therapy by leucovorin. Leucovorin (5-formyltetrahydrofolate, also known as citrovorum or folinic acid) is converted into 5-methyltetrahydrofolate, the predominant folate in the body and, if present in sufficient concentration, can overcome methotrexate toxicity in a competitive manner by replenishment of the intracellular reduced folate pool. There are two different high-dose methotrexate regimens.[17] In the first regimen methotrexate is given at doses up to 30 g/m^2 as a bolus over 4–6 h with the intent of maximising intracellular drug concentrations in poorly perfused or transport-resistant tumours. In the second regimen methotrexate is infused over a 20–42-h period with the dual intent of encompassing more of the cell cycle of the tumour cells and to reach an equilibrium between the concentration of drug in plasma and in interstitial and intracellular compartments. With both regimens lethal toxicity, predominantly to the gastrointestinal epithelium and bone marrow, would occur if leucovorin rescue was not employed. The severity of methotrexate toxicity is proportional to the duration of methotrexate exposure beyond a time threshold rather than the magnitude of methotrexate elevation. For gastrointestinal epithelium and bone marrow, the plasma concentration and time threshold are around 2×10^{-8} M and 42 h respectively. Leucovorin rescue has to be started sometime before 42 h of exposure to suprathreshold methotrexate plasma concentrations and continued until methotrexate plasma concentrations have fallen to below 10^{-7}–10^{-8} M, which is the threshold level for lung and liver toxicity.

11.6.5 Combined Modality Therapy and Adjuvant Chemotherapy

Combined modality therapy refers to combinations of surgery, chemotherapy and radiotherapy for the treatment of cancer, either all three or just two of the modalities. Adjuvant chemotherapy usually refers to the more limited use of chemotherapy after surgery in patients with minimal residual disease. The effectiveness of chemotherapy generally diminishes with increasing body burden of tumour cells. The lower limit of palpability or visibility of a tumour on X-ray is about 1 cm^3, or 10^9 cells. Anticancer drugs probably kill cancer cells by first-order kinetics, that is a given dose of drug will kill a constant fraction of cells regardless of the size of the tumour. If a drug or drug combination kills 99.99 per cent of tumour cells (that is, a fourfold log kill) this will leave 10^5 viable tumour cells. This is below the level of detectability and clinically the patient may appear to be in remission. It is possible that by immunological or other mechanisms the body could overcome this number of tumour cells and a cure might be effected. Alternatively a slowly growing tumour might be held in check so that the patient could live out a normal life-span. The lethal tumour body burden for humans is 10^{12} cells. The same drug treatment for a tumour of this size with a fourfold log kill would leave 10^8 viable tumour cells, almost certainly more than could be disposed of by the body's own defence mechanism. Because of this chemotherapy is frequently given after the body burden of tumour cells has been decreased, either by surgery or by radiotherapy. An example is the use of doxorubicin as an adjuvant to surgical treatment of patients with osteogenic sarcoma. Conversely, prior chemotherapy may be sufficiently effective in treating disseminated disease to permit successful surgical or radiation treatment of remaining localised tumour. The principles for use of drugs with surgery or radiotherapy are similar to those used in combination regimens. Clinical evidence suggests that not only very large, but also very small tumours, are less sensitive to treatment than tumours of intermediate size. This is because most human tumours grow in a Gompertzian fashion and the growth rate is lowest for both very large and very small tumours, and is maximum at an intermediary inflection point.[18] For adjuvant chemotherapy, where the tumour-cell population has been markedly reduced by surgery or radiotherapy, with only microfoci of disease remaining, it is considered by some, based on Gompertzian consideration and on the possibility that tumour may be sequestered in areas relatively inaccessible to drug, such as brain or testes, or that tumour cell resistance may have developed as a consequence of prior chemotherapy, that more intensive schedules involving higher doses of more prolonged therapy than is frequently employed in the adjuvant setting would be most effective in obtaining a cure. However, an additional factor to be considered and weighed against potential benefit of adjuvant chemotherapy is the risk of delayed drug toxicity, such as induction of secondary malignancies as with the leukaemogenic effect of methyl-CCNU, as a significant number of patients may remain free of disease with surgery or radiotherapy alone.

11.6.6 Combination of Anticancer Drugs With Other Drugs

Anticancer drugs are usually given to cancer patients who are also receiving one or more of a variety of non-cytotoxic drugs and there exists the possibility of interaction between the anticancer drugs and the other drugs. Non-cytotoxic drugs a cancer patient may receive include antiemetic agents to reduce the incidence or severity of nausea and vomiting caused by the anticancer drugs, pain killers, antibiotic therapy to prevent bacterial infection in patients with marked neutropenia due to a suppressed bone-marrow function and allopurinol. Allopurinol is a drug that inhibits xanthine oxidase, an enzyme that converts hypoxanthine into xanthine and xanthine into uric acid. Uric acid is sparingly soluble in urine and can precipitate in the distal renal tubules causing damage to the kidneys. Allopurinol will prevent the transient rise in uric acid formed from purines secondary to rapid lysis of a tumour. Xanthine oxidase is also a major route for elimination of 6-mercaptopurine, converting it into the inactive metabolite 6-thiouric acid. In the presence of allopurinol the elimination of 6-mercaptopurine is decreased. This was one of the major developments leading to the clinical use of allopurinol before its uric acid-lowering properties were clinically recognised. If allopurinol is given the dose of 6-mercaptopurine must be reduced to avoid excessive toxicity. However, not all drug interactions are of clinical significance. Aspirin has been shown to displace methotrexate from binding to albumin but this does not appear to affect methotrexate toxicity.

11.7 Newer Methods of Drug Delivery

The toxicities of anticancer drugs have prompted searches for new ways of administering these agents which would provide greater selectivity toward the tumour cell. The lysomotrophic hypothesis has featured prominently in attempts to develop new dosage forms of anticancer drugs. This hypothesis is based on an observation that pinocytotic activity is increased in tumour cells compared with that in normal cells and it has been proposed that if a cytotoxic drug could be induced to be taken into cells by pinocytosis it should render tumour cells more susceptible to the cytotoxic effects of the drug than normal cells. The only clinical trials of the lysomotrophic hypothesis so far have employed complexes of the anthracycline daunomycin bound to DNA in an attempt to stimulate selective uptake. Daunomycin binds strongly but reversibly to DNA. Pharmacokinetic studies suggest, however, that the complex dissociates fairly readily *in vivo* and that daunomycin disappears more rapidly from plasma than the DNA. This makes a lysomotrophic mechanism less appealing and the beneficial effects of a daunomycin–DNA complex that have been reported, possibly result from the preparation acting as an intravascular depot for slow release of the drug. Encapsulation of a water-soluble anticancer drug within liposomes, lipid vesicles containing an aqueous core formed when phospholipid is mixed with aqueous medium, has been suggested as a means for controlling the toxicity for some anticancer drugs, particularly the anthracyclines. Although a lysomotrophic mechanism has been advanced to account for improved tumour control with a decrease in toxicity of anthracyclines in animals, there is little evidence to suggest significant uptake of intact vesicles by cells. Doxorubicin-loaded liposomes have been shown to be less cardiotoxic in animals but to date there have been no human studies. Any improvement in therapeutic index appears to derive from slow release of doxorubicin from the liposomes into the systemic circulation. The use of liposomes should be distinguished from the use of lipid emulsions for administering poorly water-soluble anticancer drugs. Hexamethylmelamine and some newer investigational drugs have been dissolved in Intralpid®, a parenteral nutritional lipid formulation, to facilitate intravenous administration. Animal evidence suggests that this mode of administration does not affect the pharmacokinetics of the anticancer drug. An intriguing new method for delivering anticancer drug is with iron-loaded micro-aggregates of albumin. Animal studies have shown that the micro-aggregates can be guided by an externally applied magnetic field to a specific organ or tissue in the body. The technique has not yet been applied in humans. Another novel method for delivering anticancer drug is to load autologous erythrocyte ghosts with drug. This technique has been shown to offer some advantages over free drug in animals but needs more work before it can be applied to humans. A major problem with many anticancer drugs is their inability to penetrate the blood–brain barrier and to act on tumours localised within the central nervous system. Attempts have been made to improve penetration of systemically administered anticancer drugs into the brain by simultaneous administration of dimethylsulphoxide, a water miscible organic solvent which temporarily disrupts the blood–drain barrier, with anticancer drugs such as doxorubicin or cyclophosphamide. The therapeutic benefit of this approach remains to be established.

11.8 Factors Affecting Human Response to Anticancer Drug

Anticancer drugs possess a smaller therapeutic index than any other class of drugs in use today. This means that toxicity is a complication of practically all chemotherapy regimens. Animal studies have shown wide variability in tumour-regression rates and drug toxicity even in inbred animals with a transplanted tumour derived from the same single source, at the same state of growth, treated on the same day with the same dose of drug. Some animals show complete tumour regression, others only partial regression and some no tumour regression at all. It is small wonder that variability is seen in the response of human patients to chemotherapy. Many chemotherapy regimens employ the highest dose of anticancer drug that can reasonably be tolerated by patients. Because of inevitable variability in the way individuals respond to drugs some patients will experience more, in a few cases even life-threatening toxicity than other patients receiving the same dose of drug. The clinician constantly has to monitor the status of the patient, modifying the dose of the drug and providing supportive measures as necessary to avoid excessive toxicity, while providing the best chance of therapeutic benefit. In the light of the serious consequences of anticancer drug toxicity an understanding of factors that might result in individual variability in response to anticancer drugs can be of enormous benefit in planning appropriate individualised dosage regimens.

11.8.1 Status of the Patient

An important factor in determining human response to chemotherapy is the performance status of the patient. Clinical experience has shown that individuals who are not severely debilitated are better able to tolerate, and have a greater chance of responding to, chemotherapy than anorectic, bedridden or uncomfortable patients. Prior radiotherapy or treatment with nitrosourea anticancer drugs can leave a patient much more sensitive to the bone-marrow depressing effects of subsequent chemotherapy with other drugs. This is thought in both cases to be due to a reduction in the functional haemopoietic reserve of bone marrow.

11.8.2 Pharmacokinetic and Biochemical Basis for Variable Response

For many anticancer drugs it is not unusual to see a wide variation in the rate of plasma drug elimination in patients with no other obvious difference in personal or clinical characteristics. These differences are presumably determined by genetic variations among individuals and could contribute to variability in drug response. It is not surprising that patients with poor hepatic or renal function often experience more toxicity, since many anticancer drugs are eliminated by these organs. Almost one-third of patients with solid tumours exhibit some form of hepatic dysfunction due, primarily, to tumour metastases in the liver, autoimmune-type infiltration of the liver with leucocytes or drug-induced liver damage. Doses of doxorubicin and vincristine, two anticancer drugs eliminated by biliary excretion, are frequently reduced in patients with hepatic dysfunction to avoid the risk of increased, even fatal, toxicity. Doses of many other anticancer drugs are often decreased in patients with poor liver function. Streptozotocin has toxicity to the liver and has been shown to lead to delayed elimination of concomitantly administered doxorubicin and an increase in the incidence of doxorubicin-related side-effects. Renal function is often compromised in cancer patients, due in part to the advanced age of cancer patients (renal function decreases progressively with advancing age), to tumour infiltration of the kidney and to drug-induced renal damage. It is generally recommended that drug dosage be reduced progressively for any drug that is more than 30 per cent eliminated by the kidney, after renal function has decreased to about one-half the value for a normal young adult. The dose of methotrexate, an anticancer drug eliminated predominantly by the kidneys, is reduced in patients with renal dysfunction. Long-term therapy with *cis*-diamminedichloroplatinum, a drug which is toxic to the kidneys, can decrease the elimination of other drugs that are excreted by the kidneys, for example bleomycin, resulting in an increase in the toxicity of bleomycin.

Sometimes an understanding of the mechanism of response to drug treatment can be of help in selecting patients who are most likely to respond to a particular form of therapy. Selection of patients for endocrine therapy on the basis of hormone-receptor analysis provides a good example of this approach (see Chapter 8).

11.9 Non-cytotoxic Therapy

Cytotoxic chemotherapy has established itself as the third important modality, along with surgery and radiotherapy, for treating cancer. Although cytotoxic chemotherapy has led to some dramatic advances in the treatment of cancer and is responsible for our ability to cure some types of cancer, many types of cancer are still incurable. Non-cytotoxic chemotherapy derives from recent advances in our understanding of basic immunological processes and in the molecular biology of cancer causation and the oncogene. Non-cytotoxic chemotherapy has the appeal of newness and the potential of lesser toxicity than cytotoxic chemotherapy but in most cases has yet to be shown to be of increased therapeutic benefit. The promise of non-cytotoxic chemotherapy may only be realised when the basic mechanisms underlying the causation of cancer and the host immunological response are fully understood. There are several approaches to non-cytotoxic chemotherapy that are dealt with below.

11.9.1 Biological-response Modifiers

One of the most promising areas of non-cytotoxic chemotherapy is that of biological-response modification. The approach is based on the hypothesis that cancer elicits an immunologic reaction from the host which, if stimulated or modulated to a sufficient degree, will be able to attack the neoplastic cell at a level of intensity to be of clinical benefit to the patient. It is now known that the cellular and humoral-based response to tumour-associated antigens is a highly complex one with interplay of helper, suppressor and effector mechanisms. With our present immunological knowledge it is not difficult to understand the failure of past approaches

Table 11.6 Biological-response Modifiers

Substrate	Properties	Clinical results
Non-specific immune modulators	BCG, C-parvum	Little benefit
Interferon	Group of single-chain polypeptides, inhibit tumour-cell growth, stimulate immune system, inhibit tumour viruses; species-specific. Recombinant material available	Some measurable responses; side-effects fever, fatigue, neutropenia
Thymosin	Family of small acidic polypeptides, important for normal maintenance of immune system, accelerate development of precursor T-cells	Prolong survival of lung-cancer patients receiving chemotherapy
Thymic humoral factor	Acidic protein from animal thymus, important for maintenance of immune system	Some clinical improvement in leukaemia
Prostaglandins	Acidic lipids derived from C-20 essential fatty acids	No demonstrated clinical activity
Chalones	Proteins or polypeptides, non-cytoxic mitotic inhibitors, tissue-specific species non-specific	Inconclusive due to impure preparation
Interleukins	Promote expansion of populations of T-cell precursor	Only partially purified preparations so far used, side-effects, pyrexia, hypoglycaemia, lymphocytopenia
Lymphotoxin	Cytolytic factor released by cytotoxic T-cells	Not studied
Transfer factor	Polypeptide-polynucleotide, species-specific, transfers donor antigen sensitivity to host	Some activity, but needs further study

to non-specific immune stimulation with crude vaccines like bacillus Calmette-Guérin (BCG) and *Corynebacterium*. A list of some newer biological-response modifiers and their properties is given in Table 11.6.

Interferons are a group of single-chain glycosylated polypeptides that exhibit relatively similar biological effects, but differ in their stability. Interferon was first identified in 1957 as a mediator of viral interference, whereby infection with one strain of virus protects the host organism from further infection with the same viral strain. Since 1957, a great deal of information concerning interferon has been compiled. There are several classes of interferons (α, β, γ), they are species-specific and in addition to anti-viral activity interferon has been shown to stimulate natural killer-cell activity, inhibit tumour-cell growth, and increase the expression of cell-surface antigens, thus making the cells more vulnerable to the attack of killer cells. Much of the initial clinical work with interferon was conducted with material with a purity of less than 1 per cent. Current clinical studies are utilising more highly purified material from leucocytes, fibroblasts or lymphoblastoid cells. Recombinant interferon that can be produced in much larger quantities through recombinant DNA technology by cloning human leucocyte or fibroblast genes in *Escherichia coli* is also available. The anti-tumour effect of interferon appears to be threefold: first, growth of virally induced tumours may be inhibited indirectly by restraining multiplication of the cell-borne virus; secondly, growth of the tumour cell may be directly inhibited; thirdly, interferon may enhance the host's immune response against the tumour. Interferons have not led to miraculous cures of cancer but there have been a few individual cases where interferon has been of clinical benefit. In most cases, however, the responses were less, or no greater than could be obtained with other forms of therapy. Side-effects of interferon therapy are fever, chills, loss of appetite and fatigue but while they resemble the side-effects of chemotherapy, except for fatigue, they are generally less severe.

Thymosin is a factor that can be isolated from thymic tissue and consists of relatively small acidic polypeptides with molecular-weight ranges from 1000 to 5000. These thymic peptides seem to play a central role in the normal development and maintenance of the immune system. Based on the premise of immune surveillance whereby deficiencies in the immune response are thought responsible for tumour development, thymosin has been given with some success to patients with lung cancer as an adjuvant to chemotherapy. It is probable that thymosin does not induce a hyperimmune state but rather acts to accelerate development of thymosin-sensitive precursor T-cells. Prostaglandins are a series of naturally occurring acidic lipids derived from arachidonic acid or dihomo-α-linolenic acid. They have been shown to possess tumour growth-modifying properties in animals but clinical results have so far been disappointing.

Several other biological-response modifiers are waiting in the wings for clinical development. Chalones are the oldest of these and have been known for 20 years or more and the name now encompasses a variety of biological proteins or polypeptides that are tissue-specific, species non-specific, non-cytotoxic inhibitors of mitosis. Chalones are present at very low levels in normal tissue. The attraction of chalones is that they provide reversible tissue-specific mitotic inhibition and could be used to protect normal tissues, which are more sensitive to chalones than tumour cells, from the cytotoxic effect of chemotherapy and radiation. Problems associated with purification and with physical and chemical characterisation have hampered chalone research and recurrent criticism of work with chalones stems from the use of impure preparations. There have so far been only a few clinical trials in humans with partially purified chalones.

Another class of biological-response modifiers that could eventually find application in the treatment of cancer is the interleukins. Current theories suggest that antigen- or mitogen-stimulated macrophages produce interleukin-1 which in turn promotes production of interleukin-2 by the helper cell T-lymphocyte subset that promotes expansion of functional T-cell populations including the cytotoxic T-cell percursor. Preliminary studies have shown that systemic injection of purified preparations of interleukin-2 could be a feasible approach to induce cytotoxic T-cells in humans. Lymphotoxin, the cytolytic factor released by cytotoxic T-cell lymphocytes, is a candidate for future development as also is Transfer Factor, a species-specific polypeptide-polynucleotide derived from leucocytes of a donor sensitive to a particular antigen which when injected into a recipient insensitive to that antigen will bestow donor-sensitivity on the recipient. Interest in Transfer Factor stems from its putative role in the abrogation of the immunodeficient state characteristic of cancer.

11.9.2 Other Forms of Non-cytotoxic Therapy

Forms of non-cytotoxic therapy already in clinical use for treatment of cancer include endocrine therapy (see Chapter 8) and the use of agents that modify the response of tissues to radiation, the so-called radiosensitisers and radioprotectors (see Chapter 9). Other concepts of non-cytotoxic chemotherapy offering future promise will be discussed briefly. The concept of redifferentiation has recently become popular as a potential drug-development strategy. The concept states that tumour cells of an established malignant disease can be reverted to a state where they no longer exert a negative effect on the host. Theoretically, a redifferentiated tumour could exist in the host for an indefinite period, without being harmful to the host. Although antimetastatic drugs might not cure the primary tumour, they could inhibit the spread of tumour cells to form secondary foci of disease and would still be very useful drugs, since it is much easier to treat localised tumour than disseminated disease. Unfortunately, it is probable that most tumours have metastasised before they are diagnosed clinically. The oncogene concept is based on recent discoveries that a single point mutation can change a normal gene into a cancer gene (oncogene, see Chapter 4). Oncogenes have been found in some human tumours that are identical with oncogenes responsible for the tumorigenic properties of an important group of RNA viruses called retroviruses. This similarity has made it possible to visualise carcinogenesis by mutagens and viruses in the same conceptual framework. The effect of an oncogene in causing malignant transformation may be mediated by a protein product made by the oncogene. It remains to be established whether the protein product of the oncogene is a unique product or an altered form of a normal protein. Potential targets for therapeutic intervention are the oncogene itself, the gene product or the mechanism by which the oncogene protein disturbs cellular metabolism in a way that leads to a transformed phenotype.

11.10 Conclusions

The therapeutic index of most cytotoxic anticancer drugs is low and consequently drug 'cure', that is removal of a large enough fraction of the viable tumour-cell population so that tumour cells surviving treatment can never re-establish the clinically recognisable disease, has not been accomplished for most clinically evident tumours. There are eleven types of clinical neoplasia for which chemotherapy plays a major role in eliciting long-term disease-free survival or 'cures'. These responsive tumours comprise only about 25 per cent of cancer incidence and 8 per cent of cancer deaths. In the case where cures can be obtained we are able to recognise critical patterns in the kinetics of the tumour cells to which the therapeutic successes can be attributed, for example high growth fraction and marked drug sensitivity as with Burkitt's lymphoma, choriocarcinoma and childhood acute leukaemia. The majority of tumour types does not exhibit such factors beneficial to therapeutic intervention. It will require continuing effort in the development of new agents and treatment modalities and their integration with knowledge of tumour-cell growth kinetics and drug pharmacodynamics to be able to achieve 'cures' of most tumours which afflict man.

References

1. DeVita, V.T., Oliverio, V.T., Muggia, F.M., Wiernik, D.W., Ziegler, J., Goldin, A., Rubin, D., Honney, J. & Shepartz, S. (1979) The drug development and clinical trials programs of the Division of Cancer Treatment National Cancer Instititue. *Cancer Clin. Trials*, *2*, 195-216
2. Grein, A., Spalla, DiMarco, A. & Canevazzi, G. (1963) Descrizione e classificazione di un attinomicete (*Streptomyces peucetius* Sp. Nova) Produttore di una sostanza ad attivita antitumorale: Ladaunomica. *J. Gen. Microbiol.*, *11*, 109-18
3. Rosenberg, B., Van Camp, L. & Krigas, T. (1965) Inhibition of cell division in *Escherichia coli* by electrolysis products from a platinum electrode. *Nature (Lond.)*, *205*, 698-9

4. DeVita, V.T. (1982) Principles of chemotherapy. In V.T. DeVita, S. Hellman & S.A. Rosenberg (eds) *Cancer: Principles and Practice of Oncology*, Lippincott, Philadelphia, pp. 132-55
5. Gellhorn, A. & Hirschberg, E. (1955) Investigation of diverse systems for cancer chemotherapy screening. *Cancer Res.*, *3*, (Suppl.) 1-125
6. Hamburger, A.W. & Salmon, S.E. (1977) Primary bioassay of human tumor stem cells. *Science*, *197*, 461-3
7. Goldin, A. & Venditti, J.M. (1962) A manual on qualitative drug evaluation in experimental tumor systems. Part II, Quantitative assessment of various classes of agents employing advanced leukaemia L1210 in mice. *Cancer Chemother. Rep.*, *17*, 145-52
8. Ovejera, A.A. & Houchens, D.P. (1981) Human tumor xenografts in athymic mice as a preclinical screen for anticancer agents, *Sem. Oncol.*, *8*, 386-93
9. Skipper, H.E. (1965) The effects of chemotherapy on the kinetics of leukaemic cell behaviour. *Cancer Res.*, *25*, 1544-50
10. Guarino, A.M. (1979) Pharmacologic and toxicologic studies of anticancer drugs: of sharks, mice and men (and dogs and monkeys). *Methods Cancer Res.*, *17*, 91-174
11. Freireich, E.J., Gehan, E.A., Rall, D.P., Schmidt, L.H. & Skipper, H.E. (1966) Quantitative comparison of toxicity of anticancer agents in mouse, rat, hamster, dog, monkey, and man. *Cancer Chemother. Rep.*, *50*, 219-44
12. Goldin, A. & Venditti, J.M. (1981) A prospective screening program: current screening and its status. *Recent Results Cancer Res.*, *76*, 176-91
13. Staquet, M.J., Byar, D.P., Green, S.B. & Rosencweig, M. (1983) Clinical predictivity of transplantable tumor systems in the selection of new drugs for solid tumors: rationale for a three-stage strategy. *Cancer Treat. Rep.*, *67*, 753-65
14. DeVita, V.T., Lewis, B.J., Rozencweig, M. & Muggia, F.M. (1978) The chemotherapy of Hodgkin's disease. *Cancer*, *42*, 979-90
15. Mihich, E. & Grindey, G.B. (1977) Multiple basis of combination chemotherapy. *Cancer*, *40*, 534-43
16. Blum, R.H. & Frei, E. (1979) Combination chemotherapy. *Methods Cancer Res.*, *17*, 215-57
17. Bleyer, W.A. (1978) The clinical pharmacology of methotrexate. New applications of an old drug. *Cancer*, *41*, 36-51
18. Norton, L. & Simon, R. (1977) Tumor size, sensitivity to therapy and design of treatment schedules. *Cancer Chemother. Rep.*, *61*, 1307-17

Further Reading

Carter, S.K. & Stringfellow, D. (1984) Anti-cancer drug development – A challenge for industry and government. *Cancer Treat. Rep.*, in press

DeVita, V.T. (1982) Principles of chemotherapy. In V.T. DeVita, S. Hellman & S.A. Rosenberg (eds) *Cancer: Principles and Practice of Oncology*, Lippincott, Philadelphia, pp. 132-55

DeVita, V.T., Oliverio, V.T., Muggia, F.M., Wiernik, D.W., Ziegler, J., Goldin, A., Rubin, D., Honney, J. & Shepartz, S. (1979) The drug development and clinical trials programs of the Division of Cancer Treatment National Cancer Institute. *Cancer Clin. Trials*, *2*, 195-216

Goldin, A. (1978) Rationale of chemotherapeutic adjuvant treatment. *Arch. Geschulstforsch.*, *48*, 627-35

Goldin, A. & Carter, S.K. (1980) Screening and evaluation of antitumor agents. In J.F. Holland & E. Frei (eds) *Cancer Medicine*, 2nd edn, Lea and Febiger, Philadelphia, pp. 633-62

Mihich, E. & Grindey, G.B. (1977) Multiple basis of combination chemotherapy. *Cancer*, *40*, 534-43

GLOSSARY

Ablation: the removal of tissue by surgery

ACTH: adrenocorticotropic hormone

Addisonian Crisis: deficient secretion of aldosterone and cortisol from the adrenals, causing, among other effects, lowered blood pressure, great muscular weakness and gastrointestinal upsets

Adenoma: a benign tumour, derived from and resembling glandular epithelium

ADH: antidiuretic hormone

Adjuvant Chemotherapy: the use of chemotherapy in conjunction with surgery, radiotherapy, or both, in an attempt to ensure that all the tumour is killed at the time of treatment

Aetiology: the cause, or to do with relations of, cause and effect

Alkylating Agent: an electrophilic chemical compound that can covalently alkylate (transfer an alkyl group to) nucleophilic molecules. Many alkylating agents posses anticancer activity

Alpha (α)-Particle: a particle composed of two protons and two neutrons emitted by a radionuclide

Anaplasia: a form of neoplastic growth pattern characterised by extremely atypical cells not conforming to the architectural rules of alignment, associated with a complete loss of differentiation

Aneuploid: see Ploidy

Antimetabolites: a class of anticancer agents whose structures are related to those of endogenous metabolites

Ascites: abnormal fluid collection in the peritoneal cavity of the abdomen. It may either be exudative (high protein) or transudative (low protein). Both types may arise in cancer patients. Exudative ascites usually implies direct malignant infiltration of the peritoneum and indeed malignant cells in small clumps (spheroids) may thrive suspended in the fluid

BCG (bacillus Calmette-Guérin): an attenuated strain of the bacterium *Mycobacterium bovis*, that causes bovine tuberculosis

Benign Tumours: tumours that grow only by progressive expansion, not by local invasion and infiltration, nor by spread to distant sites

Beta (β)-Particle: a particle with the mass and charge of an electron and emitted by a radionuclide

Biopsy: removal of a fragment of tissue from the living patient for microscopical examination

Burkitt's Lymphoma: a tumour of the lymphoid system that arises in facial bones, ovaries, and the abdominal lymph nodes

Cachexia: a state of wasting, weakness, malnutrition and anaemia characteristic of the final stages of cancer

Cancer: a malignant neoplasm

Carcinogen (or complete carcinogen): an agent that, when administered to animals, leads to a statistically significant increased incidence of malignant neoplasms compared with that in untreated control animals. In the two-stage model of carcinogenesis, a complete carcinogen acts as both initiator and promoter

Carcinogenesis: development of cancerous cells

Carcinoma: a malignant tumour arising from epithelial cells

Cell Cybrids: a population of cells produced by combining the cytoplasm of one cell type with the nucleus of another cell type

Cell Hybrids: a population of cells produced by the fusion of two different cell lines, e.g. normal diploid cells with malignant cells

Cell Survival Curve: plot of the logarithm of the fraction of cells surviving vs the dose of the damaging agent

Chemotherapy: treatment with pure chemical substances, usually of synthetic origin

Choriocarcinoma: a highly malignant cancer derived from trophoblast, that occurs about once in every 40 000 pregnancies (See also Trophoblastic Tumour)

Chromatin: a term used to describe the complex of DNA and protein found in the eukaryotic cell nucleus (See also Chromosome)

Chromosome: a highly condensed form of chromatin identifiable in the eukaryotic cell nucleus at the time of cell division

Clone: a group of similar cells derived from a single ancestor and therefore, barring mutations, genetically identical

Clonogenic: a cell capable of generating a clone of cells (see above); not necessarily a stem cell

Cocarcinogen: an agent that administered alone to experimental animals, does not induce tumours, but when co-administered with a carcinogen, causes a significant increase in the number of tumours compared with that seen after treatment with the carcinogen alone

Contact Inhibition: the phenomenon whereby healthy cells growing in culture stop dividing and become immobilised once they have formed a contiguous monolayer covering the surface on which they grow

Corticotropic: property of stimulating the adrenal cortex, as in ACTH (adrenocorticotropic hormone)

Cyst: a defined cavity containing fluid

Differentiation: a period of cellular development during which time the cells acquire special characteristics which fit them for the specialised tasks for which they are destined

Distant Effects: effects felt at vulnerable sites remote from the tumour, and mediated largely by active substances, released into the circulation from the tumour; synonyms: remote, paraneoplastic, systemic, non-metastatic

Dormancy: a phase, in the natural history of some neoplasms, of persistent viability without an absolute increase in the number of tumour cells. This concept is controversial

Dysgenesis: defective development

Dysplasia: a form of cellular proliferation characterised by the appearance of deranged forms which increasingly disobey the rules of mutual alignment

Ectopic: out of place; a product not normally associated with the organ or tissue producing it

Electrophile: a compound or ion containing an electron-deficient centre. In reactions it removes an electron pair from the substrate

Embryonal Tumours: malignant tumours evident at birth, or occurring in early childhood, and derived from developing tissue; they contain cells native to the area in which they arise (cf. teratoma), e.g. medulloblastoma, nephroblastoma, neuroblastoma, retinoblastoma

Endemic: present within a localised area or peculiar to persons in such an area

Endometrial Cancer: cancer of the tissue lining the uterus (endometrium)

Endoplasmic Reticulum (ER): a series of membranes found in the cytoplasm of both pro- and eu-karyotic cells, and normally subdivided into 'rough' and 'smooth' according to their appearance under the electron microscope. The rough ER have ribosomes attached to them and these ribosomes synthesise proteins for export from the cell

Epidemiology: analysis of the occurrence and development of (epidemic) diseases in a population and correlation with potential causative factors

Epigenetic: factors that influence the phenotype but which do not arise in the genotype

Episomal: existing as an independent, autonomously replicating, genetic element not associated with the cellular chromosomes

Epithelium: a cell type that covers internal and external surfaces

Eukaryotic Cell: a cell whose genetic material is organised into a well-defined compartment (the nucleus), e.g. plant and mammalian cells are all eukaryotic (cf. Prokaryotic Cell)

Euploid: see Ploidy

Eutopic: literally 'in the correct place'. Applied to the production of substances but really means that the cells which produce them are in the correct location (cf. Ectopic, 'out of place')

Frameshift Mutation: an alteration in the structure of DNA that results in a change of the reading frame (triplet code) being used by the enzyme RNA polymerase. As a result of this mutation the amino acid sequence of the protein coded for by the region of DNA containing the mutation is completely altered beyond the point of mutation and normally results in the production of a 'useless' protein

Free Radical: a species with an unpaired electron in the outer shell as a result of which it is highly reactive

Fungating: term describing the overall growth pattern of a neoplasm. Fungating neoplasms project out of the surface (internal or external) and the exposed parts are often ulcerated and eroded. (cf. Sessile neoplasms)

Gamma (γ)-Ray: a quantity of energy (photon) without mass or charge that is propagated as a wave

Genome: the quantity of DNA that carries a complete set of genetic instructions for an organism

Germ-cell Cancers: cancers derived from the reproductive cells

Gonadotropins: any hormone that stimulates the male or female sex glands (gonads)

Gynaecomastia: enlargement of the male mammary glands

Haemopoetic Tissues: blood-forming tissues

Hamartoma: a developmental malformation with excess tissue; a localised error involving cells native to the area in which they occur; possessing no power of progressive encroachment or distant spread, e.g. a port wine stain. Distinct from a neoplasm

HCG: human chorionic gonadotropin

Hepatoma: a malignant tumour of the liver

Histogenesis: the tissue of origin of a tumour

Hyper-: an excess of. If suffixed by 'aemia' it refers to an increase per unit volume of a soluble substance, e.g. hypercalcaemia, an excess of calcium in the blood

Hyperplasia: a proliferation of cells, active only while an external cause is operating; usually but not always physiological or protective; may be pathological

Hyperthermia: treatment of tumours by heat

Hypertrophy: increase in size (but not number) of individual cells

Hypo-: a deficiency of. If suffixed by 'aemia' it refers to decrease per unit volume of a soluble substance, e.g. hyponatraemea, a deficiency of sodium in the blood

Hypophysectomy: the removal by surgery of the pituitary gland

Hypoxic Cells: cells existing at a reduced oxygen tension (<10–15 mmHg)

Inappropriate Production: used with respect to substances which may be produced by tumours not in accordance with the body's physiological needs (See also Ectopic)

Initiator: an agent, a single dose of which irreversibly initiates tumours in experimental animals. Tumours will often only develop if the treatment with the initiator is followed by multiple treatments with a promoter

Involution: normal shrinkage of an organ after fulfilling its functional purpose, e.g. uterus after labour

Ion: electrically charged atom or group of atoms

Ionising Radiation: radiation with sufficient energy to remove electrons from atoms or molecules. These may be electromagnetic (e.g. X-rays, γ-rays) or particulate (e.g. neutrons, heavy charged particles)

Karyotype: chromosomal content of a cell. It may be abnormal in both overall quantity and in the form of individual chromosomes. Determined at the level of light microscopy

Lectin: a protein that has two or more combining sites for carbohydrates and that can agglutinate cells or isolated glycoproteins which have the appropriate sugar residues

Leukaemia: a systemic cancer in which abnormal leucocytes accumulate in the blood and bone marrow. There are four main types of leukaemia: acute and chronic lymphocytic leukaemia, and acute and chronic myelocytic leukaemia

Linear Energy Transfer (LET): rate of energy loss along the track of an ionising particle. High LET radiations (e.g. α-particles and neutrons) are usually more effective per unit dose than low LET radiations (X- and γ-rays and electrons)

Lymphatic System: third compartment of the circulatory system after the arteries and veins, responsible for drainage of a significant proportion of the transcapillary filtrate (and chyle from the gut). It is interrupted by lymph nodes containing many of the cellular components of the immune systems

Lymphoma: a malignant tumour of lymphocytes

Malignant Tumour: a tumour that grows by local invasion and infiltration as well as by spread to discontinuous sites; invariably fatal unless treated effectively

Medulloblastoma: a tumour of the supporting cells of the central nervous system of children

Meiosis: form of cell division that occurs in the gonads to generate spermatozoa or oocytes which are haploid, i.e. half the normal dual chromosomal complement is carried

Mesothelioma: cancer of the pleura (lining of the lung) usually induced by asbestos

Metaplasia: the orderly substitution of one type of differentiated adult cell by another, in response to a long-standing stimulus

Metastasis: the process of spread to distant discontinuous sites; local metastases refers to invasive growth in continuity with the primary tumour; also used as a concrete noun to denote the colonies so-formed themselves

Mitosis: common form of cellular division in which the normal dual complement of chromosomes (the diploid state) is generated in each of the daughter cells

Mitotic Inhibitor: anticancer agent that acts by inhibiting cells at, or just before, mitosis

Monoclonal Antibody: homogeneous antibody product of a single, clonally expanded, B-lymphocyte

Multipotential: capable of differentiating into one of several different cellular types

Mutation: a chemical change in nuclear DNA. Mutations in sperm or egg cells or their precursors may cause heritable effects. Mutations in body cells (somatic mutations) may lead to effects in the individual

Myeloma: a malignant tumour of plasma cells; infiltrates and replaces bone marrow diffusely

Necrosis: death of cells

Neoplasia: a state in which the control mechanism becomes deficient and an excessive proliferation of cells continues indefinitely without relation to normal growth-control mechanisms

Neoplasm: literally 'new growth'

Neutron: an elementary particle with unit atomic mass and no charge

Neutropenia: a condition showing a deficiency of neutrophil granular leucocytes in the circulating blood

Nitrogen Mustards: compounds containing the bis(chloroethylamino) function $[N(CH_2CH_2Cl)_2]$

Nucleophile: a compound or ion containing an electron-rich centre. In reactions it brings an electron pair to the substrate

Nude Mouse: a strain of mouse characterised by lack of hair and immunological incompetence. Tumours from other species (xenografts) including man can be grown in these animals without rejection

Oncofetal Antigen: an antigen found in the fetus and in malignant disease that is absent from or much reduced in quantity in the adult e.g. α-fetoprotein

Oncogene: a genetic element responsible for or contributing to the formation of a tumour

Oncogenesis: development of a tumour

Oncologist: specialist in the treatment of cancer

Oncology: theoretically, the study of tumours ('ONKOS', Greek for tumour or mass), but in effect the practice of cancer medicine

Oophorectomy: surgical removal of the ovaries

Orchidectomy: surgical removal of the testicles

Organotropism: affinity of chemicals for particular tissues or organs of the body

Palliation: relief of symptoms without attempting cure
Papilloma: a benign tumour of surface epithelium
Paraneoplastic: see Distant Effects
Pathogenesis: origin and development of diseased conditions
Permissive/Non-permissive: terms that refer to cells which will or will not support the full replication of a particular virus. Non-permissive cells are usually derived from species or tissues which are not the natural host for the virus
Phenotype: physical characteristics or behaviour of a cell or organism
Pinocytosis: process of engulfment of external solid and liquid matter by a cell
Pleiotropic: refers to a single change or activity that influences more than one characteristic of a cellular phenotype
Pleomorphism: possessing different shapes (and sizes)
Ploidy: the number of chromosomes in a cell. Similar to karyotype. Euploid, the correct number; aneuploid, an abnormally high or low number; polyploid, an increased number
Point Mutation: an alteration in the base sequence of DNA that results in a single base being replaced by another. As a result of this mutation a single amino acid is changed in the protein sequence that is coded for by the region of DNA containing the mutation
Polyploid: see Ploidy
Precarcinogen: carcinogenic agent that requires metabolic activation to exert its biological effect
Procarcinogen (proximate carcinogen): a metabolite of a carcinogen (precarcinogen) that exhibits carcinogenic activity equal to, or greater than, that of the precarcinogen. An intermediate in the pathway of metabolic activation of the precarcinogen to its ultimate carcinogen
Progestational Agents: substances that have a similar biochemical effect to the steroid hormone progesterone
Prognosis: (knowledge of) what will happen, and when
Progression: evolution of a tumour into a more malignant form
Prokaryotic Cell: a cell whose genetic material is present throughout the cell and *not* organised into a discrete nucleus, e.g. bacteria are all prokaryotes (cf. Eukaryotic Cell)
Promoter (carcinogenesis): an agent that, although incapable of inducing tumours when administered alone, will significantly increase the yield of tumours when administered after an initiator or carcinogen
Promoter (DNA structure): an element of DNA sequence directing the initiation of RNA synthesis
Provirus: the double-stranded DNA form of a retrovirus genome that has become covalently integrated into the chromosomal DNA of the host cell
PTH: parathormone (parathyroid hormone)

Radiation Therapy (radiotherapy): treatment of tumours by ionising radiation

Radioactivity: property of radionuclides of spontaneously emitting ionising radiation

Radionuclide: an unstable species of atom that emits ionising radiation

Regression: a decrease in the number of tumour cells induced by treatment or, very rarely, occuring without known reason

Reverse Transcriptase: RNA-directed DNA polymerase. This enzyme synthesises DNA using RNA as a template

Sarcoma: a cancer derived from connective tissue or from cells derived from it

Secondary Tumour: a tumour that arises as a result of metastasis

Seminoma: a tumour of the testis

Sessile: term describing the overall growth pattern of a neoplasm. Sessile neoplasms grow close to the surfaces from which they arise (cf. Fungating neoplasms)

Somatic Cell: any cell other than a germ (egg or sperm) cell

Squamous (epidermoid): referring to an epithelial tumour resembling squamous epithelium by virtue of keratin production and the formation of intercellular bridges

Stem Cells: proliferating cells that can both maintain their own numbers by 'self-renewal' divisions and give rise to differentiated progeny

Strand Breaks: a type of DNA damage in which one or both strands of the DNA helix are broken, by radiation or drugs

Stroma: the non-cellular material in between cells; may include non-neoplastic cells caught up in a tumour, as well as the matrix

Systemic: cancer is often said to be 'a systemic disease'. This may refer to the distant effects or to the spread of cells to other organ systems; or that many of the cells in the body are to an extent premalignant, either because of inheritance or after broad exposure to a carcinogen. Very ambiguous

Taxonomy: classification of organisms into groups based on similarities of structure, origin, etc.

Teratogen: a compound that when administered during pregnancy induces abnormalities in the offspring

Teratoma: a benign or malignant developmental neoplasm containing (variably) cells from all three germ layers; these cells are foreign to the area in which the tumour arises

Therapeutic Ratio (index): ratio of doses producing damage in tumour and damage in normal tissue

Transduction: transfer of genetic information from one cell to another by its incorporation into the genome of a virus (originally defined for bacteriophages). Retroviruses that contain oncogenes are said to have transduced parts of the cellular equivalents of these genes

Transfection: introduction of DNA into cultured cells (usually in the form of a co-precipitate with calcium phosphate)

Transformation: any alteration in the apparent growth properties, morphology or behaviour of a cell in culture, most commonly referring to changes which parallel those involved in tumorigenesis

Trophoblastic Tumour: a tumour arising from the trophoblast, the ectodermal cell layer covering the blastocyst which erodes the uterine mucosa and through which the embryo receives nourishment from the mother, e.g. Choriocarcinoma

TSH: thyroid-stimulating hormone

Tumour: synonym for neoplasm; more loosely, any swelling

Ultimate Carcinogen: an electrophilic and carcinogenic metabolite of a precarcinogen. Ultimate carcinogens are direct-acting mutagens and are the forms of precarcinogens via which DNA binding *in vivo* occurs

X-Ray: a discrete quantity of energy without mass or charge propagated as a wave

Xenograft: cells or tissues from one species grown in another species, e.g. human colon cancer cells may be grown as a solid tumour in nude mice

LIST OF CONTRIBUTORS

J.E. Coggle Department of Radiobiology, The Medical College of St Bartholomew's Hospital, Charterhouse Square, London EC1M 6BQ, UK

P.B. Farmer MRC Toxicology Unit, MRC Laboratories, Woodmansterne Road, Carshalton, Surrey SM5 4EF, UK

J.T. Gallagher Cancer Research Campaign, Department of Medical Oncology, Manchester University and Christie Hospital, Wilmslow Road, Manchester M20 9BX, UK

G. Peters Imperial Cancer Research Fund Laboratories, Dominion House, St Bartholomew's Hospital, London EC1A 7BE, UK

D.H. Phillips Chester Beatty Laboratories, Institute of Cancer Research, Royal Cancer Hospital, Fulham Road, London SW3 6JB, UK

G. Powis Division of Developmental Oncology Research, Mayo Clinic, Rochester, MN 55901, USA

M.G. Rowlands Drug Development Section, Cancer Research Campaign Laboratory, Institute of Cancer Research, Clifton Avenue, Sutton, Surrey SM2 5PX, UK

P. Thomas Mallory Gastrointestinal Research Laboratory and the Department of Medicine, Harvard Medical School, Boston City Hospital, Boston, MA 02118, USA

M.D. Vincent Royal Marsden Hospital, Downs Road, Sutton, Surrey, UK and Groote Schuur Hospital, Observatory, Cape Town, South Africa

J.M. Walker Department of Biological and Environmental Sciences, The Hatfield Polytechnic, P.O. Box 109, College Lane, Hatfield, Hertfordshire AL10 9AB, UK

P. Workman MRC Clinical Oncology and Radiotherapeutic Unit, Hills Road, Cambridge CB2 2QM, UK

INDEX